Lecture Notes in Computer Science

Lecture Notes in Artificial Intelligence   **14374**
Founding Editor

Jörg Siekmann

Series Editors

Randy Goebel, *University of Alberta, Edmonton, Canada*
Wolfgang Wahlster, *DFKI, Berlin, Germany*
Zhi-Hua Zhou, *Nanjing University, Nanjing, China*

The series Lecture Notes in Artificial Intelligence (LNAI) was established in 1988 as a topical subseries of LNCS devoted to artificial intelligence.

The series publishes state-of-the-art research results at a high level. As with the LNCS mother series, the mission of the series is to serve the international R & D community by providing an invaluable service, mainly focused on the publication of conference and workshop proceedings and postproceedings.

Jinchang Ren · Amir Hussain · Iman Yi Liao ·
Rongjun Chen · Kaizhu Huang · Huimin Zhao ·
Xiaoyong Liu · Ping Ma · Thomas Maul
Editors

Advances in Brain Inspired Cognitive Systems

13th International Conference, BICS 2023
Kuala Lumpur, Malaysia, August 5–6, 2023
Proceedings

Editors
Jinchang Ren
Robert Gordon University
Aberdeen, UK

Iman Yi Liao
University of Nottingham
Semenyih, Malaysia

Kaizhu Huang
Duke Kunshan University
Kunshan, China

Xiaoyong Liu
Guangdong Polytechnic Normal University
Heyuang, China

Thomas Maul
University of Nottingham
Semenyih, Malaysia

Amir Hussain (iD)
School of Computing
Edinburgh Napier University
Edinburgh, UK

Rongjun Chen
School of Computer Science
Guangdong Polytechnic Normal University
Guangzhou, China

Huimin Zhao
School of Computer Science
Guangdong Polytechnic Normal University
Guangzhou, China

Ping Ma
Robert Gordon University
Aberdeen, UK

ISSN 0302-9743 ISSN 1611-3349 (electronic)
Lecture Notes in Artificial Intelligence
ISBN 978-981-97-1416-2 ISBN 978-981-97-1417-9 (eBook)
https://doi.org/10.1007/978-981-97-1417-9

LNCS Sublibrary: SL7 – Artificial Intelligence

This Springer imprint is published by the registered company Springer Nature Singapore Pte Ltd.
The registered company address is: 152 Beach Road, #21-01/04 Gateway East, Singapore 189721, Singapore

Paper in this product is recyclable.

Preface

Welcome to the proceedings of BICS 2023 – the 13th International Conference on Brain-Inspired Cognitive Systems. BICS has now become a well-established conference series on brain-inspired cognitive systems around the world, with growing popularity and increasing quality. BICS 2023 followed on from BICS 2004 (Stirling, UK), BICS 2006 (Island of Lesvos, Greece), BICS 2008 (Sao Luis, Brazil), BICS 2010 (Madrid, Spain), BICS 2012 (Shenyang, China), BICS 2013 (Beijing, China), BICS 2015 (Hefei, China), BICS 2016 (Beijing, China), BICS 2018 (Xi'an, China), BICS 2019 (Guangzhou, China), BICS 2020 (Hefei, China), and BICS 2022 (Chengdu, China).

Kuala Lumpur, Malaysia's vibrant heart, covers 243 square kilometers with 145 years of rich history. Once a hub for tin mining, it's now the capital and cultural nucleus. The iconic Petronas Towers showcase its economic prowess. With a diverse population harmoniously coexisting, Kuala Lumpur is a Southeast Asian key player, celebrated for its global gastronomic appeal—from lively street markets to upscale restaurants, it entices visitors with culinary diversity.

BICS 2023 aimed to provide a high-level international forum for scientists, engineers, and innovators to present the state of the art in brain-inspired cognitive systems research and applications in diverse fields. The conference featured plenary lectures given by world-renowned scholars, regular sessions with broad coverage, and some special sessions and workshops focusing on popular and timely topics.

In BICS 2023, a single-blind review process was adopted. Each paper was reviewed by three different reviewers, and the average score was used to decide the outcome. Each reviewer was asked to handle up to five manuscripts, and the proportion of external reviewers involved was around 30%. Invited contributions and contributions co-authored by any committee members were reviewed by external reviewers for an objective decision. Among fifty-eight validated submissions, only thirty-six full papers were accepted, with an acceptance rate of 62%.

BICS 2023 was co-organized by the University of Nottingham Malaysia (UNM), the National Subsea Centre, the Robert Gordon University (RGU), the Edinburgh Napier University (ENU), and the Guangdong Polytechnic Normal University (GPNU), China. The main conference was held in Kuala Lumpur, Malaysia. A joint workshop was also organized in China, co-hosted by Guangdong Polytechnic Normal University, with a special poster session for exchanges and discussions.

We would like to sincerely thank all the committee members for their great efforts and time spent in organizing the event. Special thanks go to the Program Committee members and reviewers, whose insightful reviews, and timely feedback ensured the high quality of the accepted papers and the smooth flow of the conference. We would also like to thank the publisher, Springer. Finally, we would like to thank all the speakers, authors, and participants for their support.

August 2023

Jinchang Ren
Amir Hussain
Iman Yi Liao
Rongjun Chen
Kaizhu Huang
Huimin Zhao
Xiaoyong Liu
Ping Ma
Thomas Maul

Organization

General Co-chairs

Amir Hussain	Edinburgh Napier University, UK
Chung-Lim Law	University of Nottingham Malaysia, Malaysia
Qing-Yun Dai	Guangdong Polytechnic Normal University, China
Jinchang Ren	Robert Gordon University, UK

Honorary Co-chairs

Tariq Durrani	University of Strathclyde, UK
Tieniu Tan	Chinese Academy of Sciences, China

Program Chairs

Iman Yii Liao	University of Nottingham Malaysia, Malaysia
Kaizhu Huang	Duke Kunshan University, China
Hui-Min Zhao	Guangdong Polytechnic Normal University, China
Tomas Maul	University of Nottingham Malaysia, Malaysia
Hafeez Ullah Amin	University of Nottingham Malaysia, Malaysia
Yijun Yan	Robert Gordon University, UK

Workshop Co-chairs

Yu Tang	Guangdong Polytechnic Normal University, China
Xiao-Yong Liu	Guangdong Polytechnic Normal University, China
Rong-Jun Chen	Guangdong Polytechnic Normal University, China
M. D. Junayed Hasan	Robert Gordon University, UK

Advisory Board and Publicity Chairs

Cheng-Lin Liu	Chinese Academy of Sciences, China
Bin Luo	Anhui University, China

Gun Li	University of Electronic Science and Technology of China, China
Jin Tang	Anhui University, China
Huiyu Zhou	University of Leicester, UK

Poster Session Chairs

Genyun Sun	China University of Petroleum, China
Erfu Yang	University of Strathclyde, UK
Aihua Zheng	Chinese Academy of Sciences, China

Publication Chairs

Iman Yi Liao	University of Nottingham Malaysia, Malaysia
Ping Ma	Robert Gordon University, UK
Huihui Li	Guangdong Polytechnic Normal University, China

Local Arrangement Chairs

Iman Yi Liao	University of Nottingham Malaysia, Malaysia
Zhan-Hao Xiao	Guangdong Polytechnic Normal University, China
Yinhe Li	Robert Gordon University, UK
Jiayi He	Guangdong Polytechnic Normal University, China
Li Chen	Guangdong Polytechnic Normal University, China
Jia Ouyang	Guangdong Polytechnic Normal University, China

Program Committee

Andrew Abel	Xi'an Jiaotong Liverpool University, China
Peter Andras	Keele University, UK
Xiang Bai	Huazhong University of Science and Technology, China
Vladimir Bajic	KAUST, Saudi Arabia
Yanchao Bi	Beijing Normal University, China
Erik Cambria	Nanyang Technological University, Singapore
Lihong Cao	Communication University of China, China
Chun-I Philip Chen	California State University Fullerton, USA
Mingming Cheng	Nankai University, China

Dazheng Feng	Xidian University, China
David Yushan Fong	CITS Group, USA
Marcos Faundez-Zanuy	ESUP Tecnocampus, Spain
Fei Gao	Beihang University, China
Alexander Gelbukh	CIC IPN, Mexico
Hugo Gravato Marques	ETH Zurich, Switzerland
Claudius Gros	Goethe University of Frankfurt, Germany
Junwei Han	Northwestern Polytechnical University, China
Xiangjian He	University of Technology Sydney, Australia
Bingliang Hu	Xi'an Institute of Optics and Precision Mechanics, Chinese Academy of Sciences, China
Xiaolin Hu	Tsinghua University, China
Kaizhu Huang	Xi'an Jiaotong Liverpool University, China
Tiejun Huang	Peking University, China
Amir Hussain	Edinburgh Napier University, UK
Rongrong Ji	Xiamen University, China
Zejun Jiang	Northwestern Polytechnical University, China
Donemei Jiang	Northwestern Polytechnical University, China
Yi Jiang	Institute of Psychology, Chinese Academy of Sciences, China
Jingpeng Li	University of Stirling, UK
Yongjie Li	University of Electronic Science and Technology of China, China
Cheng-Lin Liu	Institute of Automation, Chinese Academy of Sciences, China
Huaping Liu	Tsinghua University, China
Weifeng Liu	China University of Petroleum, China
Iman Yi Liao	University of Nottingham Malaysia Campus, Malaysia
Xiaoqiang Lu	Xi'an Institute of Optics and Precision Mechanics, Chinese Academy of Sciences, China
Xuelong Li	Xi'an Institute of Optics and Precision Mechanics, Chinese Academy of Sciences, China
Bin Luo	Anhui University, China
Mufti Mahmud	University of Padova, Italy
Zeeshan Malik	University of Stirling, UK
Deyu Meng	Xi'an Jiaotong University, China
Tomas Henrique Maul	University of Nottingham Malaysia Campus, Malaysia
Junaid Qadir	National University of Sciences and Technology, Pakistan
Jinchang Ren	University of Strathclyde, UK
Simone Scardapane	Sapienza University of Rome, Italy

Bailu Si	Shenyang Institute of Automation, Chinese Academy of Sciences, China
Mingli Song	Zhejiang University, China
Genyun Sun	China University of Petroleum, China
Meijun Sun	Tianjin University, China
Walid Taha	Halmstad University, Sweden
Dacheng Tao	University of Technology Sydney, Australia
Yonghong Tian	Peking University, China
Isabel Trancoso	INESC-ID, Portugal
Stefano Vassanelli	University of Padua, Italy
Liang Wang	Institute of Psychology, Chinese Academy of Sciences, China
Zheng Wang	Tianjin University, China
Zhijiang Wang	Institute of Mental Health, Peking University, China
Qi Wang	Northwestern Polytechnical University, China
Hui Wei	Fudan University, China
Jonathan Wu	University of Windsor, Canada
Qiang Wu	University of Technology Sydney, Australia
Min Xu	University of Technology Sydney, Australia
Lei Xie	Northwestern Polytechnical University, China
Yong Xia	Northwestern Polytechnical University, China
Erfu Yang	University of Strathclyde, UK
Tianming Yang	Institute of Neuroscience, China
Zhijing Yang	Guangdong University of Technology, China
Jin Zhan	Guangdong Polytechnic Normal University, China
Aizhu Zhang	China University of Petroleum, China
Bing Zhang	Chinese Academy of Sciences, China
Daoqiang Zhang	Nanjing University of Aeronautics and Astronautics, China
Li Zhang	University of Birmingham, UK
Yanning Zhang	Northwestern Polytechnical University, China
Yifeng Zhang	Institute of Neuroscience, China
Jianbiao Zhang	Beijing University of Technology, China
Huimin Zhao	Guangdong Polytechnic Normal University, China
Xinbo Zhao	Northwestern Polytechnical University, China
Jiangbin Zheng	Northwestern Polytechnical University, China
Bing J. Zhou	Sam Houston State University, USA
Jun Zhu	Tsinghua University, China

Contents

Data Analysis and Machine Learning

Applications

Bio-inspired Systems and Neural Computation

Vision-Based Deep Q-Learning on Simple Control Problems: Stabilization via Neurogenesis Regularization

Ananto Joyoadikusumo and Tomas Maul[(⊠)]

University of Nottingham Malaysia, Semenyih, Malaysia
Tomas.Maul@nottingham.edu.my

Abstract. This paper is concerned with the problem of stabilizing the training of Deep Q-Networks applied to simple Reinforcement Learning problems based on visual inputs. In particular, the paper investigates a recently proposed bioinspired regularization technique, namely adult neurogenesis, where the weights of a random subset of nodes are periodically reset. We compare experimental conditions involving different types of inputs, neural network architectures, and regularization techniques. The experiments reveal that the proposed implementation of adult neurogenesis is capable of effectively speeding up and stabilizing the training process of Deep Q-Networks.

Keywords: Bioinspired Regularization · Adult Neurogenesis · Reinforcement Learning

1 Introduction

Reinforcement Learning (RL) is a paradigm in Deep Learning that involves sampling the interaction of agents with an environment. It differs from other Deep Learning paradigms, such as supervised learning (learning from labels) or unsupervised learning (identifying the underlying structure of input data). While the past decade was mostly dominated by the other two paradigms, Reinforcement Learning [19] has recently gained significant interest due to the emergence of highly intelligent agents that have surpassed human-level performance in specific tasks. Notable examples of such agents include Google DeepMind's AlphaGo [17] (which defeated former Go World Champion Lee Sedol), AlphaStar [21] (which achieved the highest possible rank in the complex game StarCraft II), OpenAI Five [1] (which convincingly defeated the two-time winner of international champions in Dota 2), and many more.

Furthermore, Reinforcement Learning has been extended to various domains beyond games. Examples include its application in nuclear fusion development [2] or sustainable energy research [22], autonomous driving systems [9], and healthcare [23].

One of the earliest developed Reinforcement Learning algorithms is Q-Learning, which is a method to learn the action-value functions for an environment using a Q-Table. In this paper, we will implement a modification of

J. Ren et al. (Eds.): BICS 2023, LNAI 14374, pp. 3–13, 2024.
https://doi.org/10.1007/978-981-97-1417-9_1

Q-Learning called Deep Q-Network (DQN). DQN utilizes a Neural Network to represent and learn the optimal mappings of state-action pairs to Q-Values. We will apply this approach to various simple control problems from OpenAI Gym. Moreover, we will explore a vision-based approach, typically employed in Atari environments, to solve these control problems. This vision-based approach will be compared to the traditional DQN approach that relies on observable positioning features. Vision-based approaches in DQN training are generally challenging to stabilize and often fail to surpass the performance levels of position-based DQNs in these control problems. Hence, our focus in this paper will be to address this issue. Additionally, we will investigate the depth limitations of Deep Q-Networks, specifically concerning issues of overfitting and catastrophic forgetting [5]. To mitigate these challenges, we will implement and compare various stabilization and regularization techniques, including a recent technique based on adult neurogenesis (AN), which involves the generation of new functional neurons in the adult brain. Our goal is to overcome these limitations and improve the performance of vision-based reinforcement learning problems. To the best of our knowledge, this is the first application of a bioinspired regularization technique based on adult neurogenesis in vision-based reinforcement learning problems.

It is important to note that in the control problems we will experiment on, many heuristic solutions have been discovered to solve these problems with high efficiency. However, the objective of this paper is not to obtain the most optimal solution but rather to investigate stabilization and regularization techniques in DQN. These techniques can potentially be extended to more complex problems where heuristics are not feasible. Furthermore, it should be noted that more effective and robust RL algorithms, such as Proximal Policy Optimization or Soft-Actor Critic, have been developed. These algorithms generally yield better results compared to DQN in most cases. However, given the relatively simple nature of our environments, DQN provides a straightforward training approach and will most likely perform similarly to a majority of state-of-the-art reinforcement learning approaches.

2   Background

2.1   Deep Q-Learning Stabilization

The initial implementation of Deep Q-Learning [15] was primarily focused on solving Atari games or environments, where screen captures serve as inputs to the Deep Neural Network. However, vision-based inputs are rarely utilized for solving simple control problems such as CartPole. Instead, many solutions choose to use the basic physical variables (e.g., position and velocity) provided by the environment [12]. This preference for using observable physical variables can potentially be attributed to the fact that smaller state spaces, which consist of these basic physical variables, represent the simple problems more effectively. As a result, training tends to be more stable and convergence occurs faster. We use these two extremes to investigate questions surrounding the issue of unstable training, and to study a potential stabilization mechanism. In a closely related work [6], the authors investigate stabilization mechanisms based on data augmentation, whereas in this paper, we investigate stabilization via a regularization approach based on cell death/rebirth.

2.2 Network Depth

Most studies of relatively simple Reinforcement Learning problems also make use of low-parameter models. For example, the initial Atari solution [15] utilized a 3-layer-deep convolutional neural network. Similarly, traditional control problem solutions [12] typically involve models consisting of only 2 to 4 fully-connected layers.

In contrast, the majority of Deep Learning advancements and studies conducted outside of RL in the past decade have focused on increasing the depth of models. For instance, the first deep vision model, AlexNet [11], achieved superior performance by virtue of its depth and a large number of parameters. Subsequently, VGG surpassed AlexNet's performance by further increasing the depth of the architecture and doubling the total number of parameters [18].

The trend of escalating parameters has persisted to the present day, with current state-of-the-art models such as ViT [3] and Swin Transformer [13] utilizing over 1 billion parameters. The apex of this trend can be seen in the Multilingual Large Language Model (LLM) called BLOOM, developed by BigScience, which boasts over 176 billion parameters [16].

2.3 Catastrophic Forgetting

Deep neural architectures often suffer from catastrophic forgetting, a phenomenon where the model appears to forget previously learned data points while learning new ones. This problem is particularly prominent in Reinforcement Learning due to the sequential nature of the sampled training data [10]. Several techniques have been proposed to mitigate the effects of catastrophic forgetting, such as utilizing an experience buffer and a separate target model [12]. However, fully eradicating the problem remains highly challenging, especially in vision-based approaches and deeper neural networks, as we will observe later.

2.4 Neurogenesis

Adult Neurogenesis is the process of generating new functional neurons in the brain during adulthood. It was previously believed that neuronal generation stagnates or even ceases after early postnatal life, despite the absence of convincing evidence to support this hypothesis. However, recent studies [4,14] have provided evidence confirming the existence of neuron generation in adults through Neural Precursor Cells. Although the precise functions of Adult Neurogenesis, such as memory consolidation in the hippocampus, are still subject to debate, it is well-established that it contributes directly to improved cognitive functions [7]. Moreover, it may play a role in mitigating the effects of cell death in the brain.

Consequently, various studies in the field of Deep Learning have explored the application of the Neurogenesis concept to different tasks. One study investigates the process of adding new neurons to expand classes in a pre-trained ANN model from a similar domain. Another study [20], which served as inspiration for our work, incorporates Neurogenesis as a regularizer in a classification task. This approach involves randomly resetting the weights of a subset of neurons in the model, leading to improved generalization on unseen data. It is important to note that this interpretation of adult neurogenesis - which involves the simultaneous death and re-birth of a subset of neurons - is closely related, yet distinct, from techniques such as dropout, dropconnect, weight noise injection, neuronal pruning, and adaptive structure networks.

3   Methods

3.1   Environments and Experimental Design

The environments utilized from OpenAI Gym in this study are **CartPole-v1** and **Acrobot-v1**. In CartPole, the primary objective is to balance a pole attached to a cart that moves along a frictionless track for as long as possible. The action space dimensionality in this environment is 2, representing the direction (left or right) in which the cart can be pushed. The observation space dimensionality is 4, consisting of the cart position, cart velocity, pole angle, and pole angle velocity. An episode in CartPole terminates when the pole becomes unbalanced, and the episodic reward is determined by the number of steps taken in that episode. In Acrobot-v1, the goal is to swing a 2-link chain above a specified height. The action space dimensionality in Acrobot is 3, corresponding to the application of torques (positive, negative, or no torque) to the joint between the 2 links. There are 6 observable variables in the environment, including the sines, cosines, and angular velocity of the joint, as well as the relative angle of the first link. The episodic reward in Acrobot-v1 is calculated as -1 multiplied by the number of steps taken to reach the goal. To expedite training, episode truncation is set after 500 steps. For a more detailed overview of each environment, refer to the OpenAI Gym documentation.

In our experimental design, we consider CartPole to be solved when the average reward over 20 consecutive trials reaches 195.0. For Acrobot, we set the goal for the 20-moving average reward to be -110.0. It is worth noting that most studies and the OpenAI Gym leaderboard typically employ a 100-moving average goal instead of 20. However, due to computational constraints, we lowered the solution boundary to speed up the experimental process, arguably without loss of generality.

Next, we trained models, which will be further elaborated on in the next subsection, with varying numbers of parameters and depth on these environments for 5 different runs. We recorded the number of episodes required for each model to achieve the specified 20-moving average reward. We calculated the *mean* number of episodes across the 5 trials for each model. The *standard deviation* of the 5 trials was also calculated, which serves as a metric for training stability (higher standard deviations suggest that the training is more unstable and unpredictable).

3.2 Models

Table 1. Model Parameters

Model	Num params	
	CartPole-v1	*Acrobot-v1*
PosModel	10850	11043
VisionModel3L	167266	179907
VisionModel4L	987666	1708051
VisionModel6L	993746	1776723
VisionModel8L	1020818	1782419
VisionModel10L	1145330	1797587
VisionModel13L	1463538	2038483
VisionModel15L	1612338	2247518

Numbers before L in vision model indicate the number of layers in the model (e.g. Vision-Model13L has 13 layers)

Table 1 shows all the models and the number of trainable parameters that were used in our experiments. Note that the models have a slightly different number of parameters in each environment. This discrepancy is due to the environments' different action and observation spaces. PosModel is the model that will take normal positional feature information from the environment and it is comprised of 3 fully-connected layers interleaved with a Rectified Linear Unit (ReLU) activation function. The remaining VisionModels CNNs consist of alternating convolutional layers and batch normalization layers. We also vary the convolution size of each convolutional layer to capture image features of different dimensions. For very deep models, like VisionModel15L, we also vary the kernel sizes and strides of the convolutional layers.

Note that we purposefully did not choose to use current or previous state-of-the-art vision models since they have a very high number of parameters (see Sect. 2.2) and instead construct our own vision models that are relatively shallow.

3.3 Training Stabilization Strategy

During the training of different Deep Q-Networks, we incorporated several aspects from the Atari solution [15]. These aspects include the *ε-greedy strategy*, which aims to balance the rate of exploration and exploitation. Additionally, we employed *experience replay* to facilitate the training process. Experience replay helps in better approximating $Q(s, a)$ by mitigating the high, but harmful, correlation between sequential $SARS'$ pairs (also known as *experience*). Instead of learning experiences sequentially, we saved experiences in a queue buffer and randomly sampled from that buffer during network training. This approach ensures the independence and de-correlation of training data and prevents the optimizer from being influenced in the wrong direction.

Furthermore, we utilized a separate *Target Network* that acted as a copy of the policy network. The target network was used to predict Q-Values and train the policy network. The target network's weights were updated periodically every n episodes using the policy network's weights. This method of using a separate network has been shown to enhance training stability and mitigate the effects of catastrophic forgetting in the policy network over the long run. For detailed information on the training method, refer to the original Atari solution [15].

For Vision Models, instead of taking the difference between consecutive frames, we stacked the last 2 frames in the history. Additionally, we conducted the model optimization step after the conclusion of each episode, which differs significantly from the usual method of applying optimization after each step in the environment. Through our experiments, we found that performing episodal, less frequent updates resulted in faster and more stable convergence for vision models.

In addition, we chose to use the Adam optimizer [8] instead of the originally proposed RMSProp in the Atari solution. Through our experiments, we observed that Adam slightly outperformed RMSProp in some of the trials. Nevertheless, the omission of Adam is not expected to significantly alter the overall results obtained in the next section.

3.4 Neurogenesis

Regarding the implementation of neurogenesis, we randomly reset a percentage of neurons in the network every n episodes. In our experiments, we reset the weights to an arbitrary value of 0.01. It should be noted that there are various methods for selecting the value of the weight reset, such as sampling a random value from a distribution. Further studies can be conducted to explore and determine the optimal weight value assignment strategy.

4   Results

Table 2. Comparison of positional inputs and vision inputs

Input	Env	μ	σ
vision	CartPole-v1	259.6	66.62431988
vision	Acrobot-v1	870.8	89.56394364
position	CartPole-v1	154.4	46.78461286
position	Acrobot-v1	436	69.51258879

Table 3. Comparison of networks with varying depths

Model	Env	μ	σ
VisionModel3L	**CartPole-v1**	**259.6**	**66.62431988**
VisionModel4L	CartPole-v1	326.4	62.69210477
VisionModel6L	CartPole-v1	331.8	60.22624013
VisionModel8L	CartPole-v1	339	61.34329629
VisionModel10L	CartPole-v1	313	40.01249805
VisionModel13L	CartPole-v1	410	152.9101043
VisionModel15L	CartPole-v1	592	157.2005089
VisionModel3L	**Acrobot-v1**	**870.8**	**89.56394364**
VisionModel4L	Acrobot-v1	1118.6	121.0590765
VisionModel6L	Acrobot-v1	1118.6	117.2467484
VisionModel8L	Acrobot-v1	1060.6	129.2721935
VisionModel10L	Acrobot-v1	1033.8	99.0338326
VisionModel13L	Acrobot-v1	1349.4	139.4983871
VisionModel15L	Acrobot-v1	1774.8	189.5117411

*Best results from each environment is bolded

Table 4. Comparison of various regularization techniques (BN = batch normalization; AN = adult neurogenesis)

Model	Env	Mean	Standard Dev	Regularizer
VisionModel15L	CartPole-v1	592	157.2005089	BN
VisionModel15L	CartPole-v1	466.4	121.0095038	BN + Dropout
VisionModel15L	**CartPole-v1**	**292**	**57.25818719**	**AN**
VisionModel15L	CartPole-v1	448.8	114.9399843	BN + Dropout + AN
VisionModel15L	Acrobot-v1	1774.8	189.5117411	BN
VisionModel15L	Acrobot-v1	1515.4	121.0466852	BN + Dropout
VisionModel15L	**Acrobot-v1**	**889**	**104.0552738**	**AN**
VisionModel15L	Acrobot-v1	1573.6	164.5214272	BN + Dropout + AN

*Best results for each environment are bolded

Table 2 presents a brief comparison between using positional inputs and screen capture inputs. For image-based inputs, we utilized the **VisionModel3L**, which yielded the best results as shown in Table 3. Based on these findings, we can infer that employing vision-based inputs, despite all the training stabilization efforts, remains significantly less efficient than using normal positioning features. The utilization of vision-based inputs not only doubled the solving time but also resulted in slightly less stable training, as indicated by the standard deviations. Furthermore, we can observe that Acrobot generally takes about 3 times longer to solve compared to CartPole.

In Table 3, we can clearly observe the inherent depth limitations that exist in simple Reinforcement Learning tasks. For both CartPole and Acrobot, we found this limit to be 10 layers deep or approximately 1.1 million parameters. Models with fewer than 10 layers generally produce similar solution times and stability levels (around 320 episodes and 60 σ for CartPole, and 1100 episodes and 120 σ for Acrobot), with only slight degradation in results as we add more layers. However, beyond 10 layers, the network's performance declines significantly. For instance, at 15 layers deep with over 1.6 million parameters, solution times slow down by almost three-fold, and training stability does not show any improvement.

Initially, we hypothesized that overfitting might be the main cause of this depth limitation issue. Additionally, we noticed symptoms of catastrophic forgetting that were more pronounced in the 13-layer and 15-layer networks. In Fig. 1, we can observe the phases of catastrophic forgetting, which are depicted by rapid drops in the 20-moving average reward. Such instances result in a waste of training time and efficiency. Based on this observation, we decided to apply several regularization techniques on the worst-performing model, namely Vision-Model15L.

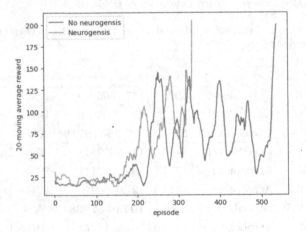

Fig. 1. Sample run of VisionModel15L in CartPole

Table 4 presents the results of applying different regularization techniques to VisionModel15L. The *BatchNorm* regularizer essentially reproduces the same results as VisionModel15L (as shown in Table 3), since the model already consists of alternating convolutional and batch normalization layers.

For *BatchNorm + Dropout*, we introduced a single dropout layer before the output linear layer, which nullifies 20% of the final batch normalization layer. The addition of dropout did, to a certain extent, reduce the solution time of the network, but not significantly enough to approach the optimal solution.

In contrast, the application of neurogenesis for 5% of the entire network every 50 episodes led to a remarkable improvement in the performance of Vision-Model15L. The network not only decreased its training time variance but also nearly achieved the optimal vision-based solution time in both CartPole (trailing by 32 episodes) and Acrobot (trailing by 18 episodes).

Unfortunately, the combination of Dropout and Neurogenesis did not yield better performance, possibly due to excessive levels of regularization, thereby interfering with the training process.

5 Discussion

Based on our experiments, we have confirmed the hypothesis that there is a limitation in viable depth when employing Deep Neural Networks in simple Reinforcement Learning problems. When coupled with the inherent instability of RL tasks, very deep networks become practically impossible to train. As evident from Fig. 1, deeper networks are more susceptible to the effects of catastrophic forgetting, leading to phases where the network seemingly "forgets" how to solve the task. In some rare cases, the training process even collapses, and the network fails to learn to maximize episodal rewards (note that we omitted these trials from the results). Although more shallow networks are not entirely immune to the phases of catastrophic forgetting, the issue is significantly exacerbated by overfitting, which occurs in much deeper networks.

With this notion, we can infer why Neurogenesis was able to remarkably stabilize the training of our deep networks. Neurogenesis, through the process of resetting weights to simulate cell death and rebirth, acts as a regularizing mechanism for deep networks, effectively stabilizing their training and preventing overfitting. This is achieved by forcing the network to partially re-learn the problem at hand. Furthermore, we can hypothesize multiple potential explanations for why Neurogenesis is able to effectively counteract the effects of catastrophic forgetting. Firstly, by applying random resetting during training, the network is forced to distribute its learning across multiple paths or nodes, thereby avoiding heavy reliance on specific neurons or connections for a particular task. Consequently, the network maintains a more distributed representation of knowledge, which reduces the risk of catastrophic forgetting when new tasks are introduced. Secondly, the stochasticity introduced by random resetting allows the network to explore various configurations of weights and activations, encouraging more stable and generalizable representations of knowledge. This stability further assists in mitigating catastrophic forgetting during the learning of new tasks. Lastly, random resetting diminishes interference between tasks by preventing individual neurons from dominating the learning process. By deactivating nodes randomly, the network becomes more resilient to interference, thus enabling the learning of new tasks without significantly overwriting or disrupting previously learned task representations. We can observe the efficacy of Neurogenesis in Fig. 1, where it successfully mitigates the decline in the 20-moving average reward, leading to faster attainment of the training goal and fewer episodes wasted on recovering from catastrophic forgetting. By resetting the weights (simulating cell death and rebirth), Neurogenesis acts as a regularizing mechanism for the network, forcing it to re-learn aspects of the problem at hand, preventing overfitting.

This work demonstrates the usefulness of adult neurogenesis as a bioinspired regularization technique for reinforcement learning problems, and suggests future research directions, including an exploration of different types of weight resetting and adaptive regimes for controlling when and what neurons are reset.

Source code: https://github.com/anantoj/rl_project.

References

1. Berner, C., et al.: Dota 2 with large scale deep reinforcement learning. arXiv preprint arXiv:1912.06680 (2019)
2. Degrave, J., et al.: Magnetic control of tokamak plasmas through deep reinforcement learning. Nature **602**(7897), 414–419 (2022)
3. Dosovitskiy, A., et al.: An image is worth 16×16 words: transformers for image recognition at scale. arXiv preprint arXiv:2010.11929 (2020)
4. Eriksson, P.S., et al.: Neurogenesis in the adult human hippocampus. Nat. Med. **4**(11), 1313–1317 (1998)
5. Fu, J., Kumar, A., Soh, M., Levine, S.: Diagnosing bottlenecks in deep Q-learning algorithms. In: International Conference on Machine Learning, pp. 2021–2030. PMLR (2019)
6. Hansen, N., Su, H., Wang, X.: Stabilizing deep q-learning with convnets and vision transformers under data augmentation. Adv. Neural. Inf. Process. Syst. **34**, 3680–3693 (2021)
7. Kempermann, G., Wiskott, L., Gage, F.H.: Functional significance of adult neurogenesis. Curr. Opin. Neurobiol. **14**(2), 186–191 (2004)
8. Kingma, D.P., Ba, J.: Adam: a method for stochastic optimization. arXiv preprint arXiv:1412.6980 (2014)
9. Kiran, B.R., et al.: Deep reinforcement learning for autonomous driving: a survey. IEEE Trans. Intell. Transp. Syst. (2021)
10. Kirkpatrick, J., et al.: Overcoming catastrophic forgetting in neural networks. Proc. Natl. Acad. Sci. **114**(13), 3521–3526 (2017)
11. Krizhevsky, A., Sutskever, I., Hinton, G.E.: ImageNet classification with deep convolutional neural networks. Commun. ACM **60**(6), 84–90 (2017)
12. Kumar, S.: Balancing a cartpole system with reinforcement learning–a tutorial. arXiv preprint arXiv:2006.04938 (2020)
13. Liu, Z., et al.: Swin transformer: hierarchical vision transformer using shifted windows. In: Proceedings of the IEEE/CVF International Conference on Computer Vision, pp. 10012–10022 (2021)
14. Ming, G.L., Song, H.: Adult neurogenesis in the mammalian brain: significant answers and significant questions. Neuron **70**(4), 687–702 (2011)
15. Mnih, V., et al.: Playing Atari with deep reinforcement learning. arXiv preprint arXiv:1312.5602 (2013)
16. Scao, T.L., et al.: BLOOM: a 176B-parameter open-access multilingual language model. arXiv preprint arXiv:2211.05100 (2022)
17. Silver, D., et al.: Mastering the game of go with deep neural networks and tree search. Nature **529**(7587), 484–489 (2016)

18. Simonyan, K., Zisserman, A.: Very deep convolutional networks for large-scale image recognition (2014). https://doi.org/10.48550/ARXIV.1409.1556, https://arxiv.org/abs/1409.1556
19. Sutton, R.S., Barto, A.G.: Reinforcement Learning: An Introduction. MIT Press, Cambridge (2018)
20. Tran, L.M., Santoro, A., Liu, L., Josselyn, S.A., Richards, B.A., Frankland, P.W.: Adult neurogenesis acts as a neural regularizer. Proc. Natl. Acad. Sci. **119**(45), e2206704119 (2022)
21. Vinyals, O., et al.: Grandmaster level in starcraft II using multi-agent reinforcement learning. Nature **575**(7782), 350–354 (2019)
22. Yang, T., Zhao, L., Li, W., Zomaya, A.Y.: Reinforcement learning in sustainable energy and electric systems: a survey. Annu. Rev. Control. **49**, 145–163 (2020)
23. Yu, C., Liu, J., Nemati, S., Yin, G.: Reinforcement learning in healthcare: a survey. ACM Comput. Surv. (CSUR) **55**(1), 1–36 (2021)

Knowledge Representation for Conceptual, Motivational, and Affective Processes in Natural Language Communication

Seng-Beng Ho[1], Zhaoxia Wang[2], Boon-Kiat Quek[1], and Erik Cambria[3][✉]

[1] Institute of High Performance Computing, Singapore, Singapore
{hosb,quekbk}@ihpc.a-star.edu.sg
[2] Singapore Management University, Singapore, Singapore
zxwang@smu.edu.sg
[3] Nanyang Technological University, Singapore, Singapore
cambria@ntu.edu.sg

Abstract. Natural language communication is an intricate and complex process. The speaker usually begins with an intention and motivation of what is to be communicated, and what outcomes are expected from the communication, while taking into consideration the listener's mental model to concoct an appropriate sentence. Likewise, the listener has to interpret the speaker's message, and respond accordingly, also with the speaker's mental model in mind. Doing this successfully entails the appropriate representation of the conceptual, motivational, and affective processes that underlie language generation and understanding. Whereas big-data approaches in language processing (such as chatbots and machine translation) have performed well, achieving natural language based communication in human-robot collaboration is non-trivial, and requires a deeper representation of the conceptual, motivational, and affective processes involved in conveying precise instructions to robots. This paper capitalizes on the UGALRS (Unified General Autonomous and Language Reasoning System) framework and the CD+ (Conceptual Dependency Plus) representational scheme to demonstrate how social communication through language can be supported by a knowledge representational scheme that handles conceptual, motivational, and affective processes in a deep and generalizable way. Through an illustrative set of concepts, motivations, and emotions, we show how these aspects are integrated into a general framework for knowledge representation and processing that could serve the purpose of natural language communication for an intelligent system.

Keywords: natural language communication · natural language understanding · knowledge representation · motivational processes · affective processes

1 Introduction

Current research in AI typically regards natural language understanding and natural language generation as separate processes. In many language processing systems, the focus is on surface level comprehension without a deep representation of meaning [1–3]. Despite the absence of deep meaning representation in these systems, conversational

J. Ren et al. (Eds.): BICS 2023, LNAI 14374, pp. 14–30, 2024.
https://doi.org/10.1007/978-981-97-1417-9_2

systems like chatbots and machine translation have achieved significant commercial success [4]. However, there are certain applications of natural language understanding that require a deeper or grounded level of representations. For instance, when using natural language instructions to guide a robot in performing certain actions, the robotic system needs to comprehend the instructions at a level where the true meaning is represented, understood, and translated into actions and behaviors. Additionally, human language communication frequently involves motivational and affective processes. For language generation, there must first be underlying motivations influenced by ongoing emotional states, before constructing and producing an utterance.

In order for language understanding to occur accurately, the listener's internal language understanding processes must include a model of the utterer's motivational and emotional states. This allows the appropriate comprehension of spoken sentences and enables the listener to respond correctly. Consider the case of sentiment analysis, which focuses on detecting emotions conveyed in sentences. While extensively researched [5–11], there has yet to be an integrated natural language communication framework that connects the processes of language generation and understanding in a systematic manner while incorporating a principled treatment of motivational and emotional processes, which are closely interwoven with language generation and understanding [1, 2]. This paper aims to address these issues by utilizing the UGALRS architectural framework proposed by Ho [12].

The remainder of this paper is organized as follows: Sect. 2 discusses motivations and background; Sect. 3 introduces the basic architectural and representational constructs; Sect. 4 elaborates on the core of the representations for language communication; finally, Sect. 5 proposes concluding remarks.

2 Motivation and Background

Natural language communication is a two-way process of generation and understanding. An utterance from a person presupposes a listener in mind and is motivated by a communicative goal and intention to achieve that goal. Then, depending on the context and emotional state of the speaker and the intended effect on the listener, an appropriate sentence is formed to communicate the intended message to the listener [13]. To do this successfully, the speaker would need a good model or representation of the mental and emotional states of the listener.

While current big data approaches to language processing, such as GPT-3 [4] are successful in many applications, they do not explicitly model these internal processes. The consequence is that certain explanations and more complex responses are not possible. For example, suppose Person A asks Person B, "Why do you say that to him?" [II.1] And Person B replies, "I want him to feel hurt." [II.2] Person A might counter suggest, "Well, I would suggest a better way to do that, which is to…" [III.3]. Utterance II.2 offers an explanation of why Person B said something, and Person B, having a model of her own mental processes including the motivation behind the earlier sentence Person A is asking her about, will be able to provide Person A with an explanation of what "causes" her to say the hurtful thing earlier.

Person A, who presumably also has an internal model of herself, Person B, and perhaps also the person whom Person B wants to hurt, will then be able to infer the various mental causalities involved and propose a different way to manage those emotions that Person B is experiencing. Whether her suggestion will be a malicious one intended to go along with and please Person B or a benign one to placate the situation for the betterment of all involved will depend on the background of their conversation and other intentions of Person A to start with. Processes such as these (i.e., attributions of intention, causality, emotions) could presumably occur when humans communicate with robots to instruct them to carry out certain tasks, or when robots are communicating with each other in collaborating to perform certain tasks. Thus, a robotic or AI system could benefit from a fuller model of the internal processes of language generation, communication, and understanding, such that humans could in turn attribute mental states and processes to them.

Language communication is inherently a form of social interaction, and autonomous systems capable of interacting and collaborating with other agents through language understanding and generation, are effectively functioning as social agents. This suggests that it might be worthwhile examining existing work involving internal operating architectures for autonomous systems or social agent, as a source of inspiration for building such agents. Among various challenges, issues relating to grounding language representations would need to be addressed. In this regard, Ho [12] has introduced the Unified General Autonomous and Language Reasoning System (UGALRS)—an architecture that not only outlines the various essential operating components of autonomous systems, but also provides a framework for grounded meaning representations. This framework, known as CD+ (an enhanced version of the conceptual dependency (CD) theory [22–24]), operates within the UGALRS framework. For the purposes of this paper, we will adopt UGALRS and CD+ as our chosen representational framework.

As mentioned earlier, the role of motivational and affective processes in language communication is often overlooked in linguistic research [14–17] computational linguistics, and natural language processing [1, 2, 18]. However, these processes play significant roles in language generation and understanding. In this regard, the UGALRS plus CD+ framework offers the necessary representational constructs to address motivational and affective processes effectively. Given that motivational and affective processes are essential components in the functioning of a social agent involved in communication and collaboration, UGALRS could serve as an architecture suited for social agent as well.

Quek et al. [19–21] have developed an architecture that integrates motivation and emotion into the functioning of an autonomous system. While primarily directed at typical robotic actions such as navigation and exploration, it aligns well with our work where these "actions" correspond to language generation within the communication process. Given the breadth of conceptual, motivational, and affective processes, for the purposes of this paper we will focus our discussion to a limited subset of each aspect, as our goal is to articulate and elucidate the intricate connections between these components within a general framework, which could be expanded upon in future research.

3 Basic Architectural and Representational Constructs

In this section we explain the construction of the basic architectural and representational constructs in terms of the UGALRS architecture and the associated CD+ representational scheme.

3.1 The CD+ Concept Representational Framework

Ho [12] developed a general representational framework that could encompass a large variety of, if not all, representations of concepts. A key idea posited in [12] is that many often-encountered concepts are functional in nature. This framework, known as CD+, is an extension of Schank's conceptual dependency (CD) representational framework [22–24]. The representations used in CD+ are cognitive, causal, and grounded.

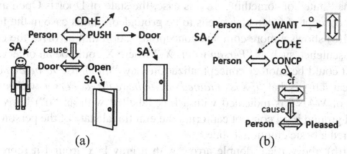

Fig. 1. Examples of CD+ representations. (a) "Person pushes the door open". (b) "Person wants something".

Figure 1(a) shows the representation of the sentence, or conceptualization, "*Person pushes the door open.*" The horizontal double arrow links the subject, Person, to the object (o), Door. This is termed a "conceptualization". In CD+, the two main constructs being enhanced over CD are the Structure Anchor (SA) and the CD+ Elaboration (CD+E) [12]. SA is a detailed structural or analogical representation of the object the symbol refers to, which provides grounding for the symbol. In this case, the concept of a Person, could encompass representational details of its bodily structure, specified to a level of detail that includes every point, limbs, and joints, as shown.

A convenient way to implement this representational model could be using the kind of analogical representation used in computer graphics in the form of a high density point cloud consisting of points corresponding to every point of the object involved, or some vectorial representations that represent the loci of these points. The various parts on the object involved (say, in the case of Person, could include the various body parts such as the head, torso, and limbs) may be movable relative to one another, and would have to be captured in the model. CD+E is used to elaborate on the symbols that represent certain actions, such as in this case, PUSH. PUSH involves a number of sub-steps, such as "*first place palm flat on Door near center of Door,*" "*then exert strength in the direction perpendicular to Door's surface,*" etc.

Here, we use English sentences to describe these steps for the ease of explanation, but each of these steps is in turn representable in CD+ form. This could carry on as we traverse a hierarchy of details until an eventual, "ground level", is reached, in which the concepts used are "ground level concepts". In [12], a comprehensive set of ground level concepts is presented, serving as a common ground for many concepts. In the two English sentences above, every concept used in their representations has to be clearly defined and further elaborated in the form of CD+E, or by using ground level concepts (which could be a SA). For instance, in the sentence "*first place palm near center of Door*", all symbols such as "*first*", "*place*", "*palm*", "*near*", "*center*", "*of*", and "*Door*" have to be explicitly defined via the CD+ framework [12].

The sentence "*Person pushes the door open*" connotes some causality that is not explicit in the sentence. The implicit concept involved is "*Person pushes the door and it causes the door to open*". This causality is shown in Fig. 1(a), as a vertical arrow with a line down in the middle. The horizontal double arrow with a line across the middle represents the "state" of something, in this case, the state of Door is Open after being pushed. The concept of Open also needs to be grounded, in this case in the form of an SA. Figure 1(b) shows a more complex conceptualization that involves the concept of WANT. In a sentence such as "*Person wants X*," where X could be an object (say, "*ice cream*") or it could be another conceptualization (say, "*the house to be demolished*"), the implicit causality is that "*if X is obtained or can be realized, Person will be pleased*". (The *object* of WANT is indicated with a link labelled with an "o") "*Pleased*" is a fundamental ground level concept capturing the emotional state of the person involved. It is considered a basic emotional state.

In Fig. 1(b), the vertical double arrow with a gray box around it represents any conceptualization (such as "*House be Demolished*"), and when Person "WANT" that, the concept of WANT is elaborated by CD+E into a causal connection between, say, "*House be Demolished*" (the same conceptualization of the object of WANT) and Person being in the state of "*Pleased*". The "c" above the horizontal gray box represents the conditional ("if"), and "f" represents the future tense – that is, "*if the house is demolished, it will cause Person to be Pleased*". Now, this entire causation is in turn a conceptualization created in Person's internal mental processes. Therefore, this entire conceptualization is the object of what Person conceptualizes - CONCP (CONCePtualize). There are other more complex CD+ constructs discussed in [12] but these examples would suffice for our subsequent discussion.

3.2 The UGALRS Architecture

As mentioned above, CD+ representations operate within a general autonomous system (or social agent) architecture, the UGALRS, for the representation of concepts [12]. Figure 2 shows part of the full UGALRS architecture that focuses on the language aspects. The focus of our attention is on the LANGUAGE COMMUNICATION REASONING CORE (LCRC) but it is by no means the only module that is important. The reason why this module occupies a larger space in this figure and has its details – the sub-modules – illustrated is that the CD+ representations that we will be using for illustrating the concepts involved in language communication reference these sub-modules in LCRC. The submodules in LCRC are PROBLEM SOLVING (PS), SIMULATION

(SM), BUFFER (BF), and CONTROL (CT) modules. In the full UGALRS [12] there is a corresponding module, REASONING CORE (REAC) to LCRC, whose major input and output are the PERCEPTION and ACTION systems respectively (called the SENSORY AND ACTION CORE – SAAC) that are non-language related, but involve the prototypical perception and action processes. The basic idea behind UGALRS views the language communication process as similar to the usual "perception and action" processes related to vision and robotic limb action, but in the sphere of language, such that "language understanding" in this context corresponds to "perception", while "language generation" is the analogous "action". These are done through the LANGUAGE SENSORY AND ACTION CORE (LSAC) here. To initiate an utterance, an intelligent autonomous system (IAS) would begin with some motivation, thus, the MOTIVATION CORE (MOTC). (On the REAC side, the MOTC would drive its reasoning to either understand the perceived information or to generate physical actions).

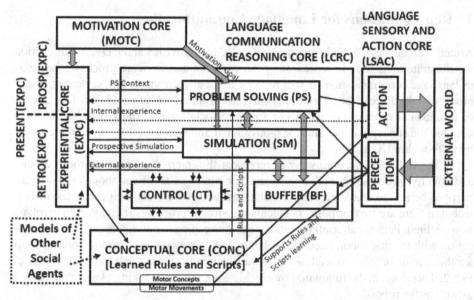

Fig. 2. The UGALRS architecture for an Intelligent Autonomous System (IAS), a robot, or a social agent. Based on [12]. See text for explanation.

PS directs a problem-solving process to concoct an appropriate sentence to prospectively satisfy the motivation involved (like concocting an appropriate physical action through PS on the REAC side). SM is used to anticipate, based on some earlier learned language communication rules, what language responses from the intended recipient of the current utterance are expected. BF contains the concepts that are currently being operated on. Finally, CT directs and control all other modules.

The EXPERIENTIAL CORE (EXPC) records all experiences, linguistic or otherwise. Therefore, there is a path from the PERCEPTION module to EXPC. EXPC also records internal experiences such as what takes place in PS, SM, or what actions (utterances) are (were) emitted. EXPC also provides the context for PS, and directs prospective simulation in SM and receives its results. EXPC can be divided into 3 portions – the present (PRESENT(EXPC)), the past (RETRO(EXPC)) and the future (PROSP(EXPC)). RETRO(EXPC) and PROSP(EXPC) can be used to ground the concepts of past and future as illustrated in [12].

The CONCEPTUAL CORE (CONC) stores knowledge as rules and scripts represented in the form of CD+. Scripts are long causal sequences of events that pertain to certain knowledge complexes, as articulated in [24] and explained in the context of CD+ in [12]. In Fig. 2, it is shown that "Models of Other Social Agents" are stored in EXPC and CONC, in EXPC in the form of un-generalized instances, and in CONC, in the form of generalized knowledge. The dotted box of "Models of Other Social Agents" is not a functioning module of UGALRS, but serves to indicate the knowledge involved and where it is located in UGALRS. The sources of external knowledge for EXPC and CONC go through the PERCEPTION module in LSAC.

4   Representations for Language Communication

Armed with the devices and constructs provided by UGALRS and CD+, in this section we illustrate their uses in representing the complex and intricate processes involved in language communication. It will be seen that between just a few sentences, many processes take place in the internal reasoning and problem-solving modules of the utterer, be it human or robotic (i.e., an IAS or a robotic social agent). These processes involve not only the conceptual, but also the motivational and affective, that can be represented by CD+ within UGALRS.

In the following section, we will represent the internal conceptual, motivational, and affective processes in both Person and Robot using CD+. Even though the primary purpose here is to elucidate the computational and representational processes in IAS (robots), there are two purposes in elucidating similar processes in the Person involved as well. First, Person can itself be another robot engaging in natural language communication with the first robot. Second, a robot or IAS can model the "mental" processes of another person or robot as well, which is the block indicated in the bottom left corner of Fig. 2. Therefore, in the following, we elucidate the processes taking place in the person as well as the robot.

4.1   Motivation and Sentence Concoction/Generation

Before an utterance is made, the utterer must begin with an idea in mind. This idea could be just a thought to be shared, or a want to be conveyed. Suppose a person (Person) thinks of asking a robot (Robot) to bring her a tool, Tool(X), from the table. This would be her "WANT," which if the robot could succeed in listening to her and satisfying it, she would be *Pleased*. At the very top part of Fig. 3, this is represented after the same fashion as the representation of Fig. 1(b). Firstly, here are SAs associated with Person and Robot, and SAs such as these will be omitted in subsequent figures to avoid clutter. The PTRANS concept is Physical TRANSfer, used and explained in [22] and [12]. There is a "from" location (Loc(Table)) and a "to" location (Loc(Person)) and a Direction (D) of transfer. So, in this case, Person conceptualizes (CONCP) that if Robot were to PTRANS Tool(X) from Table to her, she would be *Pleased*. "I" on the right most side of the PTRANS representation represent the "Instrument" that Robot might use for this purpose, such as using its legs to propel itself along the ground. The PTRANS process involves a series of steps. Suppose Robot is currently next to Person and Table is some distance away, Robot would first turn its body and face Table, viewing from far to see that Tool(X) is on Table, then mentally, through a PS process, plot a path to Table, and after executing, say, a pickup action, bring Tool(X) to Person. This sequence of events is first worked out in Robot's REAC module (i.e., the usual non-language related problem solving and planning process). This is the HOW in the CD+ Elaboration (CD+E) pointed to from PTRANS. The desired Pleased state in the WANT conceptualization is a motivational force that propels Person to proceed to look for solutions to realize the *Pleased* state. *Pleased* is a basic and ground level emotion, as discussed in [12].

Now, as discussed in [12], when a WANT is conceptualized, there may or may not be a solution to satisfy the object of the WANT. Therefore, the exact "HOW" is "IRRELEVANT" in the concept of WANT. Hence, there could be a situation in which "I want to get rich but I can't". If a solution exists, then the concept of CAN comes into play. So, if "Robot CAN bring Tool(X) from Table to Person," then it means a solution exists. This representation of CAN is given in [12] and will be shown in Fig. 4. The entire CONCP encased in a box is called a MOTIVATION CONCEPT (M-CONC).

Next, having the WANT of a certain event (namely the Robot bringing Tool(X) to the Person), Person then WANTs to communicate the concept that she WANTs this certain event to happen to Robot. There is a general rule that says if an IAS (a human is a natural IAS while a robot is an artificial one) wants something, a motivation to achieve the state of "*Pleased*" will drive the IAS to do one of three things: 1. Carry out a planning or problem solving process to achieve the state of Pleased by herself or itself; 2. Request the help of someone else to do so; and 3. Command a servile agent to do so. This is encoded as the first causal link near the top of the figure. Now, note that at the very top of the figure, the WANT conceptualization (the topmost double arrow) is encased in a gray box with a label "1". This entire conceptualization "1", that the Person wants Robot to do something, is now the object of an MTRANS (Mental TRANSfer) process intended to "mentally" transfer the WANT conceptualization "1" from Person to Robot, that will make Person *Pleased*. (I.e., Person WANTs her WANT, currently in her mind, to be MTRANS to Robot's mind. In computational terms, "mind" is simply the internal memory and processing mechanisms of the human or robot involved).

In order to realize this communication, Person concocts a sentence by Mentally BUILDing (MBUILDing) the sentence from considering the conceptualization involved (labeled "1"), together with the grammar of the language involved, the intended tone of the sentence, the emotional state of Person, etc. (the intended tone is dependent on the existing context of communication). The sentence constructed is "Robot, please bring me Tool(X) from the table," as shown in the figure as comceptualization "2". (The fact that Person WANTs Robot to do something is not explicitly stated in the sentence, but it is implied. Person could also have stated more explicitly, "Robot, I want you to bring me Tool(X) from the table".) The MBUILD concept is discussed in [22] and [12]. The MBUILDing of the sentence takes place in BF(LCRC(PERSON)). The precise process of converting an internal meaning representation in CD+ form to a grammatical surface sentence for communication is relegated to future work.

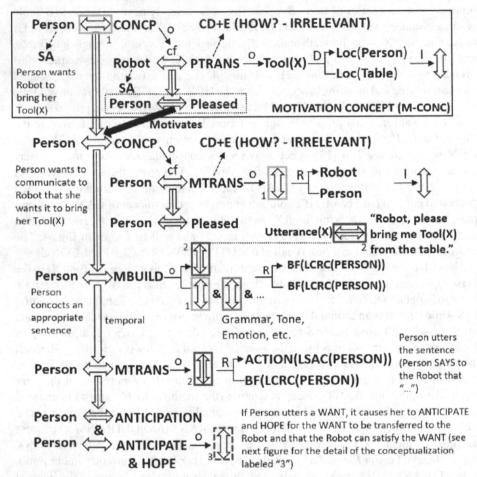

Fig. 3. The conceptual, motivational, and affective processes involved in the utterance "Robot, please bring me Tool(X) from the table." See text for explanation. The black "Motivates" arrow is not part of the representation, but an indication of the source of the causal link involved.

After the sentence is concocted in BF(LCRC(PERSON)), it it MTRANSed to ACTION(LSAC(PERSON)) to be emitted as an utterance. The link between MBUILD and MTRANS is a "temporal" one, not a causal one, as the second step simply follows the first step as part of the process (temporal links are indicated as a thickened arrow without a line running down its middle).

When a sentence, whether one that is a command or request, or just a factual statement, is uttered toward a recipient, a state of ANTICIPATION is entered (which is part of the implication of command or request) and the utterer then ANTICIPATEs something, as shown in the bottom of Fig. 3. This is the first affective state that emerges in the present communication process and will be discussed in detail in the next section. This ANTICIPATION is accompanied by HOPE as the prospect is positive [25].

4.2   Affective States and Illocutionary Forces

In the non-linguistic sphere, when a certain action is emitted by an IAS, it is expected to cause certain effects. Similarly, in the linguistic sphere, an emitted utterance is expected to cause some effects. This has been investigated by speech act theorists [26]. If an utterance is meant to communicate certain information to the recipient, there may be no immediate overt actions or responses expected in the recipient, but the information conveyed may cause future actions or responses, or in the least, it may cause certain changes in the beliefs of the recipient. If the utterance is in the form of a command or request, immediate actions and responses are expected. The utterance is said to have an "illocutionary force" [26].

At the bottom of Fig. 3 we show that Person enters am affective state of ANTICIPA-TION and she ANTICIPATEs something. In Fig. 4(a) we show the functional representation of ANTICIPATE, an action that accompanies the affective state ANTICIPATION. If an Agent ANTICIPATEs a certain conceptualization, she conceptualizes that the conceptualization involved will happen in the future. It is shown in Fig. 4(a) that the object of Agent's CONCP is labeled with an "f", which means it resides in the prospective part of the EXPC (Fig. 2), PROSP(EXPC). This formulation of an affective state and its associated causal consequences is in consonant with the cognitive appraisal theory of emotion [25]. The same approach will be adopted with the other affective states in subsequent discussions. Specifically, in the situation depicted in Fig. 3 in which Person asks Robot to bring her Tool(X) from Table, she ANTICIPATEs both the facts that "Robot WANTs to PTRANS Tool(X) from Table to Person so that Person is *Pleased*" and "Robot CAN PTRANS Tool(X) from Table to Person so that Person is *Pleased*", as shown in Fig. 4(b).

First, let us consider the representation for "Robot WANTs to PTRANS Tool(X) from Table to Person so that Person is *Pleased*". This is a transfer of what Person WANTs to what Robot WANTs. Now, for Person to reasonably assume that Robot would WANT to *Please* her, it must be assumed that Robot has either a SERVILE or an ALTRUISTIC attitude. In situations in which Robot or other recipient(s) of the utterance is REBELLIOUS or UNCOOPERATIVE, then this situation does not obtain.

In Fig. 5 we depict that Robot is indeed SERVILE or ALTRUISTIC and hence in Fig. 4(b) Person ANTICIPATES that "Robot WANTs to PTRANS Tool(X) from Table to Person so that Person is *Pleased*." Hence, Robot being *Pleased* is caused by Person being *Pleased*. CD+ can be used to represent the connections between attitudes such as being SERVILE, ALTRUISTIC, COOPERATIVE, REBELLIOUS, or UNCOOPERATIVE and whether the entity/IAS involves WANTs to do certain things. The details of these are left to future work.

The locus of the illocutionary force is this. In any IAS, ultimately it will do whatever *Please*s it. The arrow labeled PS shows the flow of Robot's actions: in order to please itself, it has to please Person, and in order to please Person, it has to PTRANS Tool(X) from Table to Person, if it CAN.

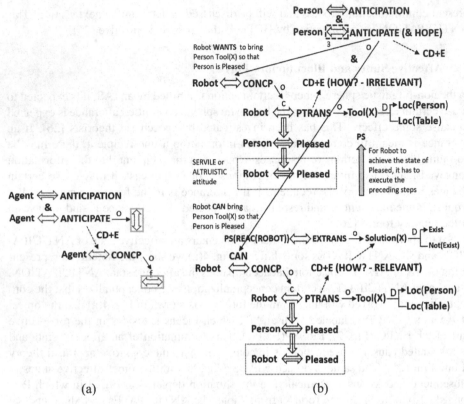

(a)   (b)

Fig. 4. (a) The general representation of ANTICIPATION. There is a state of ANTICIPATION, and an object referred to by the ANTICIPATE action. (b) The specific case of ANTICIPATION at the bottom of Fig. 3.

As mentioned above, there is a difference between WANT and CAN [12]. The primary difference is that WANTing something to happen (e.g., PTRANSing something from one place to another) does not imply that a solution exists for the thing to happen, but CAN implies that the solution exists.

Therefore, there could be a situation that "I want to go from here to there but I can't". Hence, the representation for "Robot CAN PTRANS Tool(X) from Table to Person so that Person is Pleased" shown in Fig. 4(b) is that PS(REAC(ROBOT)) returns a solution (Solution(X)) for Robot to PTRANS Tool(X) from Table to Person. (EXTRANS stands for EXistential TRANSformation in which something goes from non-existence to existence or vice versa – which is used to represent the existence of a Solution(X) – see [12]).

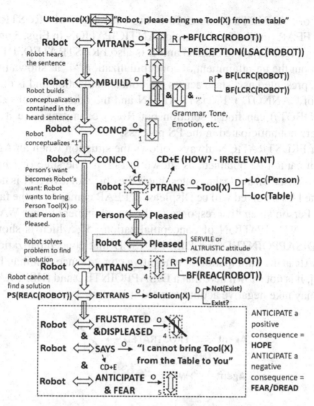

Fig. 5. Robot's internal processing in response to Person's utterance in Fig. 4. See text for explanation. Unlike in Fig. 3, ANTICIPATE here is accompanied by FEAR as it is anticipating a negative prospect [25].

4.3 Sentence Understanding, Actions, and Affective States

Now that the utterance has been made and presumably received by Robot, the first step of the process on Robot's side is to MTRANS the received utterance into BF(LCRC(ROBOT)) for further processing as shown in Fig. 5. This causes Robot to MBUILD the conceptualization corresponding to the received utterance, based on the grammar of the language, the tone present in the sentence, the current emotional state of Robot and the perceived emotional state of Person, etc. This MBUILDed conceptualization is labeled "1", which is the same as conceptualization "1" in Fig. 3. Assuming Robot has either a SERVILE, COOPERATIVE, or ALTRUISTIC attitude, this causes it to create the next conceptualization capturing the fact that Robot will be *Pleased* if Person is *Pleased* due to Robot carrying out a certain task. This then motivates Robot to seek a solution to conceptualization "4", which is "Robot PTRANS Tool(X) from Table to Person". This conceptualization is MTRANS from BF(REAC(ROBOT)) to PS(REAC(ROBOT)).

Suppose PS(REAC(ROBOT)) cannot find a solution subsequently. This situation is represented in CD+ using EXTRANS showing that a solution does not exist (see [12] for the concept of CANNOT). It causes Robot to enter states of FRUSTRATED, DIS-PLEASED and FEAR (unlike for the case of ANTICIPATION in Figs. 3 and 4(a), these are not shown in Fig. 5 to avoid clutter) and it is also FRUSTRATED, DISPLEASED, and FEARful about the un-attainment of conceptualization "4" as shown in Fig. 5 ("un-attainment' is represented as a slash across the conceptualization and is directly related to the concept of CANNOT"). FRUSTRATION and the other emotions can also arise if PS(REAC(ROBOT)) can find a solution but Robot cannot execute it due to other situations that are not anticipated in the PS process.

The state of FRUSTRATION always follows the situation when an Agent WANTs something but it cannot be obtained, as shown in Fig. 6. The state of *Displeased* also accompanies this based on a rule that states that if the object of a WANT is not achievable or satisfiable, the IAS involved will be Displeased. FEAR comes from the fact that Robot has a model of Person's negative response to the un-attainment of her WANT, and it is reflected in its ANTICIPATION of conceptualization "5", which is shown in Fig. 7 as "Person is DISAPPOINTED and DISPLEASED that conceptualization "4" is not attainable." FEAR is the anticipation of a negative consequence that may happen to the agent itself [25]. It is not shown here that a DISAPPOINTED and DISPLEASED Person toward Robot may take negative actions toward it in some way.

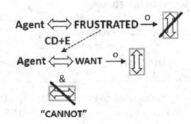

Fig. 6. The meaning and representation of FRUSTRATED

Other than the consequences above, another response to not being able to find a solution includes Robot communicating this to Person by uttering "I cannot bring Tool(X) from the table to you" as shown in Fig. 5. The symbol "SAY" has a CD+E that is the same processes in Fig. 3 when Person concocts and utters a sentence, but we omit the detailed CD+E here. If instead Robot is able to find a solution, it would go ahead to carry out the task and say "Here is Tool(X)" when handing it to Person.

What motives Robot to explain its failure is the communication rule that states: "if others are displeased with your failure to do something on request or command, do communicate about it, including explaining the reason involved, because this will placate the other person, which in turn should reduce your own frustration, displeasure, and fear". This entire rule could be stated in CD+ for the system to interpret and execute.

4.4 Continuing Communication

Following Robot reporting that it is not able to bring Tool(X) to Person, Person enters the state of being DISAPPOINTED and DISPLEASED and the object of the DISAP-POINTment and DISPLEASure is the un-attainment of conceptualization "4", as shown in Fig. 7. Instead of just keeping quiet, which is a possible response on the part of Person if she is no longer concerned about the un-attainment of "4" or she is taking some time to ponder her response, a typical immediate response on the part of Person is to try and understand the cause of the un-attainment of "4". To this end, Person asks "Why can't you bring Tool(X) to me?" as shown in Fig. 7.

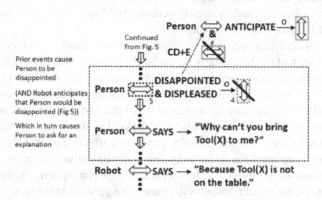

Fig. 7. Possible communication continued from Fig. 5.

As the UGALRS and CD+ representational framework articulated in [12] is a fully explainable framework, when in the problem-solving process, PS(REAC(ROBOT)) fails to return a solution, because the steps of processing everywhere in UGALRS using CD+ representations are explicit, the cause(s) of the PS failure is easily identified. Hence, the Robot would respond with "Because Tool(X) is not on the table". What causes Robot to respond is the illocutionary force present in Person asking the Why question (i.e., it is in Robot's CONC, where general knowledge is stored – Fig. 2 – that Person would be *Pleased* if her Why question is answered to, and would be *Displeased* if this is not so. This knowledge is also represented in CD+ form in CONC. These are the representations of the illocutionary forces involved.) Robot may feel further RELIEVED from being FRUSTRATED, DISPLEASED, and FEARful after providing this explanation, because providing an explanation may cause Person to be more Pleased.

5 Discussion and Conclusion

Even a small set of relatively simple utterances by an intelligent system involves many complex and intricate processes encompassing conceptual, motivational, and affective aspects. In this regard, we used UGALRS and CD+ to elucidate some of these processes. For the sake of clarity and simplicity, some processes have been omitted in these diagrams, but the texts have discussed some of them, particularly the communication rules underlying the generation of sentences. Future research could delve further into explicitly representing these rules. It is worth noting that CD+ is comprehensive enough to represent these rules and the situations under which they are triggered – i.e., the reasoning processes themselves are also representable using CD+, as has been amply illustrated in [12].

Psychologists have identified up to 161 types of motivations in humans [27]. For robots, the number of motivations could be simpler and smaller in number [19]–[21]. However, for an IAS or robot to understand humans and hence be able to interact with them effectively, it could benefit from having a model of the humans' motivations, as shown in Fig. 2 and discussed in this paper. To contain the discussion, in this paper we have only dealt with a small number of motivations; future work certainly calls for expansion in this direction, and as already demonstrated in [12], it is possible to do this within a UGALRS plus CD+ framework.

It should be noted that the topic of emotions would entail a broader selection of plausible emotions that could similarly be useful for characterizing and communicating about internal states in robots and humans than what has been discussed in this paper [10, 28–30]. For instance, while Ortony's cognitive appraisal theory of emotion [25], which is amenable to computational treatment, has been partially capitalized here, there is a fuller set of emotions that has not been addressed in this paper. In future, a more comprehensive selection of emotions and treatment of affective processes could enable intelligent systems to handle a wider range of communicative scenarios. Despite the fact that this paper only covers a subset of these vast conceptual, motivational, and affective spaces, its main contribution is to articulate a general framework of knowledge representation and processing to link these together and elucidate the respective functions they serve in the complex process underlying natural language communication.

Other important future work includes: 1. The transformations between the surface sentences illustrated in many places in this paper and their corresponding deep level "meaning" representations (Section IV(A)). This has also not been fully developed in Schank's original CD work [22–24]. The transformation must take into consideration grammar, tone, emotion, etc. 2. The roles, representations and causal consequences of various attitudes such as SERVILE, UNCOOPERATIVE, etc. (Section IV(B)). 3. Extension of the current framework and paradigm to cover a wider range of communication. 4. A computational implementation of the representation and processes involved. 5. The learning of the various representations illustrated in this paper.

Even though a computational implementation of CD+ is not reported in this paper, its viability as a computationally tractable representational scheme is reflected in the original work of Schank and colleagues [22–24], which has demonstrated that a computational implementation of CD could handle natural language question-answering and communication processes that benefit from the deep meaning representations of CD. It follows then that CD+, as an enhancement of CD, will share similar prospects for computational tractability.

As a final discussion point, the aspect of learning will be an important consideration. Given the complexity of the representations discussed in this paper, a system that lacks the ability to learn could not be scaled up nor become a practical system. While Ho [12] has discussed how learning could be done in the framework of CD+, we must first understand the kind of representations that are needed for intelligent processes, in this case, language communication processes, before we understand what it is that is to be learned. This paper hopes that by contributing to the elucidation of the intricate and complex conceptual, motivational, and affective processes involved in natural language communication between social agents, it could, when appropriately extended, hopefully bring about a fuller characterization of language communication in general.

References

1. Clark, A., Fox, C., Lappin, S. (eds.): The Handbook of Computational Linguistics and Natural Language Processing. Wiley-Blackwell, Hoboken (2012)
2. Mitkov, R. (ed.): The Oxford Handbook of Computational Linguistics. Oxford University Press, Oxford (2005)
3. Wang, Z., Hu, Z., Ho, S.-B., Cambria, E., Tan, A.-H.: MiMuSa-mimicking human language understanding for fine-grained multi-class sentiment analysis. Neural Comput. Appl. **35**, 15907–15921 (2023)
4. Brown, T.B., et al.: Language models are few-shots learners (2020). https://doi.org/10.48550/arXiv.2005.14165
5. Cambria, E., Hussain, A., Havasi, C., Eckl, C.: Sentic computing: exploitation of common sense for the development of emotion-sensitive systems. In: Esposito, A., Campbell, N., Vogel, C., Hussain, A., Nijholt, A. (eds.) Development of Multimodal Interfaces: Active Listening and Synchrony. Lecture Notes in Computer Science, vol. 5967, pp. 148–156. Springer, Heidelberg (2010). https://doi.org/10.1007/978-3-642-12397-9_12
6. Cambria, E., Liu, Q., Decherchi, S., Xing, F., Kwok, K.: SenticNet 7: a commonsense-based neurosymbolic AI framework for explainable sentiment analysis. In: Proceedings of LREC, pp. 3829–3839 (2022)
7. Mao, R., Liu, Q., He, K., Li, W., Cambria, E.: The biases of pre-trained language models: an empirical study on prompt-based sentiment analysis and emotion detection. IEEE Trans. Affect. Comput. (2023)
8. He, K., Mao, R., Gong, T., Li, C., Cambria, E.: Meta-based self-training and re-weighting for aspect-based sentiment analysis. IEEE Trans. Affect. Comput. (2023)
9. Kumar, A., Trueman, T., Cambria, E.: Gender-based multi-aspect sentiment detection using multilabel learning. Inf. Sci. **606**, 453–468 (2022)
10. Wang, Z., Ho, S.-B., Cambria, E.: A review of emotion sensing: categorization models and algorithms. Multimed Tools Appl **79**, 35553–35582 (2020)
11. Wang, Z., Ho, S.-B., Cambria, E.: Multi-level fine-scaled sentiment analysis with ambivalence handling. Int. J. Uncertain. Fuzziness Knowl.-Based Syst. **28**(4), 683–697 (2020)
12. Ho, S.-B.: A general framework for the representation of function and affordance: a cognitive, causal, and grounded approach, and a step toward AGI (2022). https://doi.org/10.48550/arXiv.2206.05273
13. Howard, N., Cambria, E.: Intention awareness: improving upon situation awareness in human-centric environments. Hum.-Cent. Comput. Inf. Sci. **3**(9) (2013)
14. Talmy, L.: Toward a Cognitive Semantics Volume I and II. The MIT Press, Cambridge (2000)
15. Evans, V., Green, M.: Cognitive Linguistics: An Introduction. Lawrence Erlbaum Associates, Mahwah (2006)
16. Langacker, R.W.: Cognitive Grammar: A Basic Introduction. Oxford University Press, Oxford (2008)
17. Geeraerts, D.: Theories of Lexical Semantics. Oxford University Press, Oxford (2010)
18. van Eijck, J., Unger, C.: Computational Semantics with Functional Programming. Cambridge University Press, Cambridge (2010)
19. Quek, B.-K.: A survivability framework for autonomous systems. Ph.D. thesis, National University of Singapore (2008)

20. Quek, K., Ibañez-Guzmán, J., Lim, K.-W.: Attaining operational survivability in an autonomous unmanned ground surveillance vehicle. In: 32nd Annual Conference on IEEE Industrial Electronics, pp. 3969–3974 (2006). https://doi.org/10.1109/IECON.2006.348001

21. Quek, K., Ibañez-Guzmán, J., Lim, K.-W.: A survivability framework for the development of autonomous unmanned systems. In: 9th International Conference on Control, Automation, Robotics and Vision, pp. 1–6 (2006). https://doi.org/10.1109/ICARCV.2006.345336

22. Schank, R.C.: Identification of conceptualizations underlying natural language. In: Schank, R.C., Colby, K.M. (eds.) Computer Models of Thought and Language, pp. 187–247. WH Freemann & Company, San Francisco (1973)

23. Schank, R.C.: Conceptual Information Processing. North-Holland Publishing Company, Amsterdam (1975)

24. Schank, R.C., Abelson, R.P.: Scripts, Plans, Goals, and Understanding: An Inquiry into Human Knowledge Structure. Lawrence Erlbaum Associates, Mahwah (1977)

25. Ortony, A., Clore, G.L., Collins, A.: The Cognitive Structure of Emotions. Cambridge University Press, Cambridge (1990)

26. Searle, J.R.: Speech Acts: An Essay in the Philosophy of Language. Cambridge University Press, Cambridge (1970)

27. Talevich, J., Read, S., Walsh, D., Iyer, R., Chopra, G.: Toward a comprehensive taxonomy of human motives. PLoS One 12(2) (2017)

28. Plutchik, R.: Emotions and Life: Perspectives from Psychology, Biology, and Evolution. American Psychological Association (2002)

29. Susanto, Y., Livingstone, A., Ng, B., Cambria, E.: The hourglass model revisited. IEEE Intell. Syst. 35(5), 96–102 (2020)

30. Amin, M., Cambria, E., Schullerl, B.: Will affective computing emerge from foundation models and general artificial intelligence? A first evaluation of ChatGPT. IEEE Intell. Syst. 38(2), 15–23 (2023)

GSA-UBS: A Novel Medical Hyperspectral Band Selection Based on Gravitational Search Algorithm

Chenglong Zhang[1], Xiaoli Yang[1], Aizhu Zhang[2], Dexin Yu[3], Nian Liu[4], and Xiaopeng Ma[1(✉)]

[1] School of Control Science and Engineering, Shandong University, Jinan 250061, China
xiaopeng.ma@sdu.edu.cn
[2] College of Oceanography and Space Informatics, China University of Petroleum (East China), Qingdao 266580, China
[3] Radiology Department, Qilu Hospital of Shandong University, Jinan 250000, China
[4] PET Center, Department of Nuclear Medicine, the First Affiliated Hospital, Zhejiang University School of Medicine, Hangzhou 310003, China

Abstract. Medical hyperspectral images (MHSIs) provide the possibility of non-invasive disease diagnosis. However, due to the sparsity of MHSIs data in high-dimensional space, the "curse of dimensionality" arises, which reduces the efficiency and accuracy of data processing. Therefore, spectral dimensionality reduction has become a necessary step for MHSIs data analysis and application. To preserve the inherent properties of spectral bands, an unsupervised band selection algorithm based on Gravitational Search Algorithm, called GSA-UBS, is proposed in this paper to search for the best subset of bands. Considering the amount of information and redundancy of the candidate bands, we define an evaluation criterion consisting of a band distance matrix and an information entropy vector. Additionally, we design a simple discrete search strategy that enables GSA to directly obtain the original serial number of the selected bands instead of weighting the bands 0–1. Extensive experiments are conducted on two publicly available *in vivo* brain cancer MHSIs datasets, and the results demonstrate that GSA-UBS outperforms several state-of-the-art methods.

Keywords: Gravitational Search Algorithm · Medical hyperspectral images · band selection

1 Introduction

Hyperspectral imaging (HSI) can capture spectral information containing dozens to hundreds of bands and has been widely applied in fields such as agriculture, environmental monitoring, and food safety [1]. Recently, the application of HSI in the medical field has also attracted attention [2]. Medical hyperspectral images (MHSIs) can provide more detailed information on the biological tissue and chemical composition, which can effectively assist medical diagnosis and treatment. However, the data dimension of

J. Ren et al. (Eds.): BICS 2023, LNAI 14374, pp. 31–40, 2024.
https://doi.org/10.1007/978-981-97-1417-9_3

MHSIs is considerably large, and the inevitable redundant information and bad bands (e.g. noisy bands) limit the efficiency and accuracy of target recognition. Therefore, spectral reduction is a crucial preprocessing step in the processing of MHSIs.

In the dimensionality reduction process of HSI, two main methods are commonly used: feature extraction and band selection [3]. Feature extraction involves mapping high-dimensional data to a low-dimensional space to achieve maximum feature separability. Classical methods include principal component analysis (PCA), independent component analysis (ICA), and minimum noise fraction (MNF) [4]. However, the low-dimensional data obtained by these methods may damage the original data structure and lack interpretable physical meaning. In contrast, band selection involves selecting representative bands to achieve redundancy reduction and noise reduction. This method produces a subset of the original data, thus preserving the spectral response characteristics within the selected bands.

Band selection methods can be categorized into supervised and unsupervised approaches depending on whether they utilize class labels or not. Supervised band selection methods involve training models using labeled samples to identify the bands that maximize class separability [5]. A common strategy is to apply a suitable search algorithm to assign a binary 0–1 weight to each band [6], to optimize classification accuracy. However, obtaining labeled samples in practical applications can be challenging, and the model's robustness is often affected by sparse samples. Therefore, this paper focuses on discussing unsupervised band selection methods that do not require labeled samples.

Unsupervised band selection methods can be broadly categorized into three main approaches: ranking, clustering, and searching. Ranking-based methods typically rely on expert knowledge to design a criterion for evaluating the importance of bands, and then rank the bands based on maximizing or minimizing the criterion. Examples of such methods include maximum variance principal component analysis (MVPCA), information entropy-based (IE) or mutual information (MI) [7] ranking. However, these ranking-based methods often overlook the correlations between bands, leading to redundant subsets of selected bands. Clustering-based strategies first group the bands and then select representative bands from each group to form a new subset. Typical methods include fast neighborhood grouping for band selection (FNGBS) [8], and optimal neighborhood reconstruction (ONR) [9]. However, individual noisy bands within a cluster can be easily selected, leading to poor performance. Search-based algorithms iteratively optimize to obtain the optimal subset of bands, such as the volume-gradient-based band selection (VGBS) [10]. In addition, deep learning-based methods have become popular, such as the recently proposed autoencoder-based unsupervised band selection framework (CAE-UBS) algorithm [11], which searches for the optimal subset of bands that can best reconstruct the original image. However, using the reconstruction error as the loss function is not reasonable because bad bands in the original image can guide the network to select low-information bands. Furthermore, the computationally intensive nature of the objective function can significantly reduce the efficiency of the algorithm, particularly during a large number of iterations.

To address the shortcomings of previous methods, this study proposes an unsupervised band selection algorithm based on the Gravitational Search Algorithm (GSA-USA). GSA is a nature-inspired optimization technique that features simplicity, fast convergence, and global optimization capabilities, making it a promising solution for a wide range of optimization problems [12]. To construct a lightweight objective function for GSA, we define a local distance matrix and an information entropy vector for each subset of bands, reflecting both the information content and redundancy of the subset. By standardizing the distance matrix using hyperbolic tangent transformation, we mitigate the interference from irrelevant bands. Additionally, we pre-calculate the global distance matrix and information entropy to avoid redundant computations during the iterative process. Furthermore, we adopt a simple discretization strategy, enabling GSA to directly obtain the optimal subset of bands and yield higher efficiency compared to the 0–1 weighted approach.

2 Related Works

In this section, we focus on the basic principles of the GSA optimization algorithm.

The Gravitational Search Algorithm (GSA) proposed by [12] is a metaheuristic optimization algorithm that is inspired by the law of gravity and motion in physics. In GSA, each candidate solution in the search space is considered a celestial body, and the search process is modeled as the interaction between these celestial bodies.

GSA is designed to solve optimization problems in D-dimensional space by setting S search particles. Each search particle i is represented by its position $X_i = [x_{i1}, x_{i2}, \ldots, x_{iD}](i = 1, 2, \ldots, S)$ in space and its velocity $V_i = [v_{i1}, v_{i2}, \ldots, v_{iD}]$, which is initially set to 0. By calculating the total gravitational force and acceleration of each particle, the individual positions and velocities can be updated to achieve optimization problem solving.

According to Newton's second law of universal gravitation, the resultant force acting on an object is equal to its mass multiplied by its acceleration. Therefore, the calculation of mass is crucial, and in GSA, it is represented by its fitness value as follows:

$$q_i^t = \frac{fit_i^t - worst^t}{best^t - worst^t},\tag{1}$$

$$M_i^t = \frac{q_i^t}{\sum_{j=1}^{S} q_j^t},\tag{2}$$

q_i^t and M_i^t are the fitness value and mass of particle X_i at the t-th iteration, respectively. $worst^t$ and $best^t$ represent the worst and best fitness values among all particles in the t-th iteration.

For the resultant force and acceleration of each particle, the calculation is as follows:

$$F_{id}^t = \sum_{j \in Kbest, j \neq i} rand_j G^t \frac{M_i^t M_j^t}{R_{ij}^t + \varepsilon} \left(x_{jd}^t - x_{id}^t\right),\tag{3}$$

$$a_{id}^t = \frac{F_{id}^t}{M_i^t} = \sum_{j \in Kbest, j \neq i} rand_j G^t \frac{M_j^t}{R_{ij}^t + \varepsilon} \left(x_{jd}^t - x_{id}^t\right),\tag{4}$$

the gravitational constant G is decreased with the increase of iteration number in order to enhance the global search ability of the GSA algorithm. The distance R_{ij}^t between particles is calculated using the Euclidean distance, and to avoid division by zero, a small constant ε is often added to the denominator. *Kbest* is a collection of the top K particles, where K decreases over time from the initial value of S to 1. This setting ensures that the algorithm can fully utilize the existing information during the search process and converge to the global optimal solution faster.

In fact, the resultant force acting on particle X_i is a random weighted sum of forces from the *Kbest* particles, where rand is a random number between 0 and 1. This random weighting method helps the algorithm to avoid getting trapped in local optimal solutions and improve the global search ability. Since *Kbest* is composed of particles with higher fitness values, this random weighting method also ensures that the particles can better utilize the existing information and converge to the global optimal solution faster.

Therefore, the update of particle velocity and position is as follows:

$$v_{id}^{t+1} = rand_i \times v_i^t + a_i^t, \tag{5}$$

$$x_{id}^{t+1} = x_{id}^t + v_{id}^{t+1}, \tag{6}$$

where $rand_i$ is a random value in the interval $[0, 1]$.

3 The Proposed GSA-UBS

The workflow of the proposed method is illustrated in Fig. 1, which primarily employs a discretized GSA algorithm to search for subsets of bands that are both highly informative and non-redundant. In this section, we provide a detailed introduction to the discretization process and objective function of GSA.

In this article, a raw hyperspectral image with L bands is represented as $E = [e_1, e_2, \ldots, e_B] \in \mathbb{R}^{L \times P}$, P is the number of pixels.

3.1 Discretization of GSA

Originally, GSA was designed to address continuous optimization problems. In this paper, we have employed a simple discrete strategy to enable GSA to effectively tackle the problem of band combination optimization.

To obtain a subset containing l bands, we aim for the particles in GSA to find an optimal solution in an l-dimensional space. In this work, the space boundary is set to $[0, 1]$. When calculating the fitness value, each particle $X_i = [x_{i1}, x_{i2}, \ldots, x_{il}]$ is discretely encoded as the serial number of the candidate bands:

$$Y_i = round(X_i(L-1)+1), \tag{7}$$

where, $round(\omega)$ means rounding ω to the nearest integer, so that the elements of Y_i are integers in the range of 1 to L. Once GSA converges, the subset of bands with the optimal performance, denoted as $E_{Y_{best}} \in \mathbb{R}^{l \times P}$, is output.

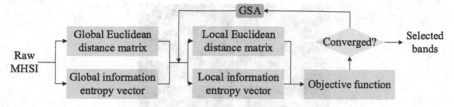

Fig. 1. The workflow of GSA-UBS

3.2 Objective Function

Information entropy has been established as a dependable metric for characterizing image information content, and has thus been extensively employed in band selection tasks [8]. As such, this paper employs information entropy as a measure for quantifying the information of bands. Calculated as follows:

$$IE_i = -\sum P_i(v) \log P_i(v), \tag{8}$$

where, $P_i(v)$ denotes the probability of the pixel value v occurring in the spectral band i.

Euclidean distance is a prevalent method for assessing the similarity between bands, The smaller the distance between bands, the higher the degree of aggregation between them. Therefore, to avoid redundancy, maximizing the distance between bands is used as an optimization objective. The Euclidean distance between two bands is as follows:

$$ED_{ij} = \left\| e_i - e_j \right\|_2. \tag{9}$$

Therefore, this paper proposes a criterion that integrates both information entropy and Euclidean distance to evaluate candidate bands. Specifically, the objective function can be follows:

$$arg \max \left(\frac{1}{\sqrt{l(l-1)}} \| ED_{local} \|_F + \frac{1}{l} \sum IE_{local} \right), \tag{10}$$

here, $ED_{local} \in \mathbb{R}^{l \times l}$ represents the Euclidean distance matrix between l ($l \leq L$) candidate bands, while $IE_{local} \in \mathbb{R}^{l \times 1}$ denotes the information entropy vector of these bands. Prior to GSA optimization, we pre-calculated the global distance matrix $ED \in \mathbb{R}^{L \times L}$ and information vector $IE \in \mathbb{R}^{L \times 1}$. This reduced the need for repeated calculations during the iterative process by indexing local data. To ensure that the distance and information entropy are on the same scale, we need to normalize ED and IE. The normalization formula for IE is:

$$IE = (IE - IE_{\min}) \cdot / (IE_{\max} - IE_{\min}), \tag{11}$$

where $IE_{\max}$ and $IE_{\min}$ are the maximum and minimum values in IE, respectively. As bad bands generally exhibit larger distances from other bands, to avoid selecting them, this paper normalizes ED using the hyperbolic tangent function to weaken the distances between bad bands and others. The normalization formula for ED is as follows:

$$ED = \tanh(ED) = \left(e^{ED} - e^{-ED} \right) / \left(e^{ED} + e^{-ED} \right). \tag{12}$$

36   C. Zhang et al.

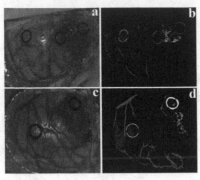

Fig. 2. Color images of *in vivo* human brain tissue from three patients, together with the corresponding groundtruth images. (a, b) Image from patient 8-01. (c, d) Image from patient 12-01.

By dividing $\|ED_{local}\|_F$ and $\sum IE_{local}$ by their theoretical maximum values $\sqrt{l(l-1)}$ and l respectively, the weights of distance and information entropy are made equal.

4 Experiment

In this section, we conduct experiments on two publicly available *in-vivo* hyperspectral human brain images to evaluate the performance of the proposed method.

4.1 Datasets

The experiment utilized MHSIs, which were collected from the European project "HypErspectraL Imaging Cancer Detection" (HELICoiD) and captured at the University of Las Palmas de Gran Canaria in Las Palmas, Spain [13]. The MHSIs were produced utilizing the Hyperspec VNIR A series camera, which operates within a spectral range of 400–1000 nm and captures data at an interval of 0.73 nm. The resulting MHSIs contained 826 spectral bands with a spatial resolution of 128.7 μm. Two MHSIs, denoted as 8-01 and 12-01, were collected from patients with grade IV glioblastoma (GBM) tumors, and the pixels in the MHSIs were labeled as *normal tissue, tumor tissue, blood vessels*, or *background* (Fig. 2 and Table 1), forming the ground truth with the support of experience and knowledge from neurosurgeons. Each band of the image was normalized to eliminate the influence of the magnitude difference between bands.

Table 1. Numbers of total pixels and labeled pixels in two MHSIs datasets

MHSIs codes	Total pixels	Number of labeled pixels				
		Normal tissue	Tumor tissue	Blood vessel	Back-ground	Total number
8-01	460 × 594	2295	1221	1331	630	5477
12-01	443 × 479	4516	855	8697	1685	15753

4.2 Compared Methods

In this paper, we compared five state-of-the-art unsupervised band selection algorithms:

TOF [14]: which combines clustering algorithms and objective function optimization to select the most discriminative and informative bands from hyperspectral images.

ONR [9]: which utilizes neighborhood information to reconstruct the original data and selects the most discriminative bands based on reconstruction errors.

FNGBS [8]: which adopts a coarse-to-fine strategy to group bands and selects bands within each group based on two criteria: information entropy and local density.

ADBH [15]: which constructs an adaptive band hierarchy clustering framework and inserts adaptive distances into the hierarchy to avoid interference from noisy bands.

HCBC [16]: which constructs a new hyperbolic clustering-based band hierarchy structure, aimed at better representing potential spectral structures and achieving more consistent band selection.

4.3 Experimental Setup

We assessed the performance of the band selection algorithms using two classifiers: support vector machine (SVM) [17] and k-nearest neighbors (KNN). The classifier parameters were determined using 5-fold cross-validation. SVM utilized the radial basis function (RBF) kernel, with penalty coefficient and gamma parameter set to 1000 and 8. KNN used 3 nearest neighbors for classification. We measured the classification accuracy for selecting 10 to 60 bands (in increments of 5), which included overall accuracy (OA) and kappa coefficient (KAPPA) [18]. For two MHSIs, we selected 5 labeled pixels for training and used the remaining pixels for testing. We conducted 20 independent experiments and recorded the average results to minimize random errors.

4.4 Experimental Results

In this section, we report our experimental results and compare them with five other contrastive algorithms based on the classification performance of two classifiers. In addition, we also consider the classification accuracy on the raw image. Table 2 presents the OA and KAPPA evaluations of different algorithms for two MHSIs at the optimal number of bands, where the best results are highlighted in bold. Figure 3 and 4 display the accuracy curves of different algorithms for different numbers of selected bands.

It is evident from Table 2 that the proposed GSA-UBS algorithm exhibits higher OA values than other algorithms for both images, with an advantage of over 5%. Particularly,

38   C. Zhang et al.

Table 2. The accuracy and runtime of different band selection methods with different classifiers on two MHSIs.

MHSIs	Classifiers	Metrics	Raw	TOF	ONR	FNGBS	ADBH	HCBC	GSA-UBS
8-01	SVM	OA (%)	83.79 ± 2.16	88.73 ± 4.64	81.77 ± 5.50	93.44 ± 4.73	66.78 ± 4.32	75.05 ± 4.26	**96.75 ± 1.75**
		KAPPA (×100)	77.17 ± 2.94	83.95 ± 6.62	74.35 ± 7.50	90.69 ± 6.67	53.62 ± 5.56	64.80 ± 5.88	**95.37 ± 2.49**
	KNN	OA (%)	78.86 ± 3.57	77.40 ± .45	73.33 ± 3.87	82.69 ± 6.05	60.47 ± 4.61	65.15 ± 5.18	**88.94 ± 3.87**
		KAPPA (×100)	70.25 ± 4.74	67.92 ± 4.82	62.41 ± 5.42	75.70 ± 8.09	44.63 ± 5.49	50.83 ± 6.50	**84.36 ± 5.40**
		Runtime (s)	–	14.702	18.925	128.279	21.139	17.72	**2.479**
12-01	SVM	OA (%)	78.17 ± 8.14	85.91 ± 6.06	75.08 ± 8.26	69.25 ± 6.23	68.43 ± 5.50	69.99 ± 7.08	**92.06 ± 5.12**
		KAPPA (×100)	66.92 ± 10.9	78.73 ± 8.20	63.83 ± 11.0	55.43 ± 8.12	55.06 ± 6.63	56.99 ± 8.73	**87.65 ± 7.18**
	KNN	OA (%)	60.20 ± 4.90	67.90 ± 5.50	58.00 ± 5.14	54.41 ± 4.83	53.61 ± 3.57	55.56 ± 5.54	**72.03 ± 5.90**
		KAPPA (×100)	46.27 ± 6.22	56.08 ± 6.73	43.45 ± 6.26	38.66 ± 5.62	38.53 ± 3.34	40.49 ± 5.58	**60.91 ± 7.41**
		Runtime (s)	–	15.112	18.4	110.718	18.113	15.272	**2.296**

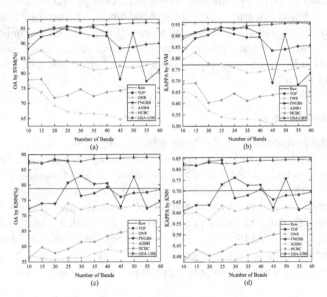

Fig. 3. OA and KAPPA values on the 8-01 for different band selection methods.

when using KNN for 8-01, the GSA-UBS algorithm demonstrates an advantage of over 10% in both OA and KAPPA measures. It is noteworthy that the GSA-UBS algorithm has the lowest precision standard deviation when using the SVM classifier, indicating its higher stability. Figure 3 and 4 show that the GSA-UBS algorithm outperforms other methods in terms of accuracy curves for most numbers of selected bands. Although there are a few cases where its accuracy is lower than TOF, GSA-UBS consistently generates the highest point. Moreover, TOF and FNGBS perform similarly to GSA-UBS when selecting a small number of bands. However, as the number of bands increases, their performance tends to decline and even fall below the classification accuracy of the raw image. These findings demonstrate the superiority and stability of the proposed GSA-UBS algorithm in hyperspectral image classification.

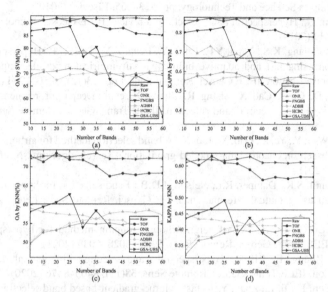

Fig. 4. OA and KAPPA values on the 12-01 for different band selection methods.

Furthermore, due to the Euclidean distance matrix and the information entropy vector are pre-calculated, the iterative process of GSA-UBS is not time-consuming. This is confirmed by the runtime shown in Table 2, which demonstrate the best efficiency of GSA-UBS.

5 Conclusion

In this work, we proposed a novel unsupervised band selection framework called GSA-UBS, which is based on the gravitational search algorithm (GSA) and aims to solve the band combination optimization problem in medical hyperspectral imaging (MHSI). To directly encode and thus obtain band serial number, we adopted a simple yet effective discretization strategy. Considering the redundancy and information content among candidate bands, we constructed an objective function consisting of a local Euclidean distance matrix and an information entropy vector. The use of hyperbolic tangent-normalized Euclidean distance helped to avoid selecting poor bands. The pre-computed Euclidean distance matrix and information entropy vector also avoided repeated calculations during the iterative process, thus improving the algorithm's efficiency. Comparative experiments with five state-of-the-art band selection algorithms demonstrated that our proposed GSA-UBS algorithm not only effectively addresses the spectral dimensionality reduction problem in MHSIs, and has potential for clinical applications.

References

1. Zhang, A., et al.: Bayesian gravitation-based classification for hyperspectral images. IEEE Trans. Geosci. Remote Sens. **60**, 1–14 (2022)
2. Fei, B.: Chapter 3.6 - hyperspectral imaging in medical applications. In: Amigo, J.M. (ed.) Data Handling in Science and Technology, pp. 523–565. Elsevier (2019)
3. Sun, W., Du, Q.: Hyperspectral band selection: a review. IEEE Geosci. Remote Sens. Mag. **7**(2), 118–139 (2019)
4. Yang, M.D., Huang, K.S., Yang, Y.F., Lu, L.Y., Feng, Z.Y., Tsai, H.P.: Hyperspectral image classification using fast and adaptive bidimensional empirical mode decomposition with minimum noise fraction. IEEE Geosci. Remote Sens. Lett. **13**(12), 1950–1954 (2016)
5. Feng, J., Li, D., Gu, J., Cao, X., Shang, R., Zhang, X., et al.: Deep reinforcement learning for semisupervised hyperspectral band selection. IEEE Trans. Geosci. Remote Sens. **60**, 1–19 (2022)
6. Wang, M., Wan, Y., Ye, Z., Gao, X., Lai, X.: A band selection method for airborne hyperspectral image based on chaotic binary coded gravitational search algorithm. Neurocomputing **273**, 57–67 (2018)
7. Guo, B., Gunn, S.R., Damper, R.I., Nelson, J.D.B.: Band selection for hyperspectral image classification using mutual information. IEEE Geosci. Remote Sens. Lett. **3**(4), 522–526 (2006)
8. Wang, Q., Li, Q., Li, X.: A fast neighborhood grouping method for hyperspectral band selection. IEEE Trans. Geosci. Remote Sens. **59**(6), 5028–5039 (2021)
9. Wang, Q., Zhang, F., Li, X.: Hyperspectral band selection via optimal neighborhood reconstruction. IEEE Trans. Geosci. Remote Sens. **58**(12), 8465–8476 (2020)
10. Geng, X., Sun, K., Ji, L., Zhao, Y.: A fast volume-gradient-based band selection method for hyperspectral image. IEEE Trans. Geosci. Remote Sens. **52**(11), 7111–7119 (2014)
11. Sun, H., Ren, J., Zhao, H., Yuen, P., Tschannerl, J.: Novel gumbel-softmax trick enabled concrete autoencoder with entropy constraints for unsupervised hyperspectral band selection. IEEE Trans. Geosci. Remote Sens. **60**, 1–13 (2022)
12. Rashedi, E., Nezamabadi-pour, H., Saryazdi, S.: GSA: a gravitational search algorithm. Inf. Sci. **179**(13), 2232–2248 (2009)
13. Fabelo, H., Ortega, S., Szolna, A., Bulters, D., Piñeiro, J.F., Kabwama, S., et al.: In-vivo hyperspectral human brain image database for brain cancer detection. IEEE Access **7**, 39098–39116 (2019)
14. Wang, Q., Zhang, F., Li, X.: Optimal clustering framework for hyperspectral band selection. IEEE Trans. Geosci. Remote Sens. **56**(10), 5910–5922 (2018)
15. Sun, H., Ren, J., Zhao, H., Sun, G., Liao, W., Fang, Z., et al.: Adaptive distance-based band hierarchy (ADBH) for effective hyperspectral band selection. IEEE Trans. Cybern. **52**(1), 215–227 (2022)
16. Sun, H., Zhang, L., Ren, J., Huang, H.: Novel hyperbolic clustering-based band hierarchy (HCBH) for effective unsupervised band selection of hyperspectral images. Pattern Recogn. **130**, 108788 (2022)
17. Ma, P., et al.: Multiscale superpixelwise prophet model for noise-robust feature extraction in hyperspectral images. IEEE Trans. Geosci. Remote Sens. **61**, 1–12 (2023)
18. Li, Y., et al.: CBANet: an end-to-end cross band 2-D attention network for hyperspectral change detection in remote sensing. IEEE Trans. Geosci. Remote Sens. **61** (2023)

Enhancing Generalizability of Deep Learning Polyp Segmentation Using Online Spatial Interpolation and Hue Transformation

Mahmood Haithami[1]([✉]) [iD], Amr Ahmed[2] [iD], and Iman Yi Liao[1] [iD]

[1] University of Nottingham Malaysia, Semenyih, Malaysia
{hcxmh1,iman.liao}@nottingham.edu.my
[2] Edge Hill University, Ormskirk, UK
ahmeda@edgehill.ac.uk

Abstract. Polyps, which are precursors to colon cancer, can be detected early to reduce mortality rates. However, the limited availability of public datasets and the variability of polyp shapes, textures, and colors restrict the generalizability of existing deep learning models. To overcome this challenge, researchers often employ data augmentation techniques or generative models to increase the number of training samples, regardless of the downstream learning task (i.e., polyp segmentation). In this study, we propose a deep learning framework that combines an image transformation layer with a segmentation model, where the transformed images serve as input for the segmentation model. The image transformation layer comprises a random hue shifting function and an autoencoder. The autoencoder removes textures while preserving key polyp features, transforming the input images. To control the intensity of the transformation, we employ a simple interpolation between the original and transformed images. During training, the image transformation layer generates multiple levels of texture and color variations for each input image in every epoch, effectively regularizing the segmentation model. By exposing the segmentation model to different texture and color levels within the same training batch, we encourage the model to update its weights based on intrinsic features present in both the original images and their corresponding transformed versions. This approach enhances the generalizability of deep learning models on unseen test sets. Experimental results using various configurations consistently demonstrate significant improvements in polyp Intersection over Union (IoU) ranging from 1.8% to 16.4% across different test sets.

Keywords: Deep learning · Polyp Segmentation · Spatial interpolation · Generalizability

© The Author(s), under exclusive license to Springer Nature Singapore Pte Ltd. 2024
J. Ren et al. (Eds.): BICS 2023, LNAI 14374, pp. 41–50, 2024.
https://doi.org/10.1007/978-981-97-1417-9_4

1   Introduction

Colorectal cancer ranks as the third leading cause of cancer-related deaths worldwide with reported mortality rate of nearly 51% [1]. If polyps are overlooked, they develop from adenomatous polyps (i.e., cancerous polyps), which initially begin as benign growths but can potentially progress into malignancy over time. Polyp identification and removal often pose significant challenges. This is primarily due to the intricate and complex anatomical structure of the colon and rectum, making such screening require a high level of expertise. To address these challenges, computer-assisted systems have emerged as a promising solution. These systems aim to reduce operator subjectivity and enhance adenoma detection rates.

The computer vision community has shown significant interest in deep learning models due to their ability to surpass hand-crafted classifiers commonly used in conventional machine learning [2]. However, deep learning models come with their own set of challenges. These challenges include the need for substantial training datasets, addressing the issue of overfitting, and effectively tuning hyperparameters to achieve optimal performance. Furthermore, a notable decline in performance is observed when these models are evaluated on unseen test sets. Poor generalizability renders the proposed models unsuitable for real-world clinical practice. Several types of deep learning segmentation models are found in the literature. However, they mainly focus on modifying the architecture to achieve invariant properties which enable them to achieve better segmentation results. For instance, encoder-decoder base Unet models proposed to capture multi-level contextual information [9,15,16], meanwhile, Pyramid-based architectures were proposed to enable deep learning models to acquire multi-resolution invariant property [4,11]. Sequence-bsed models such as GRU were used as well to enrich the feature maps by capturing spatial contextual information [5,6]. Moreover, recent architecture based on self-attention and Transformers were utilized in polyp segmentation as in [2,3,8].

All these discussed methods have one common subtle assumption, that is, the training images are adequate to have representative features that need to be extracted. This assumption is inherited from the nature of deep learning models which are known to be data hungry. After all, deep learning is a function that learns patterns found in the input training data. To mitigate this problem image augmentation and generative adversarial models (GANs) are used. However, such augmentation methods (i.e., rotation, flipping, ...etc.) would not significantly enhance the performance on unseen data due to some invariants properties embedded in deep learning models [13]. On the other hand, GANs based models is not a straightforward process and in some cases, it needs manual intervention [18,19]. Intuitively, adding realistic-looking polyp images to training dataset may enhance the downstream task (i.e., polyp segmentation). However, it is noticed that such approach does not enhance the performance of segmentation models except when the training data is extremely small [18]. The reason is that GANs based models sample its data from the training distribution which yield images that have similar features as the original training images.

This paper presents an alternative approach to improve generalizability without explicitly increasing the dataset size. Instead of augmenting or adding new images to the dataset, we expose the segmentation model to out-of-domain images during training epochs. The method involves gradually transforming the original images by applying spatial interpolation between an input image with fine-grained texture and a corresponding version with reduced texture details. The main objective is to expose the segmentation model to different texture and color levels, enabling it to develop invariance to these factors. By training on a range of texture and color variations, the segmentation model becomes more robust and adaptable to different conditions.

2   Methodology

In this section, we briefly introduce the dataset used in this research, the hyper-parameters of the proposed model, and the evaluation metrics. Following that, the proposed model along with the intuition for such design is discussed.

2.1   Dataset

The experiments in this study utilized four publicly available polyp datasets, which are summarized in Table 1. It is worth noting that the resolutions of the images in these datasets were not standardized. Therefore, prior to being fed into the deep learning models, all images were resized to a uniform size of 216-by-288.

2.2   Proposed Framework

Polyps do not exhibit specific shapes, colors, or sizes, and the elasticity of the colon lining can sometimes obscure or create tissue that resembles polyps. These factors significantly impact the precision and recall of deep learning models used for polyp segmentation. Moreover, when these models are tested on unseen polyp images from different medical centers, their generalizability performance tends to worsen. This raises concerns about their suitability for clinical usage. Due to the scarcity of large polyp datasets, polyp segmentation remains a challenging problem.

To address this issue, researchers often resort to techniques such as data augmentation or generative adversarial networks (GANs) to artificially increase

Table 1. The used datasets in the experiment section

Dataset name	Number of images	Resolution
CVC-ClinicDB [1]	612	384×288
Kvasir-Seg [10]	1000	332×487 to 1920×1072
ETIS-Larib [17]	196	1224×966
CVC-EndoSceneStill [20]	300	384×288 to 574×500

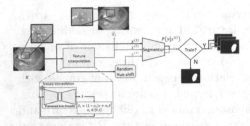

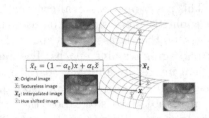

Fig. 1. Overview of the proposed deep learning framework. During the inference the model considers only the mask generated by the original image.

Fig. 2. Linear interpolation between original and textureless training manifold.

the size of available datasets. However, these approaches have not been proven to enhance generalizability significantly. In contrast, this paper proposes a novel approach that involves on-the-fly image-to-image transformation during training, as illustrated in Fig. 1.

In each training epoch, the proposed framework applies a texture interpolation transformation using the input images and their corresponding textureless versions. The textureless images are generated by a pre-trained autoencoder, which was trained separately to reconstruct the original images. The autoencoder's latent space has a lower number of dimensions compared to the encoding space, allowing it to capture essential features while omitting texture details. Additionally, the autoencoder was trained to generate corresponding masks to prevent it from becoming an identity function. The texture interpolation is defined as follows:

$$\bar{x}_t = (1 - \alpha_t)\, x + \alpha_t \check{x}, \quad \alpha_t \in [0,1] \tag{1}$$

where x, $\bar{x}_t$, and $\check{x}$ are original image, interpolated image, and textureless image, respectively. α_t is a scalar that controls the interpolation rate at training epoch t:

$$\alpha_t = 0.5 \cdot \left(1 + cos\left(\frac{cycles \cdot 2\pi \cdot t}{T}\right)\right) \tag{2}$$

where t and T are the epoch number and the total number of training epochs, respectively. The variable *cycles* is an integer number that controls the periodicity. As a result, the interpolation factor α_t will oscillate between its maximum and minimum values multiple times before reaching the final training epoch.

In each epoch, the segmentation model receives an input image with a different level of texture details compared to the previous epoch, see Fig. 2 and Fig. 3. Additionally, to enhance the model's robustness to color changes, random hue transformations are applied. The underlying idea behind this design is to enable the segmentation model to learn from both the original training manifold and a corresponding, yet distinct, training manifold, as seen in Fig. 2. This approach helps mitigate overfitting issues that may arise due to limited training samples. By including three different images within the same batch, the learnable weights of the model are updated to capture robust features, resulting in improved results, as demonstrated in the experiments section. Mathematically, we can interpret the training process as performing stochastic gradient descent on the following expectation:

$$- E_{P_{data(x)}} \left[log P_\theta \left(y \,|x \right) + E_{C(\hat{x}|x)} \left(log P_\theta \left(y \,|\hat{x} \right) \right) + E_{T_\emptyset(\bar{x}|x, \; \hat{x})} \left(log P_\theta \left(y \,|\bar{x} \right) \right) \right]$$
(3)

where $\log P_\theta \left(\cdot \,|\cdot \right)$ is the log likelihood function that should be optimized by a segmentation model. $x, \hat{x},$ and $\bar{x}$ are an original image, corresponding hue shifted version, and corresponding transformed version by $T_\emptyset$ "texture interpolation layer", respectively. $C \left(\hat{x} \,|x \right)$ is a hue-shift function that randomly shift the hue component of the input image. $P_{data(x)}$ is the training distribution.

During training, the segmentation model produces three masks for each input image. These masks are then compared to the corresponding true mask, and the loss is calculated using binary cross-entropy. However, during inference or testing, only the mask that corresponds to the original input image is used for comparison with the true mask. This helps the model learn robust features across different variations. However, during inference, when the model is applied to unseen images, only the mask generated for the original image is utilized for evaluation and comparison with the ground truth mask.

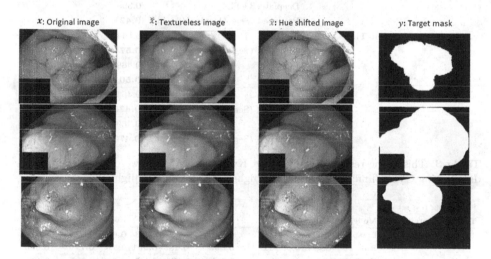

Fig. 3. Some examples for input images and their corresponding transformations. The segmentation model receives as an input $\mathbf{x}, \hat{\mathbf{x}},$ **and** $\bar{\mathbf{x}}$ as a training batch to produce a mask for each input image during training epochs.

2.3   Hyperparameters and Metrics

The number of training epochs is 300 and the learning rate lr = 0.01 is fixed for all experiments. The optimizer used is Adam [12] and the batch size is 14 images of size 216 by 288. Pytorch framework was used with NVIDIA Tesla T4. Finally, we adopt Binary Cross Entropy loss between the target and the input probabilities. In this study, we utilized widely adopted computer vision metrics to evaluate the performance of the proposed framework. These metrics include Accuracy, Precision, Recall, Intersection over Union (IoU), and mean IoU between background and polyp mask (mIoU).

3   Experiments and Results

To evaluate the performance of the proposed model, extensive experiments were conducted to compare it to several state-of-the-art models under various settings, as summarized in Table 2, Table 3, Table 4, and Table 5, respectively. The

Table 2. Validation results on the CVC-ClinicDB dataset. The training and validation split is 70% and 30%, respectively. The highest results are highlighted.

Validation set (30% of CVC-ClinicDB dataset)		
Transfer Learning	Model	Polyp IoU
TL with Coco dataset	Lraspp +TL	**0.569**
	Proposed (Seg: Lraspp+TL)	0.557
	FCN+TL	0.460
	Proposed (Seg: FCN+TL)	**0.484**
	DeeplabV3+TL	0.396
	Proposed (Seg: DeeplabV3+TL)	**0.421**
Without TL	Lraspp	0.544
	Proposed (Seg: Lraspp)	**0.578**
	FCN	0.489
	Proposed (Seg: FCN)	**0.501**
	DeeplabV3	0.451
	Proposed (Seg: DeeplabV3)	**0.470**
	Unet	0.460
	Proposed (Seg: Unet)	**0.470**

Table 3. This is the results of the test set Kvasir-Seg. Models were trained and validated using CVC-ClinicDB dataset. The highest results are highlighted.

Kvasir-Seg test set							
Transfer Learning	Model	Accuracy	IoU	Dice	Recall	Precision	mIoU
TL with Coco dataset	Lraspp +TL	0.859	0.426	0.553	**0.632**	0.607	0.635
	Proposed (Seg: Lraspp+TL)	**0.895**	**0.483**	**0.591**	0.557	**0.779**	**0.685**
	FCN+TL	0.729	0.298	0.417	**0.683**	0.393	0.497
	Proposed (Seg: FCN+TL)	**0.873**	**0.396**	**0.516**	0.519	**0.646**	**0.628**
	DeeplabV3+TL	0.820	0.192	0.280	0.291	0.375	0.501
	Proposed (Seg: DeeplabV3+TL)	**0.861**	**0.350**	**0.472**	**0.478**	**0.592**	**0.600**
Without TL	Lraspp	0.865	0.368	0.479	0.486	0.610	0.611
	Proposed (Seg: Lraspp)	**0.898**	**0.510**	**0.622**	**0.596**	**0.779**	**0.700**
	FCN	0.720	0.276	0.397	**0.635**	0.378	0.482
	Proposed (Seg: FCN)	**0.873**	**0.400**	**0.526**	0.530	**0.645**	**0.630**
	DeeplabV3	0.805	0.211	0.309	0.343	0.389	0.502
	Proposed (Seg: DeeplabV3)	**0.867**	**0.375**	**0.500**	**0.507**	**0.625**	**0.615**
	Unet	0.678	0.283	0.409	**0.767**	0.354	0.458
	Proposed (Seg: Unet)	**0.780**	**0.307**	**0.446**	0.674	**0.400**	**0.529**

objective of these experiments was to demonstrate that irrespective of the chosen model and initial weights, the proposed framework consistently outperforms the state-of-the-art models in terms of generalizability.

The deep learning models were trained and validated using the CVC-ClinicDB dataset. Subsequently, they were tested on three different datasets: Kvasir-Seg, CVC-EndoSceneStill, and ETIS-LaribPolypDB. Table 1 provides detailed information about these datasets. By evaluating the proposed model against the state-of-the-art models across these datasets, the experiments aimed to highlight its superior generalizability and performance. The best results

Table 4. This is the results of the test set CVC_EndoSceneStill. Models were trained and validated using CVC-ClinicDB dataset. The highest results are highlighted.

CVC_EndoSceneStill test set							
Transfer Learning	Model	Accuracy	IoU	Dice	Recall	Precision	mIoU
TL with Coco dataset	Lraspp +TL	0.965	0.647	0.734	**0.799**	0.746	0.805
	Proposed (Seg: Lraspp+TL)	**0.974**	**0.680**	**0.753**	0.778	**0.779**	**0.826**
	FCN+TL	0.939	0.581	0.663	**0.768**	0.664	0.758
	Proposed (Seg: FCN+TL)	**0.963**	**0.630**	**0.707**	0.746	**0.732**	**0.795**
	DeeplabV3+TL	0.948	0.549	0.625	0.702	0.620	0.747
	Proposed (Seg: DeeplabV3+TL)	**0.951**	**0.572**	**0.660**	**0.760**	**0.632**	**0.760**
Without TL	Lraspp	0.970	0.663	0.744	0.779	0.763	0.815
	Proposed (Seg: Lraspp)	**0.976**	**0.712**	**0.787**	**0.828**	**0.789**	**0.843**
	FCN	0.916	0.537	0.624	0.729	0.637	0.724
	Proposed (Seg: FCN)	**0.966**	**0.646**	**0.723**	**0.749**	**0.750**	**0.805**
	DeeplabV3	0.938	0.551	0.625	0.704	0.633	0.743
	Proposed (Seg: DeeplabV3)	**0.951**	**0.604**	**0.691**	**0.760**	**0.696**	**0.776**
	Unet	0.924	0.483	0.572	0.646	0.630	0.701
	Proposed (Seg: Unet)	**0.946**	**0.542**	**0.643**	**0.710**	**0.665**	**0.743**

Table 5. This is the results of the test set ETIS_LaribPolypDB. Models were trained and validated using CVC-ClinicDB dataset. The highest results are highlighted.

ETIS_LaribPolypDB test set							
Transfer Learning	Model	Accuracy	IoU	Dice	Recall	Precision	mIoU
TL with Coco dataset	Lraspp +TL	0.826	0.130	0.196	**0.534**	0.168	0.476
	Proposed (Seg: Lraspp+TL)	**0.955**	**0.169**	**0.227**	0.218	**0.345**	**0.561**
	FCN+TL	0.652	0.085	0.133	**0.473**	0.110	0.364
	Proposed (Seg: FCN+TL)	**0.931**	**0.130**	**0.182**	0.243	**0.207**	**0.530**
	DeeplabV3+TL	0.922	0.074	0.113	0.129	0.144	0.498
	Proposed (Seg: DeeplabV3+TL)	**0.955**	**0.116**	**0.162**	**0.158**	**0.246**	**0.535**
Without TL	Lraspp	0.941	0.136	0.188	0.220	0.215	0.538
	Proposed (Seg: Lraspp)	**0.956**	**0.234**	**0.305**	**0.329**	**0.360**	**0.594**
	FCN	0.540	0.082	0.131	**0.667**	0.098	0.304
	Proposed (Seg: FCN)	**0.935**	**0.115**	**0.164**	0.207	**0.203**	**0.524**
	DeeplabV3	0.880	0.069	0.104	0.176	0.110	0.474
	Proposed (Seg: DeeplabV3)	**0.918**	**0.143**	**0.197**	**0.263**	**0.223**	**0.529**
	Unet	0.408	0.075	0.127	**0.861**	0.090	0.231
	Proposed (Seg: Unet)	**0.698**	**0.094**	**0.154**	0.570	**0.118**	**0.392**

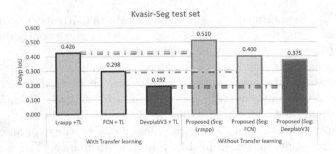

Fig. 4. The state-of-the-art segmentation model with transfer learning against the proposed framework without transfer learning. The used unseen test set is Kvasir-Seg meanwhile the training and validation is conducted using CVC-ClinicDB.

achieved on the validation set as well as on the test sets are highlighted in Table 2, Table 3, Table 4, and Table 5, respectively. Compared to other models, the proposed framework consistently outperformed in terms of Accuracy, Intersection over Union (IoU), Dice, Recall, Precision, and mean IoU (mIoU, the average of polyp and background IoU).

However, there were some exceptions observed in the case of Recall when using the Fully Convolutional Network (FCN) model [14]. FCN models exhibited a bias towards polyps, leading to higher Recall scores compared to the proposed framework. For instance, in the Kvasir-Seg test set (Table 3), FCN achieved a Recall of 0.683, while the proposed framework with the same FCN segmentation model achieved a Recall of 0.519. On the other hand, FCN had a lower Precision score of 0.393, indicating a higher rate of false positive errors. In contrast, the proposed framework with the same FCN segmentation model achieved a Precision score of 0.646, demonstrating a 24% improvement over FCN.

The biases of the FCN model towards polyp pixels in the Kvasir-Seg dataset were also evident in the IoU scores. The proposed model consistently outperformed FCN by approximately 10% to 13% in IoU, despite using the same FCN segmentation model within the proposed framework.

An interesting observation is that the Lite Reduced Atrous Spatial Pyramid Pooling (Lraspp) model [7] consistently outperformed other state-of-the-art models in general. The Lraspp model utilizes a segmentation decoder that is combined with a MobileNetsV3 encoder. MobileNets models are specifically designed to be efficient for deployment on mobile CPUs. One advantage of this design is that the Lraspp model is light compared to other models, which makes it less prune to overfitting. It allows the model to generalize well and achieve higher performance across various evaluation metrics. Despite the already high performance of the Lraspp model, the proposed framework managed to further improve its performance across various metrics in all test sets. This result clearly demonstrates that the proposed framework is effective in enhancing the performance of both large and lightweight architectures. Regardless of the model's size or complexity, the proposed framework was able to boost its performance, highlighting its versatility and effectiveness in improving segmentation models.

Transfer learning (TL) using the COCO dataset was applied to each segmentation model in the experiments. It was observed that the results of segmentation models we tested have improved when TL was employed. However, the proposed framework, even without transfer learning, outperformed the state-of-the-art models, particularly on the Kvasir-Seg test set, as illustrated in Fig. 4. These results highlight the effectiveness of the proposed framework in enhancing the generalizability performance of all tested segmentation models. Despite not relying on transfer learning, the proposed framework demonstrated superior performance, demonstrating its ability to enhance the segmentation models' robustness and adaptability.

4   Conclusions

This paper introduced a novel framework aiming at improving the generalizability of segmentation models. The limited availability of datasets poses a challenge to deep learning-based segmentation models. Whilst traditional solutions involve data augmentation or generative models to increase the training datasets, those approaches have not shown significant effectiveness in the existing literature. In contrast, our proposed approach suggests employing image-to-image transformations in conjunction with a segmentation model to enhance its performance. The transformations are texture-based interpolation and random hue shifting to variate texture and color of input images, respectively. Experimental results have shown that the proposed framework consistently enhances segmentation results with improvements in IoU ranging from approximately 1.8% to 16.4% across three different test sets when considering only the polyp mask. On the other hand, when considering the background mask in addition to the polyp mask, the improvements in mIoU were even more significant. The mIoU improvements ranged from approximately 1.2% to 21.9% across the unseen test sets. These results indicate that the proposed framework not only enhances the segmentation accuracy of polyps but also improves the overall segmentation performance, including the accurate delineation of the background region. The reported enhancements highlight the effectiveness of the proposed framework in enhancing the generalizability of segmentation models, irrespective of their internal architectural design.

References

1. Bernal, J., Sánchez, F.J., Fernández-Esparrach, G., Gil, D., Rodríguez, C., Vilariño, F.: WM-DOVA maps for accurate polyp highlighting in colonoscopy: validation vs. saliency maps from physicians. Comput. Med. Imaging Graph. **43**, 99–111 (2015)
2. Duc, N.T., Oanh, N.T., Thuy, N.T., Triet, T.M., Dinh, V.S.: ColonFormer: an efficient transformer based method for colon polyp segmentation. IEEE Access **10**, 80575–80586 (2022)

3. Fan, D.-P., et al.: PraNet: parallel reverse attention network for polyp segmentation. In: Martel, A.L., et al. (eds.) MICCAI 2020. LNCS, vol. 12266, pp. 263–273. Springer, Cham (2020). https://doi.org/10.1007/978-3-030-59725-2_26
4. Guo, X., Zhang, N., Guo, J., Zhang, H., Hao, Y., Hang, J.: Automated polyp segmentation for colonoscopy images: a method based on convolutional neural networks and ensemble learning. Med. Phys. **46**(12), 5666–5676 (2019)
5. Haithami, M., Ahmed, A., Liao, I.Y., Jalab, H.: Employing GRU to combine feature maps in DeeplabV3 for a better segmentation model. Nordic Mach. Intell. **1**(1), 29–31 (2021)
6. Haithami, M., Ahmed, A., Liao, I.Y., Jalab, H.A.: An embedded recurrent neural network-based model for endoscopic semantic segmentation. In: EndoCV@ ISBI, pp. 59–68 (2021)
7. Howard, A., et al.: Searching for MobileNetV3. In: Proceedings of the IEEE/CVF International Conference on Computer Vision, pp. 1314–1324 (2019)
8. Huang, C.H., Wu, H.Y., Lin, Y.L.: HarDNet-MSEG: a simple encoder-decoder polyp segmentation neural network that achieves over 0.9 mean dice and 86 fps. arXiv preprint arXiv:2101.07172 (2021)
9. Jha, D., Riegler, M.A., Johansen, D., Halvorsen, P., Johansen, H.D.: DoubleU-Net: a deep convolutional neural network for medical image segmentation. In: 2020 IEEE 33rd International Symposium on Computer-Based Medical Systems (CBMS), pp. 558–564. IEEE (2020)
10. Jha, D., et al.: Kvasir-SEG: a segmented polyp dataset. In: Ro, Y.M., et al. (eds.) MMM 2020. LNCS, vol. 11962, pp. 451–462. Springer, Cham (2020). https://doi.org/10.1007/978-3-030-37734-2_37
11. Jia, X., et al.: Automatic polyp recognition in colonoscopy images using deep learning and two-stage pyramidal feature prediction. IEEE Trans. Autom. Sci. Eng. **17** (2020)
12. Kingma, D.P., Ba, J.: Adam: a method for stochastic optimization. arXiv preprint arXiv:1412.6980 (2014)
13. Kvinge, H., Emerson, T.H., Jorgenson, G., Vasquez, S., Doster, T., Lew, J.D.: In what ways are deep neural networks invariant and how should we measure this? arXiv preprint arXiv:2210.03773 (2022)
14. Long, J., Shelhamer, E., Darrell, T.: Fully convolutional networks for semantic segmentation. In: Proceedings of the IEEE Conference on Computer Vision and Pattern Recognition, pp. 3431–3440 (2015)
15. Mahmud, T., Paul, B., Fattah, S.A.: PolypSegNet: a modified encoder-decoder architecture for automated polyp segmentation from colonoscopy images. Comput. Biol. Med. **128**, 104119 (2021)
16. Nguyen, N.Q., Lee, S.W.: Robust boundary segmentation in medical images using a consecutive deep encoder-decoder network. IEEE Access **7**, 33795–33808 (2019)
17. Silva, J., Histace, A., Romain, O., Dray, X., Granado, B.: Toward embedded detection of polyps in WCE images for early diagnosis of colorectal cancer. Int. J. Comput. Assist. Radiol. Surg. **9**, 283–293 (2014)
18. Thambawita, V., et al.: SinGAN-Seg: synthetic training data generation for medical image segmentation. PloS One **17**(5), e0267976 (2022)
19. Thambawita, V.L., Strümke, I., Hicks, S., Riegler, M.A., Halvorsen, P., Parasa, S.: ID: 3523524 data augmentation using generative adversarial networks for creating realistic artificial colon polyp images: validation study by endoscopists. Gastrointest. Endosc. **93**(6), AB190 (2021)
20. Vázquez, D., et al.: A benchmark for endoluminal scene segmentation of colonoscopy images. J. Healthc. Eng. **2017** (2017)

MLM-LSTM: Multi-layer Memory Learning Framework Based on LSTM for Hyperspectral Change Detection

Yinhe Li[1], Yijun Yan[1], Jinchang Ren[1(✉)], Qiaoyuan Liu[2], and Haijiang Sun[2]

[1] National Subsea Centre, Robert Gordon University, Aberdeen, UK
{y.li24,y.yan2,j.ren}@rgu.ac.uk
[2] Changchun Institute of Optics, Fine Mechanics and Physics, Chinese Academy of Sciences, Changchun, China
liuqy@ciomp.ac.cn

Abstract. Hyperspectral change detection plays a critical role in remote sensing by leveraging spectral and spatial information for accurate land cover variation identification. Long short-term memory (LSTM) has demonstrated its effectiveness in capturing dependencies and handling long sequences in hyperspectral data. Building on these strengths, a multilayer memory learning model based on LSTM for hyperspectral change detection is proposed, called MLM-LSTM for hyperspectral change detection is proposed. It incorporates shallow memory learning and deep memory learning. The deep memory learning module performs deep feature extraction of long-term and short-term memory separately. Then fully connected layers will be used to fuse the features followed by binary classification for change detection. Notably, our model has higher detection accuracy compared to other state-of-the-art deep learning-based models. Through comprehensive experiments on publicly available datasets, we have successfully validated the effectiveness and efficiency of the proposed MLM-LSTM approach.

Keywords: Hyperspectral image · Change detection · Long short-term memory

1 Introduction

Change detection (CD) is a fundamental approach for monitoring land-cover changes, involving the analysis of disparities between bi-temporal remote sensing images of the same geographic area [1]. Hyperspectral images (HSIs) provide a comprehensive representation of objects by combining pixel-wise 1-D spectral data with spatial information in the form of a standard 2-D image. The integration of spectral and spatial data in HSIs enables a more detailed and accurate assessment of changes occurring within the observed area [2]. Therefore, hyperspectral change detection (HCD) has emerged as a prominent research area in recent years.

Despite the abundance of spatial and spectral information, HSIs often plagued by highly redundancy information and various noise that primarily due to sensor limitations and atmospheric effects during the data acquisition step. As a result, processing HSIs

J. Ren et al. (Eds.): BICS 2023, LNAI 14374, pp. 51–61, 2024.
https://doi.org/10.1007/978-981-97-1417-9_5

data poses significant challenges [3]. Over the past two decades, numerous methods for hyperspectral change detection (HCD) have been proposed to address these challenges, encompassing both unsupervised and supervised approaches [4].

Recently, the application of deep learning-based methods to HCD tasks has emerged as a new trend in extracting more effective and representative spectral, spatial, and spatial-spectral features. In [5], a cross-temporal interaction symmetric attention (CSA) network was proposed that employed a Siamese module to hierarchically extract change information in a symmetric manner. A cross-temporal self-attention module was incorporated to joint spatial-spectral-temporal features and enhance the feature representation ability. In [6], a novel end-to-end 2-D CNN was introduced that utilizes a mixed affinity matrix and subpixel representation to effectively extract cross-band gradients. In addition, some deep learning-based methods based on recurrent neural networks (RNN) and Long Short-Term memory (LSTM) have been proposed, for example, in [7], Re3FCN was proposed that leverages 3-D CNN layers to extract spatial-spectral features, while LSTM is utilized for extracting comprehensive features and screening significantly changed features. In [8], a novel multilevel encoder-decoder attention network is introduced to extract hierarchical spatial-spectral features more effectively. In this approach, the extracted features are transferred to a LSTM module for analysing temporal dependencies. Although deep learning models have yielded impressive results, they often rely on a substantial amount of training data, which is difficult to acquire. Consequently, the computational cost becomes exceptionally high, thus needs further efforts to address this issue.

Although LSTM has shown good performance in hyperspectral change detection (HCD), it has limitations:

1. Difficulty handling long sequence data: HSIs have hundreds of consecutive bands, which can lead to issues like gradient explosion when processing with LSTM models, hindering the capture of long-term dependencies effectively.
2. Information omission: Traditional LSTM methods only utilize the final output result, neglecting the full utilization of memory from the last hidden state.

To address these limitations, we propose an end-to-end multi-layer memory learning framework based on LSTM. It incorporates unsupervised PCA for dimensionality reduction, retaining essential features while reducing dimensionality and mitigating the gradient explosion risk. Multi-layer LSTM modules are employed for initial feature extraction, with long-term and short-term memories integrated separately for fusion in subsequent layers. This facilitates the identification of significant change features.

The remainder of this paper is organized as follows. Section 2 describes the details of the proposed MLM-LSTM. Section 3 presents the experimental results and assessments. Finally, some remarkable conclusions are summarized in Sect. 4.

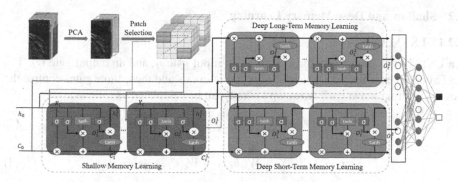

Fig. 1. The architecture of the proposed MLM-LSTM model

2   The Proposed Approach

Figure 1 shows the architecture of the proposed MLM-LSTM method, which is composed of two main steps, i.e., 1) spectral feature extraction and dimension reduction, 2) shallow memory learning and hierarchical deep memory learning.

2.1   Spectral Feature Extraction and Dimension Reduction

Given two HSIs $T^1, T^2 \in \mathfrak{R}^{W*H*B}$ acquired on the same geographical area at times t_1 and t_2, where W, H and B denote the numbers of rows, columns and spectral bands, respectively. To analyze the behaviors of spectral differences between the two images, let us compute the HS difference image T^D by subtracting bitemporal images from each other pixel by pixel, i.e.,

$$T^D = \left| T^2 - T^1 \right| \tag{1}$$

Then principal component analysis (PCA) is introduced to reduce the high-dimensional original input while retaining more spatial and spectral features. The spectral feature of T^D can be represented as $T^D(PCA) \in \mathfrak{R}^{H*W*q_{PCA}}$, where q_{PCA} is the number of bands after PCA dimensionality reduction. T^D will be feed into LSTM modules in the next stage.

2.2 Shallow and Deep Memory Learning

2.2.1 LSTM Unit

An LSTM unit comprises a forget gate f_t, an input gate i_t, and an output gate O_t. The LSTM cell memories values at arbitrary time intervals and these three gates control the flow of information at each time step t which can be calculated as follows:

$$f_t = \sigma(W_{xf}X_t + W_{hf}h_{t-1} + W_{cf}C_{t-1} + b_f) \tag{2}$$

$$i_t = \sigma(W_{xi}X_t + W_{hi}h_{t-1} + W_{ci}C_{t-1} + b_i) \tag{3}$$

$$O_t = \sigma(W_{xo}X_t + W_{ho}h_{t-1} + W_{ci}C_t + b_o) \tag{4}$$

where W, b, σ represent the coefficient matrix, bias vector and sigmoid function, respectively. These three gates are crucial parts of the LSTM unit, which is used to update the current memory state of this unit, obtain the short-term memory C_t and long-term memory h_t, that can be represented as:

$$\tilde{C}^t = \tanh(W_{\tilde{C}}h_{t-1} + W_{\tilde{C}}h_{t-1}X_t + b_{\tilde{C}}) \tag{5}$$

$$C_t = f_t C_{t-1} + i_t * \tilde{C}^t \tag{6}$$

$$h_t = O_t * \tanh(C_t) \tag{7}$$

2.2.2 MLM-LSTM: Multi-layer Memory Feature Extraction

T^D Will be divided into a group of overlapped 3-D neighboring patches denoted as $Z_{(\alpha,\beta)} \in \mathfrak{R}^{S*S*qPCA}$, where S is the patch size of Z, (α, β) denote the coordinates of the patch centre in the spatial domain where $\alpha \epsilon[1, W]$, $\beta \epsilon[1, H]$ (we set S = 3 in this study). The total number of 3-D patches from T^D will be $(W - S + 1) \times (H - S + 1)$. If we split the patched across the spectral channels, then Z can be considered as an $qPCA$-length sequence $\{(Z^1_{(\alpha,\beta)}, Z^2_{(\alpha,\beta)}, \ldots, Z^{qPCA}_{(\alpha,\beta)})| Z^q_{(\alpha,\beta)} \in \mathfrak{R}^{S*S*1}, 1 \leq q \leq qPCA\}$ The image patches in the sequence are fed into the memory feature extraction module one by one to extract the spectral feature via a recurrent operator. The proposed memory feature extraction module is composed of three LSTM layers with t hidden size and m LSTM layers. In order to fully extract all the features of the input, we set $t = 256$ and m = 2 in this study. The first LSTM is used to extract long-term memory h^1, short-term memory C^1_t and output O^1_t from the input patches, h_o and C_o are initialized to zero. After initial feature extraction by using the shallow memory learning, the extracted outputs O^1_t, h^1_t, C^1_t have the same size as Z.

Next, we used two LSTM modules for deep extraction of the long-term memory and short-term memory, respectively. The outputs O^2_t, O^3_t can be obtained by Eqs. (8–9).

$$O^2_t = \sigma(W_{xo}O^1_t + W_{ho}h^1_t + W_{ci}C_0 + b_o) \tag{8}$$

$$O_t^3 = \sigma(W_{xo}O_t^1 + W_{ho}h_0 + W_{ci}C_t^1 + b_o) \tag{9}$$

O_t^2, O_t^3 are used to extract the deep hierarchical feature from the long-term and short-term spatial-spectral memories, respectively. Then these two outputs are further concatenated together and fed into to a linear layer for feature integration.

Since change detection can be considered as a binary classification problem of distinguishing the change and non-change pixels, the cross entropy, which is commonly used for classification, is adopted as the loss function.

$$Loss_{(pred,label)} = -\frac{1}{u}\sum_{i=1}^{n}(l * \log(p) + (1 - l) * \log(1 - p)) \tag{10}$$

where u denotes the number of samples, l represents the ground truth value where 0 and 1 represent unchanged and changed regions. p represents the probability predicted by the Linear function. The selected optimizer is the adaptive momentum (Adam) with the initial learning rate of 0.0001.

3 Experiments and Results

3.1 Experimental Settings

Change detection task can be treated as a binary classification problem, therefore, three commonly used evaluation metrics for classification, including the overall accuracy (OA), average accuracy (AA), and Kappa coefficient (KP), were adopted in our experiments for quantitative performance assessment [9]. Two datasets (i.e., River and Hermiston) shown in Fig. 2 [10] are adopted in this study for performance evaluation.

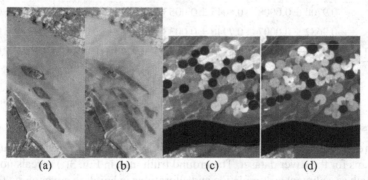

| (a) | (b) | (c) | (d) |

Fig. 2. Pseudo-colored images of the three datasets, including the River dataset captured on May 3, 2013 (a) and December 31, 2013 (b) and the Hermiston dataset captured on May 1, 2004 (c) and May 8, 2007 (d), respectively.

3.2 Results and Analysis

In this session, we evaluate the effectiveness of the proposed method by comparing it with a few start-of-the-art unsupervised methods, which include the change vector analysis (CVA) [11], principal component analysis (PCA-KM) [12] and absolute distance (AD) [13] as well as several deep-learning based methods such as 2-D-CNN [14], 3-D-CNN [15], HybridSN [16], and Traditional Long-short-term-memory (LSTM) [17]. The proposed MLM-LSTM and all other DL-based methods are trained based on the PyTorch on an NVIDIA RTX A2000, with the batch size set to 128 and the number of training epoch as 200. We randomly select 20% pixels in the changed and unchanged pixels as the training set, and the remaining for testing. To make a fairer and more reliable comparison, all DL algorithms are repeated ten times in each experiment, and the averaged results with the standard deviations are reported. In the produced change maps, false alarms and missing pixels are marked in red and green respectively for ease of comparison, white areas represent correctly detected and black area for true negatives. The quantitative assessment results of OA and KP on River and Hermiston datasets are shown in Table 1.

Table 1. Quantitative assessment of different methods on River and Hermiston datasets

	River		Hermiston	
	OA	KP	OA	KP
AD	0.9431	0.7137	0.9342	0.7904
CVA	0.9253	0.6528	0.9287	0.7705
PCA-KM	0.9517	0.7476	0.9224	0.7472
LSTM	0.9569 ± 0.0011	0.7216 ± 0.0070	0.9537 ± 0.0024	0.8580 ± 0.0009
HybridSN	0.9671 ± 0.0019	0.7826 ± 0.0087	0.9579 ± 0.0007	0.8789 ± 0.0005
3-D-CNN	0.9700 ± 0.0008	0.8045 ± 0.0053	0.9639 ± 0.0002	0.8966 ± 0.0047
2-D-CNN	0.9682 ± 0.0007	0.7946 ± 0.0033	0.9585 ± 0.0013	0.8794 ± 0.0050
MLM-LSTM	**0.9723 ± 0.0005**	**0.8248 ± 0.0023**	**0.9708 ± 0.0014**	**0.9194 ± 0.0013**

3.2.1 Results on River Dataset

Table 1 presents the quantitative assessment results and Fig. 3 provides visual comparison maps for the River dataset. The ground truth map in Fig. 3(i) reveals noticeable changes such as sediment accumulation and alterations in building cover along the riverbank. Figures 3(a–c) demonstrate that the unsupervised algorithms yield numerous false alarms, particularly in the lower left and upper left corners of the maps where non-changing pixels are incorrectly classified as changed. As a result, the KP values for all unsupervised methods remain below 80%. In contrast, DL-based algorithms effectively classify most of the false alarms. Among the benchmarked DL methods, traditional LSTM performs the worst with an average KP of 0.7261 and OA of 95.69%, indicating the highest number of missing pixels. The 2-D CNN and 3-D CNN produce similar detection results, with an OA of around 97% and KP of approximately 0.80, slightly outperforming the unsupervised methods. Our proposed MLM-LSTM method achieves the highest OA and KP among all compared methods, with an OA value of 0.9723 and KP value of 0.8248, surpassing the second-place method by 3%. Additionally, MLM-LSTM exhibits the smallest variance, further confirming its effectiveness.

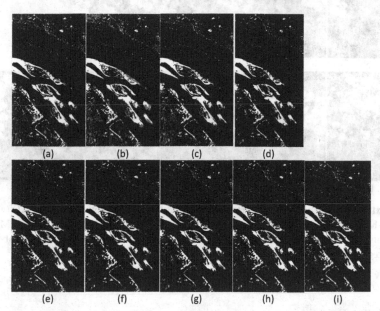

Fig. 3. Extracted change maps on the River Dataset from different methods of AD (a), CVA (b), PCA-KM (c), LSTM (d), HybridSN (e), 3-DCNN (f), 2-DCNN (g), MLM-LSTM (h) in comparison to the Ground-truth map (i), where the false alarms and missing pixels are labelled in red and green. (Color figure online)

3.2.2 Results on Hermiston Dataset

The Hermiston dataset's quantitative assessment results and extracted change maps are presented in Table 1 and Fig. 4, respectively. The changing areas mainly consist of crop regions characterized by simple round shapes. Among the unsupervised methods, there is a significant number of undetected pixels, resulting in OA values below 94% and KP values lower than 0.8. The CNN-based methods, including 2-D CNN, 3-D CNN, and HybridSN, exhibit similar detection results, with OA around 97% and KP values below 0.9. However, both 2-D CNN and 3-D CNN show significant variance, leading to unstable detection performance. Traditional LSTM consistently demonstrates the lowest detection results among all DL-based benchmarks. In contrast, our MLM-LSTM method shows a significant improvement in performance, achieving the highest detection accuracy among all methods. It achieves an OA value of 0.9708, which is 0.23% higher than the second-place method, and a KP value of 0.9194, which is 6.14% higher than the

KP of the traditional LSTM method. This demonstrates the effectiveness and robustness of our proposed MLM-LSTM in handling variations in different sizes of changes.

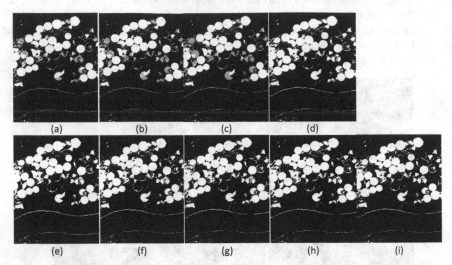

Fig. 4. Extracted change maps on the Hermiston Dataset from different methods of AD (a), CVA (b), PCA-KM (c), LSTM (d), HybridSN (e), 3-DCNN (f), 2-DCNN (g), MLM-LSTM (h) in comparison to the Ground-truth map (i), where the false alarms and missing pixels are labelled in red and green. (Color figure online)

3.3 Ablation Experiments

In our experiments, we investigated the impact of different factors on the performance of the MLM-LSTM model. Firstly, we tested patch sizes of $3 \times 3, 5 \times 5, 7 \times 7$, and 9×9, finding that increasing the patch size had minimal impact on KP and OA (Fig. 5(a)). A patch size of 5×5 showed improved performance on the River dataset but lacked robustness on the Hermiston dataset. Considering computational efficiency, a patch size of 3×3 struck a suitable balance. We also explored the optimal number of layers and found that 3 layers maximized KP on the River dataset, while 2 layers performed best on the Hermiston dataset (Fig. 5(b)). In terms of hidden size (Fig. 5(c)), increasing the size in the LSTM module resulted in higher KP values, and a hidden size of 256 was selected for optimal accuracy. Finally, we varied the training ratios from 10% to 50% and observed that increasing the number of training pixels improved detection accuracy (Fig. 5(d)). Our MLM-LSTM consistently outperformed other methods, achieving exceptional performance with a highest KP of 0.8656 on the River dataset. These results highlight the robustness and effectiveness of MLM-LSTM in hyperspectral change detection tasks.

3.4 Hyperparameter Analysis

To assess the efficiency of our proposed MLM-LSTM, we compared the hyperparameters and floating-point operations (FLOPs) of different methods, as shown in Table 2. It is

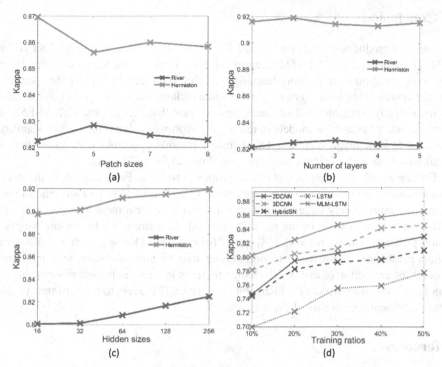

Fig. 5. Ablation experiments and results od the MLM-LSTM in different setting, including the KP values of different patch sizes (a), different number of layers (b), different hidden sizes (c) and different training ratios on the River dataset (d)

evident that the 3-D CNN and HybridSN methods have significantly more hyperparameters compared to the LSTM-based methods. Furthermore, the FLOPs associated with CNN-based methods are several dozen or even hundreds of times higher than those of LSTM-based methods. This discrepancy arises from the nature of convolutional layers used in CNN, which involve convolution operations, pooling operations, and non-linear activations. Image analysis tasks often require a large number of filters and larger input sizes, resulting in higher FLOPs for CNN. In contrast, LSTM primarily involves matrix multiplication and element-wise operations, resulting in lower FLOPs.

Table 2. Comparing the parameters of different DL-based methods on River dataset

	LSTM	HybridSN	3-D CNN	2-D CNN	Proposed
Hyperparameters (k)	213.79	5128.74	1613.03	607.43	1411.46
FLOPs	3.51	1579.24	215.35	368.21	12.77

4 Conclusion

This paper introduces a novel end-to-end DL-based network called MLM-LSTM for HCD. The proposed MLM-LSTM leverages shallow memory learning and hierarchical long-term and short-term memory learning modules to effectively capture the spectral-spatial features. This leads to more precise binary classification. Experimental results on two publicly available HCD datasets demonstrate that the proposed MLM-LSTM surpasses other benchmark models in terms of performance. It exhibits better stability compared to benchmark methods. These results provide comprehensive validation of the effectiveness and efficiency of the proposed model for HCD tasks.

There are still some limitations of our proposed method. For example, in the deep memory extraction modules of long-term memory and short-term memory, the model parameters are relatively large, and the outputs after long-term memory extraction and short-term memory extraction are not fully utilized. To address this limitation, we plan to incorporate a Siamese network [18] in our future work. This approach will allow us to share parameters during the extraction of long-term memory and short-term memory, enabling the extraction of distinctive change features in a more efficient manner. Additionally, we intend to explore the inclusion of more LSTM layers to further improve the feature representation ability of the network.

References

1. Ma, P., et al.: Multiscale superpixelwise prophet model for noise-robust feature extraction in hyperspectral images. IEEE Trans. Geosci. Remote Sens. **61**, 1–12 (2023)
2. Luo, F., Zhou, T., Liu, J., Guo, T., Gong, X., Ren, J.: Multiscale diff-changed feature fusion network for hyperspectral image change detection. IEEE Trans. Geosci. Remote Sens. **6**, 1–13 (2023)
3. Yan, Y., Ren, J., Liu, Q., Zhao, H., Sun, H., Zabalza, J.: PCA-domain fused singular spectral analysis for fast and noise-robust spectral-spatial feature mining in hyperspectral classification. IEEE Geosci. Remote Sens. Lett. (2021)
4. Hasanlou, M., Seydi, S.T.: Hyperspectral change detection: an experimental comparative study. Int. J. Remote Sens. **39**, 7029–7083 (2018)
5. Song, R., Ni, W., Cheng, W., Wang, X.: CSANet: cross-temporal interaction symmetric attention network for hyperspectral image change detection. IEEE Geosci. Remote Sens. Lett. (2022)
6. Wang, Q., Yuan, Z., Du, Q., Li, X.: GETNET: a general end-to-end 2-D CNN framework for hyperspectral image change detection. IEEE Trans. Geosci. Remote Sens. **57**, 3–13 (2018)
7. Song, A., Choi, J., Han, Y., Kim, Y.: Change detection in hyperspectral images using recurrent 3D fully convolutional networks. Remote Sens. **10**, 1827 (2018)
8. Qu, J., Hou, S., Dong, W., Li, Y., Xie, W.: A multilevel encoder–decoder attention network for change detection in hyperspectral images. IEEE Trans. Geosci. Remote Sens. **60**, 1–13 (2021)
9. Yan, Y., et al.: Non-destructive testing of composite fiber materials with hyperspectral imaging—evaluative studies in the EU H2020 FibreEUse project. IEEE Trans. Instrum. Meas. **71**, 1–13 (2022)
10. Li, Y., Ren, J., Yan, Y., Petrovski, A.: CBANet: an end-to-end cross band 2-D attention network for hyperspectral change detection in remote sensing. IEEE Trans. Geosci. Remote Sens. (2023)

11. Malila, W.A.: Change vector analysis: an approach for detecting forest changes with Landsat. Presented at the LARS Symposia LARS Symposia, p. 385 (1980)
12. Li, Y., Ren, J., Yan, Y., Liu, Q., Petrovski, A., McCall, J.: Unsupervised change detection in hyperspectral images using principal components space data clustering. J. Phys.: Conf. Ser. (2022)
13. Hotelling, H.: Relations between two sets of variates. Methodol. Distrib. 162–190 (1992)
14. He, N., et al.: Feature extraction with multiscale covariance maps for hyperspectral image classification. IEEE Trans. Geosci. Remote Sens. **57**, 755–769 (2018)
15. Hamida, A.B., Benoit, A., Lambert, P., Amar, C.B.: 3-D deep learning approach for remote sensing image classification. IEEE Trans. Geosci. Remote Sens. **56**, 4420–4434 (2018)
16. Roy, S.K., Krishna, G., Dubey, S.R., Chaudhuri, B.B.: HybridSN: exploring 3-D-2-D CNN feature hierarchy for hyperspectral image classification. IEEE Geosci. Remote Sens. Lett.Geosci. Remote Sens. Lett. **17**, 277–281 (2019)
17. Liu, Q., Zhou, F., Hang, R., Yuan, X.: Bidirectional-convolutional LSTM based spectral-spatial feature learning for hyperspectral image classification. Remote Sens. **9**(12), 1330 (2017)
18. Li, X., et al.: Siamese residual neural network for musical shape evaluation in piano performance assessment. Presented at the 31st European Signal Processing Conference (2023)

A Hierarchical Geometry-to-Semantic Fusion GNN Framework for Earth Surface Anomalies Detection

Boan Chen[1(✉)], Aohan Hu[1], Mengjie Xie[1], Zhi Gao[1(✉)], Xuhui Zhao[1], and Han Yi[2]

[1] School of Remote Sensing and Information Engineering, Wuhan University, Wuhan 430079, China
{cbchen1997,whu_ahah,mengjie_xie,zhaoxuhui}@whu.edu.cn, gaozhinus@gmail.com
[2] School of Computing, National University of Singapore, Singapore 117417, Singapore
hany24@u.nus.edu

Abstract. The increasing occurrence of earth surface anomalies (ESA) underlines the importance of timely and accurate detection of such events. Therefore, researchers have utilized satellite imagery for large-scale detection and developed advanced deep learning methods. However, the performance is hindered by inadequate labeled data and the complexity of semantic information in satellite imagery. To this end, we propose a hierarchical geometry-to-semantic fusion graph neural network (GNN) framework. Specifically, our method employs two branches to extract geoentities and construct graphs at different levels. Then, a hierarchical graph attention network (GAT) is used to mine complex semantic information from graphs, facilitating accurate and rapid detection of ESA. To fill the gap of the lack of benchmark datasets, we create a composite dataset ESAD based on existing datasets for ESA detection. Extensive experiments demonstrate that the proposed method is effective for accurate ESA detection, outperforming many baseline methods.

Keywords: Earth surface anomalies · Hierarchical fusion · Graph neural network · Satellite imagery

1 Introduction

Earth surface anomalies (ESA) refers to sudden events on earth's surface caused by natural and human factors (e.g., natural disasters) [1]. In recent years, numerous ESA have caused significant loss of life and property damage, emphasizing the need for research in Humanitarian Assistance and Disaster Response (HADR) [2]. Timely detection of ESA is of great significance for early rescue and loss reduction [3]. The integration of satellite imagery and deep learning methods provides a large-scale, accurate solution for HADR challenges and has gained considerable attention. However, the scarcity of labeled data and the complexity of satellite imagery hinder the performance of these methods [4].

J. Ren et al. (Eds.): BICS 2023, LNAI 14374, pp. 62–71, 2024.
https://doi.org/10.1007/978-981-97-1417-9_6

To achieve accurate detection of ESA, several methods have attempted to introduce additional temporal or modal data [5,6]. However, these methods entail demands for data availability and preprocessing, posing challenges for rapid response. More efficient ESA detection can be achieved via single-image based methods, including deep visual feature-based supervised learning methods like Convolutional Neural Network (CNN), but the overfitting problem caused by limited labeled data is challenging to address. Hence, certain studies have sought to gain insights from the success of unsupervised learning in industrial anomaly detection [7,8]. However, satellite imagery is much more complex than industrial images, leading to unsatisfactory model performance. Meanwhile, abnormal samples can provide valuable information and should not be completely ignored. Instead, we should strive to learn more useful information from limited data [9].

In order to address the aforementioned problems, we propose a two-stage framework, i.e. Hierarchical Geometry-to-Semantic Fusion Graph Neural Network (GNN), which solely utilizes a single image for detecting ESA, thereby reducing data requirements and decreasing preprocessing and inference time. In the first stage, the graph is constructed through a two-branch module that obtains geoentities and their relationships from a single satellite image. In the second stage, a hierarchical graph attention network (GAT) and attention-based fusion module are used to extract the graph-level embedding. Inspired by how human interpret satellite imagery, proposed method aims at simulating the human brain's ability to learn higher-level information from objects and their relationships. Meanwhile, existing datasets mostly focus on single or a few classes of ESA, e.g., flooding, hurricane, and landslide, while a large-scale, multi-class dataset is still lacking. To bridge this gap, we create a composite dataset ESAD, based on existing distributable datasets, xBD [10], Multi^{3}Net [11], and Sichuan Landslide and Debrisflow [12]. Extensive experiments demonstrate that our method is effective for accurate ESA detection, outperforming many baseline methods.

2 Related Work

2.1 Earth Surface Anomalies Detection and Diagnosis

The increasing frequency of ESA has attracted attention. Current methods primarily utilize satellite imagery for large-scale research, and according to the focus of research, these methods can be divided into post-hoc analysis and rapid response. The former aims to study the process, intensity, impact, and other factors of ESA and conducts comprehensive analysis to aid the authorities in better evaluation, planning and action. Particularly, the release of the large-scale building damage assessment dataset xBD [10] has led to many related studies, such as change detection and multimodal fusion [5,13,14]. The rapid response task focuses on timeliness, striving to obtain detection results as early as possible during or after a disaster [2,15,16]. To this end, EmergencyNet [15] utilizes single UAV images for rapid detection of several classes of ESA. However, it is better suited for on-site applications and challenging to scale up. In [16], the

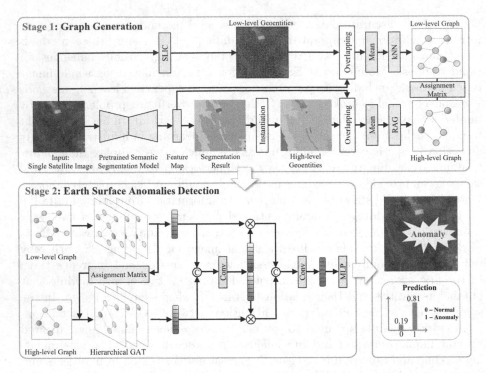

Fig. 1. Overview of the hierarchical geometry-to-Semantic fusion graph neural network.

author proposes a lightweight network to rapidly extract landslide areas from single satellite imagery. In [2], flood areas were extracted with limited resources for rapid response under on-orbit conditions. Nonetheless, these methods are only applicable to a single type of ESA. Given the aforementioned problems, our method aims to rapidly detect ESA using binary classification on a single satellite image.

2.2   Graph Neural Network in Satellite Imagery

GNN can explicitly capture relationships between nodes, edges, and graphs to learn higher-order semantic information, such as semantic relationships and constraints, enabling further exploitation of images. Consequently, some studies incorporate it into satellite imagery, especially hyperspectral image semantic segmentation. CEGCN [17] utilizes superpixels to construct the graph and combines CNN and GNN to achieve accurate land cover classification. In [18,19], various strategies are employed to construct multi-scale graphs for extracting high-order semantic information, resulting in a significant improvement in semantic segmentation. Recently, GNN has also been introduced in satellite imagery classification. H-GCN [20] uses GNN to leverage relationships between objects such as forests and ponds for remote sensing scene classification, while MLRSSC-CNN-GNN [21] combined object detection with GNN for complex scene understanding. By incorporating GNN based on deep visual features, these methods have achieved better performance in complex visual tasks. GNN can simulate the human ability of image interpretation and knowledge formation to some extent, thus facilitating the use of complex satellite imagery for high-level semantic reasoning. Inspired by this, GNN is integrated into our method for ESA detection. To the best of our knowledge, it is the first exploration of GNN in this field, which can provide a reference and benchmark for future research.

3 Our Method

The framework of our method is illustrated in Fig. 1. The proposed method consists of two stages: graph generation and earth surface anomalies detection, which will be detailed in the following sections. For an input of high spatial resolution satellite image comprised of RGB bands, the first stage generates hierarchical graphs and assignment matrix through two segmentation branches. Subsequently, a hierarchical GAT is utilized to update the node features and extract graph embeddings for each level. An attention-based feature fusion module then combines them to yield the graph-level feature vector, which is finally processed through a Multi-Layer Perceptron (MLP) for ESA detection.

3.1 Notations

In the first stage, the task is to obtain an attributed, undirected graph $G :=(V, E, H)$ from an input single satellite image X. Here, V, E, and H represent nodes, edges, and node features, respectively. Each node $v \in V$ is characterized by a feature vector $h(v) \in \mathbb{R}^d$, thus, $H \in \mathbb{R}^{|V| \times d}$, with d indicating the length of feature vectors and $|\cdot|$ denoting the cardinality of a set. An edge between two nodes $u, v \in V$ is represented by $e_{u,v}$. The graph structure is determined by a symmetric adjacency matrix $A \in \mathbb{R}^{|V| \times |V|}$, where $A_{u,v}$ equals 1 if $e_{u,v} \in E$. The neighborhood of a node $v \in V$ is defined as $N(v) := \{u \in V | e_{u,v} \in E\}$.

3.2 Graph Generation

In order to fully leverage the advantages of graphs and GNN, it is crucial to convert an image into graph. Most existing methods employ the Simple Linear Iterative Clustering (SLIC) [22] algorithm to generate superpixels as nodes, reducing noise and computational efficiency. However, SLIC may only capture a fraction of semantic a entity, such as dividing a road into segments, making it challenging to represent higher-level semantic objects and their relationships. To this end, we introduce a semantic segmentation branch that utilizes larger geoentities as nodes to construct a graph. The combination of generated graphs from different levels can enable more comprehensive graph representations.

Low-Level Graph. The construction of a low-level graph involves three steps: superpixel segmentation, node feature extraction, and topology configuration, as shown in Fig. 1. Firstly, for an input X, we employ SLIC algorithm to obtain a superpixel map S_1 that containing $|V_{low}|$ geoentites serve as nodes. Secondly, by superimposing S_1 onto X, we calculate channel-wise mean values of superpixels, which are used to serve as feature vectors $h(v) \in \mathbb{R}^{d_1}$ for each node. Lastly, we build the initial topology with k-Nearest Neighbor(kNN) algorithm. Meanwhile, we consider the Euclidean distance between superpixels, and the graph is pruned by removing edges longer than the threshold τ_{dist}. Formally, for node v, an edge $e_{v,u}$ is built if

$$u \in \{w | D(v,w) \leq d_k \wedge D(v,w) < \tau_{dist}, \forall w, v \in V_{low}\} \tag{1}$$

where d_k denotes the k-th smallest in the distance matrix $D(v,w)$. Following this process, the low-graph topology is constructed and the low-graph obtained can be formulated as $G_{low} := \{V_{low}, E_{low}, H_{low}\}$.

High-Level Graph. To emulate the human brain's ability to interpret images with high-level semantic understanding, we introduce a semantic segmentation branch. Similar to low-level graph generation, we use a pre-trained semantic segmentation model to obtain the preliminary segmentation map S_2 and feature map F_s. Then, S_2 is instantiated so that each connected component is independent to obtain $\hat{S}_2$. Secondly, we overlay $\hat{S}_2$ onto F_s and calculate the mean values of connected components to obtain a feature vector $h(v) \in \mathbb{R}^{d_2}$ for each node. Finally, we employ Region Adjacency Graph (RAG) algorithm to construct a high-level graph, which can be formulated as $G_{high} := \{V_{high}, E_{high}, H_{high}\}$.

Assignment Matrix. From a semantic perspective, objects in satellite imagery can be considered as hierarchical geoentities, ranging from low-level geometry, such as trees, to high-level semantic, such as forests. Inspired by [23], we utilize an assignment matrix to achieve a jointly representation of hierarchical graph. Intra-level topological structures already captured in the previous graph generation phase, i.e., G_{low} and G_{high}. To exploit the inter-level information, we use an assignment matrix $A_{low \rightarrow high} \in \mathbb{R}^{|V_{low}| \times |V_{high}|}$ to represent the topological relationships between geoentities of different levels. Specifically, for the ith low-level geoentity and the jth high-level geoentity, the corresponding assignment can be formulated as

$$A_{low \rightarrow high}[i,j] = \begin{cases} 1 & \text{if } i^{th} \text{ low-level geoentity} \in j^{th} \text{ high-level geoentity} \\ 0 & \text{otherwise} \end{cases} \tag{2}$$

Finally, each low-level geoentity is assigned to one and only one high-level geoentity. Formally, the final hierarchical graph of each input X can be formulated as $G_{joint} := \{G_{low}, G_{high}, A_{low \rightarrow high}\}$.

3.3 Earth Surface Anomalies Detection

Hierarchical GAT. In the second stage, we aim to leverage G_{joint} to extract high-order semantic information for ESA detection. The relationships among geoentities are often diverse, and assuming that all neighboring nodes make equal contributions during message aggregation is inappropriate. Therefore, we adopt the GAT [24] as backbone. Specifically, the updation of node features can be formulated as Eq. (3) and Eq. (4):

$$h_i^{(l)} = \sigma\Big(\sum_{j\in N_i} \alpha_{ij}^{(l)} W^{(l)} h_j^{(l-1)} + b^{(l)} \Big), \tag{3}$$

$$\alpha_{ij}^{(l)} = \frac{\exp\Big(\text{LeakyReLU}\Big(\boldsymbol{a}^{(l)^T}\Big[W^{(l)}h_i^{(l-1)}||W^{(l)}h_j^{(l-1)}\Big]\Big)\Big)}{\sum_{k\in N_i} \exp\Big(\text{LeakyReLU}\Big(\boldsymbol{a}^{(l)^T}\Big[W^{(l)}h_i^{(l-1)}||W^{(l)}h_k^{(l-1)}\Big]\Big)\Big)} \tag{4}$$

where $h_i^{(l)}$ denotes the feature vector of node v_i in the lth layer, $\sigma(\cdot)$ is the sigmoid function, N_i represents some neighborhood of node v_i, $\alpha_{ij}^{(l)}$ is the attention coefficient, $W^{(l)}$ and $b^{(l)}$ are the weight matrix and bias term of the lth layer, $\boldsymbol{a}^{(l)}$ is the learnable parameter vector, and $||$ denotes the concatenation operation.

Given the varying lengths of node features ($d_1 \neq d_2$), we employ two GATs with different input layers, namely f_{GAT}^{low} and f_{GAT}^{high}, to update node features. Following the $\hat{G}_{low} = f_{GAT}^{low}(G_{low})$, the low-level node embeddings, $\hat{h}_{low}(v)|v \in V_{low}$, and $A_{low\rightarrow high}$ are used to initialize the node features of high-level graph, which can complement the loss of low-level geometric information caused by the large areas and mean operations of high-level geoentities, i.e.,

$$h_{\text{high}}(w) = \Big[H_{\text{high}}(w)|| \sum_{v\in M(w)} \hat{h}_{low}(v)\Big] \tag{5}$$

where $M(w) := v \in V_{low}|A_{low\rightarrow high}(v,w) = 1$ is the set of nodes in G_{low} mapping to a node $w \in V_{high}$.

After that, G_{high} is likewise processed by f_{GAT}^{high}. Finally, readout operation and MLP are utilized to extract feature vectors $\mathbf{x}_{low}$ and $\mathbf{x}_{high}$ of different levels from $\hat{G}_{low}$ and $\hat{G}_{high}$, where $\mathbf{x}_{low}, \mathbf{x}_{high} \in \mathbb{R}^{d_3}$.

Attention-Based Feature Fusion. We propose an attention-based feature fusion module to adaptively merge $\mathbf{x}_{low}$ and $\mathbf{x}_{high}$ into a feature vector $\mathbf{z}$, which serves as a representation of input satellite imagery X. In the end, $\mathbf{z}$ will be fed into an MLP for ESA detection. Formally in Eq. (6)

$$\mathbf{z} = f\Big(\Big[x_{low} \odot a_{low}||a_{high} \odot a_{high}\Big]\Big) \tag{6}$$

$$[a_{low}, a_{high}] = [\sigma(f([x_{low}, x_{high}])), 1 - \sigma(f([x_{low}, x_{high}]))] \tag{7}$$

where a_{low} and a_{high} are attention weights calculated according to Eq. (7), and $\odot$ denotes element-wise product. The function $f(\cdot)$ represents a convolutional layer that maps a vector of length d_3 to m.

Table 1. Statistics of ESAD

Class		Num
Anomaly	Flood	647
	Landslide	59
	Debrisflow	48
	Hurricane	1296
	Wildfire	1201
	Earthquake	14
	Volcano	217
	Tornado	254
	Tsunami	107
	Fire	1548
	Bushfire	996
Normal		6671

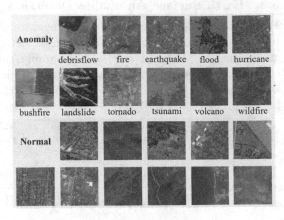

Fig. 2. Examples of ESAD Dataset

4 Experiments

4.1 Dataset

The existing ESA datasets mainly focus on single-class detection or post-hoc analysis, which is insufficient for our research needs. We propose ESAD, a composite dataset to fill the gap of large-scale multi-class datasets. Specifically, ESAD is composed of three publicly datasets: xBD, Multi³Net and Sichuan Landslide and Debrisflow. xBD comprises paired pre- and post-disaster images and is currently the largest dataset for building damage assessment. Multi³Net includes post-disaster high spatial resolution images after Hurricane Harvey, while Sichuan Landslide and Debrisflow contains images of four geohazards. The ESAD contains 13,058 ESA samples classified into 11 image-level classes as shown in Fig. 2. ESAD is integrated by carefully filtering through visual interpretation. Additionally, all the samples are preprocessed to have a resolution ranging from 0.2m to 1m and a size of 1024 × 1024. The detailed statistics of ESAD is illustrated in Table 1.

4.2 Baseline Methods and Implementation Details

To verify the performance of the proposed method, we use ResNet-50, MobileNetV3 and ViT-B/32 as visual feature-based baseline methods. As for graph classification, HGP-SL [25] is one of the state-of-the-art method, while HACT-Net [23] is an advanced method in medical image analysis that also uses hierarchical graph representations. Moreover, to verify the effectiveness of two branches, GAT-Low and GAT-High only use feature vector generated by their respective branches. We evaluate the attention-based feature fusion module using

Concat-GAT, which only uses concatenation to obtain the graph-level embedding.

We divided the dataset into training, testing, and validation sets in a 6:2:2 ratio. The semantic segmentation branch uses HRNet pre-trained on the Deep-Globe Land Cover dataset. Each method is trained from scratch for 200 epochs, with an initial learning rate of 0.0001 and Adam optimizer. The loss function is set to BCELoss. All experiments are conducted using the PyTorch framework in Ubuntu 18.04 and trained on a NVIDIA 3090 GPU with 24 GB memory.

4.3   Experimental Results

The quantitative results of comparison methods are illustrated in Table 2, where Overall Accuracy (OA), Recall, Average Inference Time (AIT) and Parameters are used to evaluate the comprehensive performance.

As shown in Table 2, the visual feature-based methods achieve the best accuracy, with only our framework from the GNN-based methods is competitive. Notably, GNN-based methods generally have fewer parameters and faster inference speed, even outperforming MobileNetV3, which is specifically designed for limited resource conditions. In comparison to advanced GNN based methods, our method achieved significant performance improvements with similar parameters, with accuracy and recall increasing by 9.36% and 8.44%, respectively, without a significant increase in the number of parameters. By fusing features of different levels, the proposed method achieved better performance compared to GAT-Low and GAT-High. Additionally, our attention-based feature fusion module better incorporates hierarchical information than Concat-GAT, resulting in accuracy and recall improvements of 5.48% and 6.61%. Overall, our method strikes a good balance between accuracy and efficiency, making it more suitable for ESA detection while saving valuable time and resources for downstream tasks.

Table 2. Quantitative result of comparison methods

Method	OA	Recall	AIT	Params
ResNet-50	91.05	90.42	16.63 ms	25.61M
MobileNetV3	88.40	88.08	14.98 ms	3.8M
ViT-B/32	93.71	93.40	16.67 ms	88.21M
HGP-SL-Low	66.85	66.87	2.04 ms	0.07M
HGP-SL-High	61.64	61.58	0.28 ms	0.14M
HACT-Net	74.53	75.42	2.47 ms	0.79M
GAT-Low	65.15	66.80	2.88 ms	0.09M
GAT-High	62.32	68.06	3.42 ms	0.21M
Concat-GAT	78.41	77.25	6.92 ms	1.02M
Our method	83.89	83.86	6.04 ms	1.01M

5 Conclusion

We propose a framework called the Hierarchical Geometry-to-Semantic Fusion Graph Neural Network for ESA detection. The proposed framework is designed based on the human brain's ability in comprehending and interpreting images. It leverages GNN to learn high-order semantic information from the satellite imagery. To fill the gap of a lack of benchmark datasets, we created the ESAD dataset based on existing related datasets. Extensive experiments demonstrate that our method achieves a good balance between accuracy and efficiency, which is more suitable for ESA detection with high timeliness requirements. In future work, we will further explore brain-inspired models for better performance and extend our method to on-orbit real-time ESA detection task.

Acknowledgements. This research is partially supported by the National Natural Science Foundation of China Major Program (Grant No. 42192580, 42192583), Hubei Province Natural Science Foundation (Grant No. 2021CFA088 and 2020CFA003), the Science and Technology Major Project (Grant No. 2021-AAA010), and Wuhan University - Huawei Geoinformatics Innovation Laboratory. The numerical calculations in this paper had been supported by the super-computing system in the Supercomputing Center of Wuhan University.

References

1. Qiao, W.: Research framework of remote sensing monitoring and real-time diagnosis of earth surface anomalies. Acta Geodaetica et Cartographica Sinica **51**(7), 1141 (2022)
2. Mateo-Garcia, G., et al.: Flood detection on low cost orbital hardware. In: Artificial Intelligence for Humanitarian Assistance and Disaster Response Workshop at NeurIPS (2019)
3. Weber, E., Papadopoulos, D.P., Lapedriza, A., Ofli, F., Imran, M., Torralba, A.: Incidents1M: a large-scale dataset of images with natural disasters, damage, and incidents. IEEE Trans. Pattern Anal. Mach. Intell. **45**, 4768–4781 (2022)
4. Rui, X., Cao, Y., Yuan, X., Kang, Y., Song, W.: DisasterGAN: generative adversarial networks for remote sensing disaster image generation. Remote Sens. **13**(21), 4284 (2021)
5. Weber, E., Kané, H.: Building disaster damage assessment in satellite imagery with multi-temporal fusion. arXiv preprint arXiv:2004.05525 (2020)
6. Saha, S., Shahzad, M., Ebel, P., Zhu, X.X.: Supervised change detection using prechange optical-SAR and postchange SAR data. IEEE J. Sel. Top. Appl. Earth Obs. Remote Sens. **15**, 8170–8178 (2022)
7. Tilon, S., Nex, F., Kerle, N., Vosselman, G.: Post-disaster building damage detection from earth observation imagery using unsupervised and transferable anomaly detecting generative adversarial networks. Remote Sens. **12**(24), 4193 (2020)
8. Liu, Q., et al.: Unsupervised detection of contextual anomaly in remotely sensed data. Remote Sens. Environ. **202**, 75–87 (2017)
9. Ding, C., Pang, G., Shen, C.: Catching both gray and black swans: open-set supervised anomaly detection. In: Proceedings of the IEEE/CVF Conference on Computer Vision and Pattern Recognition, pp. 7388–7398 (2022)

10. Gupta, R., et al.: Creating XBD: a dataset for assessing building damage from satellite imagery. In: Proceedings of the IEEE/CVF Conference on Computer Vision and Pattern Recognition Workshops, pp. 10–17 (2019)
11. Rudner, T.G., Rußwurm, M., Fil, J., Pelich, R., Bischke, B., Kopačková, V., Biliński, P.: Multi3net: segmenting flooded buildings via fusion of multiresolution, multisensor, and multitemporal satellite imagery. In: Proceedings of the AAAI Conference on Artificial Intelligence, pp. 702–709 (2019)
12. Zeng, C., Cao, Z., Su, F., Zeng, Z., Changxi, Y.: A dataset of high-precision aerial imagery and interpretation of landslide and debris flow disaster in Sichuan and surrounding areas between 2008 and 2020. China Sci. Data **7**(2) (2022)
13. Xu, J.Z., Lu, W., Li, Z., Khaitan, P., Zaytseva, V.: Building damage detection in satellite imagery using convolutional neural networks. arXiv preprint arXiv:1910.06444 (2019)
14. Lee, J., et al.: Assessing post-disaster damage from satellite imagery using semi-supervised learning techniques. In: Artificial Intelligence for Humanitarian Assistance and Disaster Response Workshop at NeurIPS (2020)
15. Kyrkou, C., Theocharides, T.: EmergencyNet: efficient aerial image classification for drone-based emergency monitoring using atrous convolutional feature fusion. IEEE J. Sel. Top. Appl. Earth Obs. Remote Sens. **13**, 1687–1699 (2020)
16. Niu, C., Gao, O., Lu, W., Liu, W., Lai, T.: Reg-SA-UNet++: a lightweight landslide detection network based on single-temporal images captured postlandslide. IEEE J. Sel. Top. Appl. Earth Obs. Remote Sens. **15**, 9746–9759 (2022)
17. Liu, Q., Xiao, L., Yang, J., Wei, Z.: CNN-enhanced graph convolutional network with pixel-and superpixel-level feature fusion for hyperspectral image classification. IEEE Trans. Geosci. Remote Sens. **59**(10), 8657–8671 (2020)
18. Jiang, J., Ma, J., Liu, X.: Multilayer spectral-spatial graphs for label noisy robust hyperspectral image classification. IEEE Trans. Neural Netw. Learn. Syst. **33**(2), 839–852 (2020)
19. Xi, B., et al.: Semisupervised cross-scale graph prototypical network for hyperspectral image classification. IEEE Trans. Neural Netw. Learn. Syst. (2022)
20. Gao, Y., Shi, J., Li, J., Wang, R.: Remote sensing scene classification based on high-order graph convolutional network. Eur. J. Remote Sens. **54**(sup1), 141–155 (2021)
21. Liang, J., Deng, Y., Zeng, D.: A deep neural network combined CNN and GCN for remote sensing scene classification. IEEE J. Sel. Top. Appl. Earth Obs. Remote Sens. **13**, 4325–4338 (2020)
22. Achanta, R., Shaji, A., Smith, K., Lucchi, A., Fua, P., Süsstrunk, S.: SLIC superpixels compared to state-of-the-art superpixel methods. IEEE Trans. Pattern Anal. Mach. Intell. **34**(11), 2274–2282 (2012)
23. Pati, P., et al.: Hierarchical graph representations in digital pathology. Med. Image Anal. **75**, 102264 (2022)
24. Veličković, P., Cucurull, G., Casanova, A., Romero, A., Lio, P., Bengio, Y.: Graph attention networks. arXiv preprint arXiv:1710.10903 (2017)
25. Zhang, Z., et al.: Hierarchical graph pooling with structure learning. arXiv preprint arXiv:1911.05954 (2019)

Zero-Shot Incremental Learning Algorithm Based on Bi-alignment Mechanism

Yang Zhao[1] , Jie Ren[1] , and Weichuan Zhang[2]([✉])

[1] Xi'an Polytechnic University, Xi'an 710600, Shaanxi, China
renjie@xpu.edu.cn
[2] Griffith University, Brisbane, QLD 4702, Australia
zwc2003@163.com

Abstract. Zero-shot incremental learning aims to enable the model to generalize to new classes without forgetting previously learned classes. However, the problem of the semantic gap between old and new sample classes has been puzzling to researchers. Therefore, this paper proposes a zero-shot incremental learning algorithm based on a bi-alignment mechanism, called BANet, which is mainly divided into an intra-class alignment module Intra-CA and an inter-class alignment module Inter-CA. The model can better extract image features, improve image classification accuracy, and fundamentally alleviate the catastrophic forgetting of the network. Extensive experiments have been conducted on two basic datasets, CUB-200-2011 and CIFAR100, and the experimental results show that our proposed algorithm outperforms the current state-of-the-art incremental learning algorithms.

Keywords: catastrophic forgetting · incremental learning · zero-shot learning · attention mechanism · alignment mechanism

1 Introduction

In recent years, with the continuous development of science and technology, machine learning [1–7] has made remarkable achievements. Machine learning is a science that studies how to use computers to simulate or realize human learning activities. However, there has always been a difficult problem in this field: catastrophic forgetting [8]. Different from the ability of human beings to continuously learn new knowledge without forgetting old knowledge, after a machine learning model is trained with a new dataset, the feature distribution will be biased towards the feature distribution of the new dataset. As a result, the performance on old dataset will drop significantly. In order to enable machines to have the same ability as humans to continuously acquire, adjust and transfer knowledge. Scholars have proposed incremental learning [9]. It enables the network to continuously process continuous information flows in the real world. Retain or even integrate and optimize old knowledge while absorbing new knowledge. It is widely used in computer vision, natural language processing, and other fields.

J. Ren et al. (Eds.): BICS 2023, LNAI 14374, pp. 72–81, 2024.
https://doi.org/10.1007/978-981-97-1417-9_7

The existing incremental learning techniques can be divided into three categories: incremental learning based on regularization, incremental learning based on replay, and incremental learning based on parameter isolation. The current research on incremental learning is mainly based on regularization-based incremental learning because the incremental learning method based on regularization is closer to the real goal of incremental learning. Li et al. [10] proposed the learning without forgetting (LWF) model which is a milestone in incremental learning technique. Jung et al. [11] presented less-forgetting learning (LFL) model which uses the stochastic gradient descent method for effectively reducing the forgetting of old sample information. Based on the LWF model, James et al. [12] introduced a parameter-related regular loss and proposed the elastic weight consolidation (EWC) model. Rahaf et al. [13] proposed a memory aware synapses (MAS) model to calculate the importance of each parameter in the model for the task.

However, the study found that, unlike the traditional classification network, the main cause of catastrophic forgetting is the semantic gap [14] between the two embedding spaces, which is caused by the different categories of two adjacent tasks. In order to solve the problem of semantic gap between old and new samples, the distance between old and new samples is reduced. Yu et al. [15] proposed the semantic drift compensation (SDC) model for semantic compensation. And inspired by the SDC model, in order to more effectively solve the semantic gap between two adjacent categories, Wei et al. [16] combined zero-shot learning [17] with incremental learning and proposed a zero-shot translation class-incremental (ZSTCI) model. Among them, zero-shot learning refers to the process in which the model can use the knowledge of seen class to identify unseen class without training samples.

Currently, when applying zero-shot incremental learning to image classification tasks, researchers only focus on building the connection between old knowledge and new knowledge. The root cause of the rapid forgetting of the network model during the incremental learning process is ignored. Therefore, Prithviraj et al. [18] proposed the learning without memorizing (LWM) model, which proved the importance of attention in the incremental learning process. The so-called attention [19] is to make the network model focus on local information, thereby strengthening useful information and suppressing useless information.

However, the single-channel attention mechanism often cannot capture the complex global information of the sample image, which makes it difficult to express the rich features of the sample, thereby inhibiting the diversity of the sample. At the same time, for the same new sample category, its intra-class distance needs to be considered to make the feature information of the new sample more compact. For different sample categories, the inter-class distance needs to be considered so that the distance between old and new samples is smaller. In order to solve the above problems, this paper proposes a zero-shot incremental learning method based on the bi-alignment mechanism as shown in Fig. 1, called BANet, which is mainly divided into the intra-class alignment module Intra-CA and the inter-class alignment module Inter-CA. The intra-class alignment module makes the network pay more attention to the region of interest, and obtains the global information of the input sample, making the feature information of the new sample more compact. The inter-class alignment module enables a unified representation between old and new tasks, making the distance between old and new samples smaller. When combined with the underlying zero-shot incremental learning method, our method can effectively improve the image classification accuracy and alleviate the catastrophic forgetting of the network.

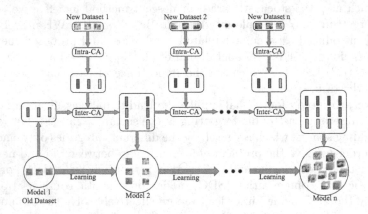

Fig. 1. BANet architecture.

2   Method

2.1   Feature Extraction Module Based on Intra-class Alignment: Intra-CA

For the same image, the main way to make it pass the spatial attention mechanism SENet [20] is global average pooling. The global average pooling is mainly to average the two-dimensional images of each channel, that is, to calculate the mean value corresponding to each channel. Although this can reduce the amount of calculation and inhibit the occurrence of overfitting, it only uses the low-frequency information of the image. It ignores some useful high-frequency information in the image, making it difficult to express the rich information of the image. The frequency domain attention mechanism FcaNet [21] was proposed. Although it is a good solution to the lack of information in SENet, the two-dimensional DCT transform he used is also a DCT transform for all channels, which increases the amount of calculation between the invisible. And there is no good adjustment of the convergence speed of the model training.

Therefore, this paper is based on the spatial attention mechanism SENet and the frequency domain attention mechanism FcaNet. Through the alignment mechanism, feature extraction module based on intra-class alignment: Intra-CA shown in Fig. 2 is designed. It makes the network focus more useful points of interest. And the global information of the input sample is obtained, so that the feature information of the new sample is more compact. The main design idea of Intra-CA is to send the image features extracted by the convolutional neural network into the spatial domain attention mechanism SENet and the frequency domain attention mechanism FcaNet in parallel. The global average pooling adopted in SENet can effectively extract the low-frequency information of the image. At the same time, the high-frequency information after DCT transformation of each channel in FcaNet is selected. That is, the high-frequency information and low-frequency information of the image are sent to the alignment mechanism at the same time, and the output high- and low-frequency information is subjected to feature fusion. Finally, image feature information is output. This information combines the useful high and low frequency components of the image, possesses the global information of the input sample, and expresses its rich image features.

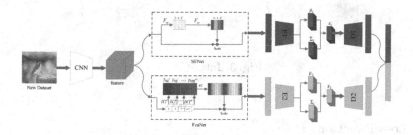

Fig. 2. Intra-CA.

The alignment mechanism used in this paper is cross and distribution aligned VAE (CADA-VAE) [22]. It is mainly used to achieve intra-class alignment and inter-class alignment. Intra-class alignment refers to aligning different features of the same sample. Inter-class alignment refers to aligning different features of old and new samples. The features are optimized by cross-aligned and distributed-aligned variational auto-encoders (VAE). So as to learn the complex feature structure of the image. A feature fusion network is constructed to classify new samples.

2.2 Feature Extraction Module Based on Inter-class Alignment: Inter-CA

For incremental learning, the addition of new samples will directly affect the performance of the old model. Because the old and new samples belong to different categories, the semantic gap between samples is too large. The problem of semantic drift will arise, which will fundamentally cause catastrophic forgetting of the network. Therefore, this paper designs a feature extraction module based on inter-class alignment: Inter-CA as shown in Fig. 3, so that a unified representation can be obtained between the old and new samples. And the distance between the old and new samples can be reduced. The main design idea of Inter-CA is to align the features of the old samples with the features of the new samples, where the features of the new samples are the features extracted by Intra-CA. Finally, the aligned new sample features are sent to the Nearest Class Mean (NCM) classifier for classification, and the most suitable category is selected.

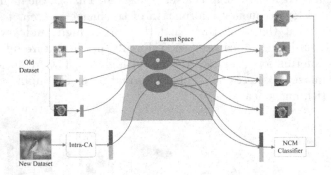

Fig. 3. Inter-CA.

2.3   Loss Function

Because in the embedded network, the distance between related images is much smaller than the distance between unrelated images. Therefore, in order to ensure that the anchor point is close to the positive sample and away from the negative sample. The objective function is as follows:

$$\mathcal{L}_{tri} = \max(0, d_+ - d_- + m), \tag{1}$$

where d_+ represents the L2-distance between the anchor point z_a and the positive sample z_p, d_- represents the L2-distance between the anchor point z_a and the negative sample z_n. And $z_i = F(x_i)$, where x_i represents the input image, z_i represents the potential features of the input image in the embedding space. The m is the margin value.

In the alignment mechanism, the VAE loss function is:

$$\mathcal{L}_{VAE} = \sum_i^M \mathbb{E}_{q_\phi(z|x)}[\log p_\theta(x^{(i)}|z)] - \varepsilon D_{KL}(q_\phi(z|x^{(i)})||p_\theta(z)), \tag{2}$$

where M represents the input image features, and ε represents the weight of KL divergence. The CA loss function is:

$$\mathcal{L}_{CA} = \sum_i^M \sum_{j \neq i}^M |x^{(j)} - D_j(E_i(x^{(i)}))|, \tag{3}$$

where D_j is the jth decoder and E_i is the ith encoder. The DA loss function is:

$$\mathcal{L}_{DA} = \sum_i^M \sum_{j \neq i}^M W_{ij}, \tag{4}$$

$$W_{ij} = (\|\mu_i - \mu_j\|_2^2 + \|\Sigma_i^{\frac{1}{2}} - \Sigma_j^{\frac{1}{2}}\|_{Frobenius}^2)^{\frac{1}{2}}, \tag{5}$$

where wasserstein distance is used to describe the difference between two distributions. So the total loss function consists of four parts:

$$\mathcal{L}_{SF} = \mathcal{L}_{tri} + \mathcal{L}_{VAE} + \alpha \mathcal{L}_{CA} + \beta \mathcal{L}_{DA}, \tag{6}$$

where α and β represent the weighting factors of cross alignment loss and distributed alignment loss, respectively.

3 Experiments

In this paper, the dataset CUB-200-2011 [23] and CIFAR100 [24] are used for experimental verification. All models are implemented by Pytorch and optimized by Adam optimizer. CUB-200-2011 used ResNet-18 as the backbone network, while CIFAR100 used ResNet-32 as the backbone network without pre-training. The image size of CUB-200-2011 was adjusted to 256×256, and the image size of CIFAR100 was adjusted to 32×32. Then they are randomly cut and flipped. Among them, the epoch size is set to 50. The batch size is set to 64. The learning rates on CUB-200-2011 and CIFAR100 are set to 1e-5 and 1e-6, respectively. In this paper, the average incremental accuracy and average forgetting are selected as the evaluation indexes of the experiment. The average incremental accuracy reflects the overall performance of the test set in the whole task. The average forgetting is to estimate how much learned knowledge has been forgotten.

3.1 Results and Analysis

We summarize the average incremental accuracy results of the baseline methods E-FT, E-LWF, E-EWC and E-MAS and our method on the two basic datasets CUB-200-2011 and CIFAR100, as shown in Table 1. It can be seen that the method equipped with BANet achieves relatively best results in all tasks on both datasets. Furthermore, all experimental results in the table demonstrate the existence of catastrophic forgetting in embedded networks. At the same time, it is also proved that adding SDC, ZSTCI, and BANet to the baseline method can improve the performance of the embedded network, but the method we propose is more conducive to alleviating the catastrophic forgetting of the network.

Table 1. The average incremental accuracy

Dataset	CUB					CIFAR100				
Method	T2	T4	T6	T8	T10	T2	T4	T6	T8	T10
E-FT	80.2	62.9	54.3	40.8	42.4	79.8	73.0	59.0	47.9	38.8
E-FT+SDC	85.4	70.2	63.6	58.3	55.7	79.7	73.3	60.5	49.5	41.2
E-FT+ZSTCI	85.5	70.3	62.8	57.2	53.1	80.1	73.8	62.8	53.3	46.4
E-FT+BANet	**84.7**	**72.2**	**64.9**	**60.0**	**57.4**	**88.0**	60.3	55.9	51.8	**47.5**
E-LWF	81.0	66.7	56.9	46.4	43.1	79.6	72.7	58.5	47.4	37.5
E-LWF+SDC	84.5	71.2	63.8	57.8	54.5	79.6	73.4	60.2	49.7	40. 9
E-LWF+ZSTCI	84.2	72.5	64.5	59.0	55.8	79.6	74.0	63.3	53.8	46.6
E-LWF+BANet	**82.1**	**71.8**	**65.3**	**60.5**	**58.3**	**80.1**	**74.9**	**66.0**	**55.6**	**48.9**
E-EWC	82.1	65.0	53.2	42.7	42.2	79.1	73.1	59.6	48.0	39.5
E-EWC+SDC	82.0	68.9	62.0	56.7	53.7	79.4	73.7	60.7	44.9	41.9
E-EWC+ZSTCI	83.2	72.6	64.3	58.0	54.9	79.6	73.8	63.0	53.4	46.7
E-EWC+BANet	**84.7**	**72.1**	**64.6**	**60.5**	**57.4**	**80.7**	**74.8**	**65.9**	**55.9**	**49.1**
E-MAS	80.5	62.8	53.8	39.1	41.0	79.7	73.2	59.1	47.5	38.1
E-MAS+SDC	85.4	69.1	63.3	57.7	55.5	79.9	73.7	60.6	49.7	41.3
E-MAS+ZSTCI	86.4	71.3	65.1	58.2	53.9	79.8	73.9	63.1	53.6	46.8
E-MAS+BANet	**84.6**	**70.7**	**65.6**	**60.1**	**57.7**	**80.4**	**74.7**	**66.0**	**55.7**	**48.7**

As can be seen from the data in the table, compared with the current state-of-the-art method, the proposal of BANet can better improve the semantic gap problem in the two tasks. And it can be more easily combined with existing baseline methods to effectively improve model performance.

Figure 4 shows the average forgetting results on the two basic data sets of CUB-200-2011 and CIFAR100. It can be seen from Fig. 4 the average forgetting of all methods becomes evident as the categories increase. Thus demonstrating the existence of catastrophic forgetting in embedded networks. However, after combining our proposed BANet method, the forgetting effect becomes less. It is proved again that our proposed method can effectively mitigate catastrophic forgetting.

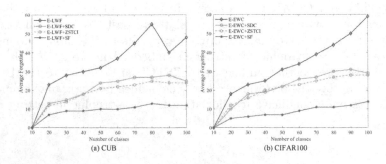

Fig. 4. Average Forgetting.

3.2   Ablation Study

We conduct a set of ablation experiments to investigate the effectiveness of our method. That is, the results of adding different modules in the basic model are shown in Table 2. The basic models include E-FT, E-LWF, E-EWC and E-MAS. On the basis of the above models, the spatial attention mechanism SENet and the frequency domain attention mechanism FcaNet are respectively added. When adding a certain attention mechanism alone, although the effect is improved, it is not the best. When adding BANet to the model, the effect has been significantly improved, and the catastrophic forgetting of the network has also become soothing.

Table 2. The average incremental accuracy

Dataset	CUB				CIFAR100			
Method	E-FT	E-LWF	E-EWC	E-MAS	E-FT	E-LWF	E-EWC	E-MAS
Base	42.4	43.1	42.2	41.0	38.8	37.5	39.5	38.1
+SE	50.4	50.4	51.2	50.3	40.9	40.7	40.9	41.1
+Fca	49.9	49.9	50.6	50.6	41.5	41.3	41.0	41.4
+BANet	**57.4**	**58.3**	**57.4**	**57.7**	**47.5**	**48.9**	**49.1**	**48.7**

When only "SENet" is added to the base model, the CUB dataset can be improved by up to 9.3%, and the CIFAR100 dataset by up to 3.2% compared to the baseline network. When only "FcaNet" is added to the base model, the CUB dataset can be improved by up to 9.6%, and the CIFAR100 dataset by up to 3.8% compared to the baseline network. When "BANet" is added to the base model, the CUB dataset can be improved by up to 16.7%, and the CIFAR100 dataset by up to 11.4% compared to the baseline network.

3.3   Visualization

As shown in Fig. 5, they are the Grad-CAM [25] visualization effect diagrams with the addition of SENet, FcaNet, and BANet three attention mechanisms, in which the baseline network is Resnet. It can be seen from the figure that the attention mechanism can effectively make the network focus on more interesting regions. The bi-alignment mechanism BANet proposed by us can more effectively capture the key information in the image and suppress the interference of background information.

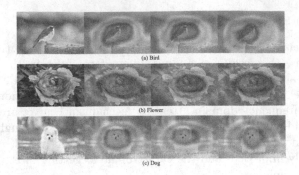

Fig. 5. Grad-CAM visualization results. From left to right is the original image, adding SENet, adding FcaNet, and adding SFNet.

4   Conclusion

In this paper, we propose a novel zero-shot incremental learning method based on double alignment mechanism to alleviate the catastrophic forgetting problem in embedding networks. While extracting features, reduce the intra-class distance and inter-class distance. Intra-class alignment enables the network to pay more attention to the region of interest and obtain compact global information of the input samples. The inter-class alignment enables a unified representation between old and new tasks and reduces the gap between old and new samples. Moreover, our proposed method can be flexibly combined with other regularization-based incremental learning methods to further improve the model performance. Experiments show that our method substantially outperforms previous methods on two benchmark datasets.

Acknowledgements. This work was partially supported by the Shaanxi Provincial Natural Science Basic Research Program under Grant NO. 2022JM-394.

References

1. Jing, J., Gao, T., Zhang, W., Gao, Y., Sun, C.: Image feature information extraction for interest point detection: a comprehensive review. IEEE Trans. Pattern Anal. Mach. Intell. **45**, 4694–4712 (2022)
2. Zhang, W., Sun, C., Gao, Y.: Image intensity variation information for interest point detection. IEEE Trans. Pattern Anal. Mach. Intell. (2023)
3. Zhang, W., Sun, C.: Corner detection using second-order generalized Gaussian directional derivative representations. IEEE Trans. Pattern Anal. Mach. Intell. **43**(4), 1213–1224 (2019)
4. Jing, J., Liu, S., Wang, G., Zhang, W., Sun, C.: Recent advances on image edge detection: a comprehensive review. Neurocomputing (2022)
5. Lu, J., Zhang, W., Zhao, Y., Sun, C.: Image local structure information learning for fine-grained visual classification. Sci. Rep. **12**(1), 19205 (2022)
6. Zhang, W., Sun, C.: Corner detection using multi-directional structure tensor with multiple scales. Int. J. Comput. Vis. **128**(2), 438–459 (2020)

7. Zhang, W., Sun, C., Breckon, T., Alshammari, N.: Discrete curvature representations for noise robust image corner detection. IEEE Trans. Image Process. **28**(9), 4444–4459 (2019)

8. Robins, A.: Catastrophic forgetting, rehearsal and pseudorehearsal. Connect. Sci. **7**(2), 123–146 (1995)

9. Mccloskey, M., Cohen, N.J.: Catastrophic interference in connectionist networks: the sequential learning problem. Psychol. Learn. Motiv. **24**, 109–165 (1989)

10. Li, Z., Hoiem, D.: Learning without forgetting. IEEE Trans. Pattern Anal. Mach. Intell. **40**(12), 2935–2947 (2017)

11. Jung, H., Ju, J., Jung, M., Kim, J.: Less-forgetting learning in deep neural networks. arXiv preprint arXiv:1607.00122 (2016)

12. Kirkpatrick, J., et al.: Overcoming catastrophic forgetting in neural networks. Proc. Natl. Acad. Sci. **114**(13), 3521–3526 (2017)

13. Aljundi, R., Babiloni, F., Elhoseiny, M., Rohrbach, M., Tuytelaars, T.: Memory aware synapses: learning what (not) to forget. In: Proceedings of the European Conference on Computer Vision, pp. 139–154 (2018)

14. Li, F., Perona, P.: A Bayesian hierarchical model for learning natural scene categories. In: Proceedings of the IEEE Computer Society Conference on Computer Vision and Pattern Recognition, vol. 2, pp. 524–531 (2005)

15. Yu, L., et al.: Semantic drift compensation for class-incremental learning. In: Proceedings of the Conference on Computer Vision and Pattern Recognition, pp. 6982–6991 (2020)

16. Wei, K., Deng, C., Yang, X., Li, M.: Incremental embedding learning via zero-shot translation. In: Proceedings of the AAAI Conference on Artificial Intelligence, vol. 35, pp. 10254–10262 (2021)

17. Lampert, C.H., Nickisch, H., Harmeling, S.: Learning to detect unseen object classes by between-class attribute transfer. In: IEEE Conference on Computer Vision and Pattern Recognition, pp. 951–958 (2009)

18. Dhar, P., Singh, R.V., Peng, K.C., Wu, Z., Chellappa, R.: Learning without memorizing. In: Proceedings of the Conference on Computer Vision and Pattern Recognition, pp. 5138–5146 (2019)

19. Bahdanau, D., Cho, K., Bengio, Y.: Neural machine translation by jointly learning to align and translate. arXiv preprint arXiv:1409.0473 (2014)

20. Hu, J., Shen, L., Sun, G.: Squeeze-and-excitation networks. In: Proceedings of the IEEE Conference on Computer Vision and Pattern Recognition, pp. 7132–7141 (2018)

21. Qin, Z., Zhang, P., Wu, F., Li, X.: FcaNet: frequency channel attention networks. In: Proceedings of the International Conference on Computer Vision, pp. 783–792 (2021)

22. Schonfeld, E., Ebrahimi, S., Sinha, S., Darrell, T., Akata, Z.: Generalized zero- and few-shot learning via aligned variational autoencoders. In: Proceedings of the Conference on Computer Vision and Pattern Recognition, pp. 8247–8255 (2019)

23. Wah, C., Branson, S., Welinder, P., Perona, P., Belongie, S.: The Caltech-UCSD birds-200-2011 dataset (2011)

24. Krizhevsky, A., Hinton, G., et al.: Learning multiple layers of features from tiny images (2009)

25. Selvaraju, R.R., Cogswell, M., Das, A., Vedantam, R., Parikh, D., Batra, D.: Grad-CAM: visual explanations from deep networks via gradient-based localization. In: Proceedings of the IEEE International Conference on Computer Vision, pp. 618–626 (2017)

Cross-Modal Transformer GAN: A Brain Structure-Function Deep Fusing Framework for Alzheimer's Disease

Junren Pan[1], Changhong Jing[1], Qiankun Zuo[1], Martin Nieuwoudt[2], and Shuqiang Wang[1](✉)

[1] Shenzhen Institutes of Advanced Technology, Chinese Academy of Sciences, Shenzhen 518000, China
sq.wang@siat.ac.cn

[2] Institute for Biomedical Engineering, Stellenbosch University, Stellenbosch 7600, South Africa

Abstract. Cross-modal fusion of different types of neuroimaging data has shown great promise for predicting the progression of Alzheimer's Disease(AD). However, most existing methods applied in neuroimaging can not efficiently fuse the functional and structural information from multi-modal neuroimages. In this work, a novel cross-modal transformer generative adversarial network(CT-GAN) is proposed to fuse functional information contained in resting-state functional magnetic resonance imaging (rs-fMRI) and structural information contained in Diffusion Tensor Imaging (DTI). The developed bi-attention mechanism can match functional information to structural information efficiently and maximize the capability of extracting complementary information from rs-fMRI and DTI. By capturing the deep complementary information between structural features and functional features, the proposed CT-GAN can detect the AD-related brain connectivity, which could be used as a bio-marker of AD. Experimental results show that the proposed model can not only improve classification performance but also detect the AD-related brain connectivity effectively.

Keywords: Cross-modal fusion · Transformer · Bi-attention Mechanism · Brain Network · Generative Adversarial Strategy

1 Introduction

Alzheimer's disease (AD), a chronic and irreversible neurodegenerative disease, is the main reason for dementia among aged people [1]. According to statistical data given by Alzheimer's Association [2], at least 50 million people worldwide are suffering from AD. AD patients will gradually lose cognitive function such as remembering or thinking, and eventually become unable to take care of themselves [3]. The widespread incidence of AD makes a severe financial burden to both patients' families and governments. With the development of artificial intelligence [4–9], researchers study AD from different angles using machine learning

J. Ren et al. (Eds.): BICS 2023, LNAI 14374, pp. 82–92, 2024.
https://doi.org/10.1007/978-981-97-1417-9_8

technology [10–12]. However, the cause of AD has not been fully revealed. One of the main reasons for the above difficulties is that brain is a highly complex network, and completing cognitions requires specific coordination between regions-of-interest (ROIs).

A brain network can be characterized as a graph. The nodes of a brain network represent ROIs of the brain. The edges of a brain network represent the interaction relationship between ROIs of the brain. There are two basic connectivity categories of brain networks: functional connectivity (FC) and structural connectivity (SC). FC is defined as the interdependence between the blood-oxygen-level-dependent (BOLD) signals of two ROIs, where BOLD signals can be extracted from rs-fMRI. SC is defined as the neural fibers connection strength among ROIs, which can be extracted from DTI. Many studies [13–15] have used FC or SC to obtain some AD-related features that can not be discovered in traditional imaging methods. This shows that brain network methods have more advantages than the traditional imaging method in AD research. However, most existing brain network studies are based on a single modal, which can only focus on one of the brain structural information and brain functional information. Since single modality data may only contain complementary cross-modal information partially, it will lose an opportunity to take advantage of complementary cross-modal information to study AD more accurately. Therefore multimodal brain network methods [16–19] are gaining more and more attention in medical imaging computing. For the structure-function deep fusing task of multimodal brain networks, the key is how to efficiently use complementary cross-modal information that is heterogeneous and hidden in different types of neuroimaging data. Most existing structure-function fusion approaches just used linear relationships between different modalities. However, changes of brain structure and function can not be fully characterized by linear relationships. Previous studies [20,21] proved that strong SC inclines to be accompanied with strong FC, but not vice versa. On the other hand, clinical studies [22–24] show that when an SC between RIOs is reduced, some regions can increase functional activity to compensate for the reduced SC.

To overcome the above problem, a novel cross-modal transformer is proposed in this work to deal with the structure-function deep fusion task based on generative adversarial networks(GANs). GANs can bee seen as variational-inference [25,26] based generative model. GANs are proved to be an efficient framework for learning complex distribution [27,28]. Currently, GANs are successfully used in many branches of medical image analysis [29–34]. Transformers [35] have shown their powerful capability for sequential analyzing in natural language processing (NLP). This is due to the self-attention mechanism which characterizes the nonlinear relationships between given inputs. Following their successful applications in the area of NLP, transformers have been adopted to image tasks very recently [36–38]. However, transformers have been few explored in the area of brain networks. In this study, a cross-modal transformer is proposed to fuse structure-function information of brain networks.

2 Method

2.1 The Architecture

The proposed CT-GAN is illustrated in Fig. 1, which consists of four components: 1) a cross-modal transformer generator that outputs multimodal connectivity brain networks; 2) two decoders that decode multimodal connectivity to the corresponding SC and FC, respectively; 3) two discriminators, one of which determines whether an SC from learned by our proposed model or output by a software template, the other discriminator for FC is similar; 4) a classifier that predicts AD stages according to multimodal connectivity.

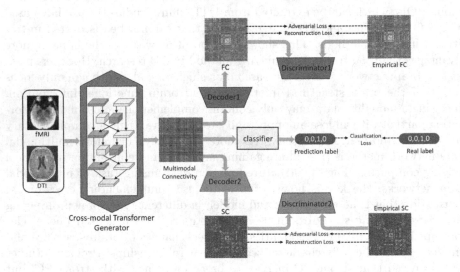

Fig. 1. The framework of the proposed CT-GAN.

2.2 Bi-attention Mechanism

The key idea of this work is to exploit the bi-attention mechanism of transformers to fuse structure-function information for rs-fMRI and DTI given their complementary nature. A transformer mapping an input feature sequence as $X = \mathbb{R}^{n \times d_X}$ to a target feature sequence as $Y = \mathbb{R}^{n \times d_Y}$, where n is the total number of ROIs, can be described as follows. Firstly, a linear projections is used to compute a set of queries matrix Q, keys matrix K, and values matrix V,

$$Q = XW^q, \quad K = XW^k, \quad V = XW^v \tag{1}$$

where $W^q \in \mathbb{R}^{d_X \times d_q}$, $W^k \in \mathbb{R}^{d_X \times d_k}$ with $d_k = d_q$, and $W^v \in \mathbb{R}^{d_X \times d_v}$ are weight matrices. The attention of Q, K, and V can be obtained as follows:

$$\text{Attention}(Q, K, V) = \text{softmax}\left(\frac{QK^T}{\sqrt{d_k}}\right)V. \tag{2}$$

And then, a fully connected layer(FC) is used to transform the attention of Q, K, and V into a feature sequence with the same dimension as the target feature sequence Y. Finally, the output of a transformer is

$$X^{out} = \text{FC}\big(\text{Attention}(XW^q, XW^k, XW^v)\big) + \lambda Y, \tag{3}$$

where λ is a hyper-parameter between 0 to 1.

There are two types of transformers in the proposed cross-modal generator. The one is used to transform functional information into structural information, abbreviated as F2S. The other is used to transform structural information into functional information, abbreviated as S2F. We can now introduce the bi-attention mechanism of transformers. BOLD signals extracted from rs-fMRI and empirical structural connectivity extracted from DTI are inputs to our cross-modal generator that uses several pairs of F2S and S2F modules for fusing intermediate features between the inputs. The intermediate features of BOLD signals and empirical structural connectivity are extracted by CNN and GCN, respectively. The detailed architecture of generator is shown in Fig. 2.

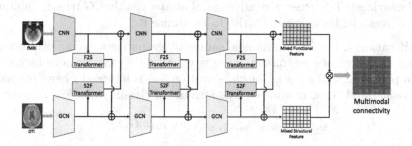

Fig. 2. The network architecture of proposed generator.

Let us denote the mixed functional feature sequence and mixed structural feature sequence in Fig. 2 to be F and S, respectively. It is worth mentioning that $F \in \mathbb{R}^{n \times d}$ represents the feature sequence of each ROI and $S \in \mathbb{R}^{n \times n}$ represents the connective feature between ROIs where n is the total number of ROIs. The multimodal connectivity matrix MC as the final output of the proposed generator is obtained by

$$\text{MC} = SFF^T S^T. \tag{4}$$

2.3 Loss Function

Success in structure-function deep fusion tasks requires semantic reasoning. Therefore a multimodal connectivity matrix learned by the proposed model should be able to decouple to be the corresponding empirical FC matrix and empirical SC matrix. A generative adversarial strategy is used to achieve the above process. To guarantee the quality of generated multimodal connectivity, a hybrid loss function is used to optimize the network, including three types of terms: the adversarial loss, the classification loss, and the pair-wise connectivity reconstruction loss. The details are given as follows.

Adversarial Loss. To make the FC matrix and SC matrix decoded by multi-modal connectivity matrix as close as possible to empirical FC matrices and SC matrices, the adversarial losses are defined as

$$\mathcal{L}_{adv}^{SC} = \mathbb{E}_x[\log D_1(SC_x)] + \mathbb{E}_x[\log(1 - D_1(Dec_1(G(x))))], \tag{5}$$

$$\mathcal{L}_{adv}^{FC} = \mathbb{E}_x[\log D_2(FC_x)] + \mathbb{E}_x[\log(1 - D_2(Dec_2(G(x))))], \tag{6}$$

where G is the generator, D_1 and D_2 are the discriminator1 and discriminator2, Dec_1 and Dec_2 are the decode1 and decode2 in Fig. 1. SC_x and FC_x represent the empirical SC matrix and empirical FC matrix of subject x. The generator G generates a multimodal connectivity matrix $G(x)$ from subject x's rs-fMRI and DTI. The decoders Dec_1 and Dec_2 decode $G(x)$ into an FC matrix and an SC matrix, respectively. While the discriminator D_1 attempts to distinguish between the FC matrix decoded by Dec_1 and empirical FC matrices(D_2 similarly works on SC matrices). The generative adversarial strategy is that G tries to minimize above adversarial losses while D tries to maximize it.

Classification Loss. For multimodal connectivity matrices, an important index to judge the effect of cross-modal fusing is whether they can achieve high accuracy in predicting AD stages. The classification loss is imposed when optimizing the classifier and the generator G. The formula of classification loss is given by

$$\mathcal{L}_{cls} = \mathbb{E}_{(x,y) \sim p_{real}(x,y)}[-\log p_c(y|G(x))], \tag{7}$$

where y represents AD stages, including normal controls(NC), early mild cognitive impairment(EMCI), late mild cognitive impairment(LMCI), and Alzheimer's Disease(AD). The $p_c(y|G(x))$, output by the classifier with the input $G(x)$, represents the probability that the subject x is now under stage y. By classification loss, the generator is trained to extract and fuse features, which contains more disease information, from rs-fMRI and DTI. Meanwhile, the classifier can achieve the highest predicting accuracy through multimodal connectivity matrices.

Pair-Wise Connectivity Reconstruction Loss. The L^1 pair-wise connectivity reconstruction loss is used to impose an additional topological constraint on the generator G. It means that the generator G and the decoders $Dec1, Dec2$ are not only needed to fool the discriminators $D1$ and $D2$. In addition, they also need to minimize the sum of pair-wise connectivity difference between FC/SC matrices decoded by $Dec1/Dec2$ and empirical FC/SC matrices. The L^1 pair-wise reconstruction losses are formalized as follows:

$$\mathcal{L}_{rcs}^{FC} = \mathbb{E}_x\|FC_x - Dec_1(G(x))\|_1, \tag{8}$$

$$\mathcal{L}_{rcs}^{SC} = \mathbb{E}_x\|SC_x - Dec_2(G(x))\|_1. \tag{9}$$

3   Experiments

DTI and rs-fMRI used to validate the proposed framework are from the public dataset ADNI(Alzheimer's Disease Neuroimaging Initiative). There are 268 subjects' data we used in this study whose detailed information can be seen in Table 1. AAL 90 atlas is used in the preprocessing. A string of preprocessing steps of rs-fMRI using the DPARSF toolbox is consisted of discarding the first 20 volumes, head motion correction, band-pass filtering, Gaussian smoothing, and extracting time series of all voxels. The structural connectivity is obtained by tracking fiber bundles between ROIs. The fiber tracking stopping conditions set in PANDA, the toolbox used to preprocess DTI, are following that: (1) the crossing angle between two consecutive moving directions is more than 45°. (2) the fractional anisotropy value is not in the range of [0.2, 1.0].

Table 1. Subjects' information in this study

Group	AD(63)	LMCI(41)	EMCI(80)	NC(84)
Male/Female	39M/24F	20M/21F	48M/32F	38M/46F
Age(mean ± SD)	75.3 ± 5.5	74.9 ± 5.3	75.8 ± 6.1	74.0 ± 5.9

The decoders Dec_1 and Dec_2, the discriminators D_1 and D_2, the classifier are implemented by multilayer perceptron. The generator G consists of three blocks. Such blocks contain a CNN layer with kernel size (1,10), a GCN layer, an F2S transformer, and an S2F transformer. The hyper-parameter λ in transformers is set to be 0.1. The whole model is trained in an end-to-end manner. The Adam optimizer is chosen to update the model's parameter during the training process. The hyper-parameter of the Adam optimizer, including learning rate, weight decay, and momentum rates, are set to be 0.001, 0.01, and (0.9,0.99), respectively.

After applying the trained model to groups of subjects with the same disease stage, the averaged multimodal connectivity matrices of different disease stages can be obtained. One of the most important biomarkers of AD is crucial brain connectomes which influence the proceeding of AD stages. We characterize such connectomes by analyzing the increase/decrease connectivity of the averaged multimodal connectivity matrices of different disease stages. The visualization of averaged multimodal connectivity matrices and the increase/decrease connectivity with different thresholds are illustrated in Fig. 3. Based on multimodal connectivity matrices, the top 8 ROIs that have the most significant connectivity changing under the AD development process are 37-Hippocampus_L, 38-Hippocampus_R, 46-Cuneus_R, 50-Occipital_Sup_R, 51-Occipital_Mid_L, 60-Parietal_Sup_R, 68-Precuneus_R, and 74-Putamen_R, where the number before ROI is the corresponding AAL id. The connectivity changing between the 8 ROIs above is shown in Fig. 4.

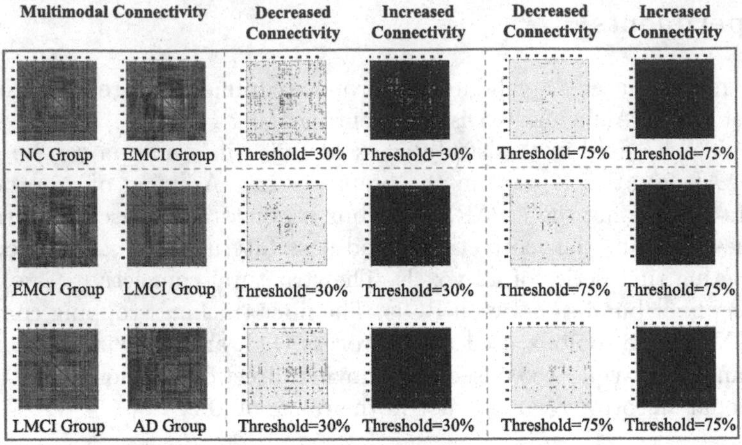

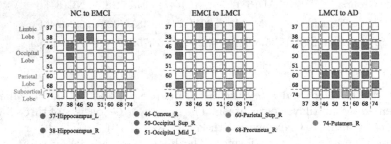

Fig. 3. Averaged multimodal connectivity of different groups with the same disease stage and the changes of such connectivity under the proceeding of AD stages.

Fig. 4. The change of multimodal connectivity between the top 8 AD-related ROIs. The number before ROI is the corresponding AAL id for this ROI. The blue represents decreased connectivity; The yellow represents increased connectivity. The red dotted lines divide the 8 ROIs into their corresponding brain lobe. (Color figure online)

To compare the ability to represent AD-related features with different fMRI-DTI fusion models, three binary classification experiments, including AD vs. NC, LMCI vs. NC, and EMCI vs. NC, are designed. Three index, detection accuracy(ACC), sensitivity(SEN), and specificity(SPEC), are used to evaluate the models performance. The experiments' result are shown in Table 2. The experiments' result shows that the proposed multimodal fusion model has the advantage of higher accuracy for predicting AD stages than other existing multimodal fusion models.

Table 2. Prediction performance in AD vs. NC, LMCI vs. NC, and EMCI vs. NC. under different fMRI-DTI fusion models

Study	Method	AD vs. NC			LMCI vs. NC			EMCI vs. NC		
		ACC	SEN	SPE	ACC	SEN	SPE	ACC	SEN	SPE
Aderghal et al. [39]	CNN+ Transfer Learning	**94.44%**	**93.33%**	**95.24%**	87.1%	87.5%	86.96%	85.37%	88.89%	82.61%
Dyrba et al. [40]	SVM+ Multi-kernel	86.11%	85.71%	86.36%	83.87%	72.73%	90.0%	82.93%	84.21%	81.82%
Lu et al. [41]	GCN+ Deep learning	91.67%	87.5%	95.0%	90.32%	88.89%	90.91%	90.24%	90.0%	90.48%
Proposed	GAN+ Bi-attention	**94.44%**	**93.33%**	**95.24%**	**93.55%**	**90.0%**	**95.24%**	**92.68%**	**90.48%**	**95.0%**

4   Conclusion

In this paper, we proposed a novel CT-GAN to fuse rs-fMRI and DTI to multimodal connectivity of brain network. The key idea of this work is that we use a bi-attention mechanism to achieve the goal of mutual conversion between structural and functional information. Therefore, the bi-attention mechanism above can help the proposed model efficiently extracts the complementary information between rs-fMRI and DTI. The experiments' results proved our multimodal connectivity has higher accuracy of AD prediction than other multimodal fusion methods. Through analyzing our multimodal connectivity matrices, some AD-related connectomes are obtained. These connectomes are highly consistent with the results of clinical AD studies. Although this work focuses only on AD, it is worth mentioning that the proposed model can be easily extended to apply to other neurodegenerative diseases.

Acknowledgement. This work was supported by the National Natural Science Foundations of China under Grant 62172403, the Distinguished Young Scholars Fund of Guangdong under Grant 2021B1515020019 and the Excellent Young Scholars of Shenzhen under Grant RCYX20200714114641211.

References

1. Dadar, M., Pascoal, T.A., Manitsirikul, S., Fonov, V.S., et al.: Validation of a regression technique for segmentation of white matter hyperintensities in Alzheimer's disease. IEEE Trans. Med. Imaging **36**(8), 1758–1768 (2017)
2. Alzheimer's Association: 2019 Alzheimer's disease facts and figures. Alzheimer's Dementia, vol. 15, no. 3, pp. 321–387 (2019)
3. Alzheimer's Association: Alzheimer's disease facts and figures. Alzheimer's Dement, vol. 14, pp. 367–429 (2018)
4. Wang, S., et al.: Skeletal maturity recognition using a fully automated system with convolutional neural networks. IEEE Access **6**, 29979–29993 (2018)
5. Wang, S., Shen, Y., Zeng, D., Hu, Y.: Bone age assessment using convolutional neural networks. In: International Conference on Artificial Intelligence and Big Data (ICAIBD), pp. 175–178 (2018)

6. Wang, S., et al.: Prediction of myelopathic level in cervical spondylotic myelopathy using diffusion tensor imaging. J. Magn. Reson. Imaging **41**(6), 1682–1688 (2015)
7. Wang, S., et al.: An ensemble-based densely-connected deep learning system for assessment of skeletal maturity. IEEE Trans. Syst. Man Cybern. Syst. **52**(1), 426–437 (2020)
8. Hu, S., Yuan, J., Wang, S.: Cross-modality synthesis from MRI to PET using adversarial U-net with different normalization. In: International Conference on Medical Imaging Physics and Engineering (ICMIPE), pp. 1–5 (2019)
9. You, S., et al.: Fine perceptive GANs for brain MR image super-resolution in wavelet domain. IEEE Trans. Neural Netw. Learn. Syst. (2022)
10. Wang, S., Wang, H., Shen, Y., Wang, X.: Automatic recognition of mild cognitive impairment and Alzheimer's disease using ensemble based 3D densely connected convolutional networks. In: 2018 17th IEEE International Conference on Machine Learning and Applications (ICMLA), pp. 517–523 (2018)
11. Wang, S., Wang, H., Cheung, A.C., Shen, Y., Gan, M.: Ensemble of 3D densely connected convolutional network for diagnosis of mild cognitive impairment and Alzheimer's disease. In: Wani, M., Kantardzic, M., Sayed-Mouchaweh, M. (eds.) Deep Learning Applications. AISC, vol. 1098, pp. 53–73. Springer, Singapore (2020). https://doi.org/10.1007/978-981-15-1816-4_4
12. Yu, S., et al.: Multi-scale enhanced graph convolutional network for early mild cognitive impairment detection. In: Martel, A.L., et al. (eds.) MICCAI 2020. LNCS, vol. 12267, pp. 228–237. Springer, Cham (2020). https://doi.org/10.1007/978-3-030-59728-3_23
13. Wang, S., Hu, Y., Shen, Y., Li, H.: Classification of diffusion tensor metrics for the diagnosis of a myelopathic cord using machine learning. Int. J. Neural Syst. **28**(02), 1750036 (2018)
14. Jeon, E., Kang, E., Lee, J., Lee, J., Kam, T.E., Suk, H.I.: Enriched representation learning in resting-state fMRI for early MCI diagnosis. In: Martel, A.L., et al. (eds.) MICCAI 2020. LNCS, vol. 12267, pp. 397–406. Springer, Cham (2020). https://doi.org/10.1007/978-3-030-59728-3_39
15. Wang, S., Shen, Y., Chen, W., Xiao, T., Hu, J.: Automatic recognition of mild cognitive impairment from MRI images using expedited convolutional neural networks. In: Lintas, A., Rovetta, S., Verschure, P., Villa, A. (eds.) ICANN 2017. LNCS, vol. 10613, pp. 373–380. Springer, Cham (2017). https://doi.org/10.1007/978-3-319-68600-4_43
16. Zhang, D., Shen, D.: Multi-modal multi-task learning for joint prediction of multiple regression and classification variables in Alzheimer's disease. Neuroimage **59**, 895–907 (2012)
17. Li, L., Kang, J., Lockhart, S., et al.: Spatially adaptive varying correlation analysis for multimodal neuroimaging data. IEEE Trans. Med. Imaging **38**(1), 113–123 (2018)
18. Hao, X., Bao, Y., Guo, Y., et al.: Multi-modal neuroimaging feature selection with consistent metric constraint for diagnosis of Alzheimer's disease. Med. Image Anal. **60**, 101625 (2020)
19. Pan, J., Lei, B., Shen, Y., Liu, Y., Feng, Z., Wang, S.: Characterization multimodal connectivity of brain network by hypergraph GAN for Alzheimer's disease analysis. In: Ma, H., et al. (eds.) PRCV 2021. LNCS, vol. 13021, pp. 467–478. Springer, Cham (2021). https://doi.org/10.1007/978-3-030-88010-1_39
20. Honey, C., Sporns, O., Cammoun, L., et al.: Predicting human resting-state functional connectivity from structural connectivity. Proc. Natl. Acad. Sci. **106**(6), 2035–2040 (2009)

21. Li, K., Guo, L., Zhu, D., et al.: Individual functional ROI optimization via maximization of group-wise consistency of structural and functional profiles. Neuroinformatics **10**(3), 225–242 (2012)
22. Daselaar, S., Iyengar, V., Davis, S.W., et al.: Less wiring, more firing: low-performing older adults compensate for impaired white matter with greater neural activity. Cerebral Cortex **25**(4), 983–990 (2015)
23. Lei, B., Cheng, N., Frangi, A.F., Tan, E.-L., Cao, J., Yang, P., et al.: Self-calibrated brain network estimation and joint non-convex multi-task learning for identification of early Alzheimer's disease. Med. Image Anal. **61**, 101652 (2020)
24. Cao, P., et al.: Generalized fused group lasso regularized multi-task feature learning for predicting cognitive outcomes in Alzheimer's disease. Comput. Meth. Programs Biomed. **162**, 19–45 (2018)
25. Wang, S.: A variational approach to nonlinear two-point boundary value problems. Comput. Math. Appl. **58**(11–12), 2452–2455 (2009)
26. Wang, S., He, J.: Variational iteration method for a nonlinear reaction-diffusion process. Int. J. Chem. Reactor Eng. **6**(1) (2008)
27. Goodfellow, I., et al.: Generative adversarial nets. In: Advances in Neural Information Processing Systems, pp. 2672–2680 (2014)
28. Wang, S., et al.: Diabetic retinopathy diagnosis using multichannel generative adversarial network with semisupervision. IEEE Trans. Autom. Sci. Eng. **18**, 574–585 (2020)
29. Yu, W., et al.: Tensorizing GAN with high-order pooling for Alzheimer's disease assessment. IEEE Trans. Neural Netw. Learn. Syst. **33**, 4945–4959 (2021)
30. Hu, S., Shen, Y., Wang, S., Lei, B.: Brain MR to PET synthesis via bidirectional generative adversarial network. In: Martel, A.L., et al. (eds.) MICCAI 2020. LNCS, vol. 12262, pp. 698–707. Springer, Cham (2020). https://doi.org/10.1007/978-3-030-59713-9_67
31. Hu, S., Yu, W., Chen, Z., Wang, S.: Medical image reconstruction using generative adversarial network for Alzheimer disease assessment with class-imbalance problem. In: 2020 IEEE 6th International Conference on Computer and Communications (ICCC), pp. 1323–1327 (2020)
32. Hu, S., et al.: Bidirectional mapping generative adversarial networks for brain MR to PET synthesis. IEEE Trans. Med. Imaging **41**(1), 145–157 (2021)
33. Yu, W., et al.: Morphological feature visualization of Alzheimer's disease via Multidirectional Perception GAN. IEEE Trans. Neural Netw. Learn. Syst. (2021)
34. Pan, J., et al.: DecGAN: decoupling generative adversarial network detecting abnormal neural circuits for Alzheimer's disease. arXiv preprint arXiv:2110.05712
35. Vaswani, A., Shazeer, N., Parmar, N., et al.: Attention is all you need. In: Advances in Neural Information Processing Systems, vol. 30 (2017)
36. Lanchantin, J., Wang, T., Ordonez, V., et al.: General multi-label image classification with transformers. In: Proceedings of the IEEE/CVF Conference on Computer Vision and Pattern Recognition, pp. 16478–16488 (2021)
37. Esser, P., Rombach, R., Ommer, B.: Taming transformers for high-resolution image synthesis. In: Proceedings of the IEEE/CVF Conference on Computer Vision and Pattern Recognition, pp. 12873–12883 (2021)
38. Hudson, D., Zitnick, L.: Generative adversarial transformers. In: International Conference on Machine Learning, pp. 4487–4499. PMLR (2021)
39. Aderghal, K., Khvostikov, A., Krylov, A., Benois-Pineau, J., Afdel, K., Catheline, G.: Classification of Alzheimer disease on imaging modalities with deep CNNs using cross-modal transfer learning. In: 2018 IEEE 31st International Symposium on Computer-Based Medical Systems (CBMS), pp. 345–350. IEEE (2018)

40. Dyrba, M., Grothe, M., Kirste, T., Teipel, S.J.: Multimodal analysis of functional and structural disconnection in Alzheimer's disease using multiple kernel SVM. Hum. Brain Mapp. **36**, 2118–2131 (2015)
41. Zhang, L., Wang, L., Gao, J., et al.: Deep fusion of brain structure-function in mild cognitive impairment. Med. Image Anal. **72**, 102082 (2021)

Image Recognition, Detection and Classification

Evaluation of Post-hoc Interpretability Methods in Breast Cancer Histopathological Image Classification

Muhammad Waqas[1,2(✉)], Tomas Maul[1], Amr Ahmed[3,4], and Iman Yi Liao[1]

[1] School of Computer Science, Faculty of Science and Engineering,
University of Nottingham Malaysia Campus,
43500 Semenyih, Selangor Dar-ul-Ehsan, Malaysia
{hcxmw1,tomas.maul,iman.liao}@nottingham.edu.my
[2] Department of Software Engineering, University of Malakand, Chakdara,
Lower Dir 18800, Khyber Pakhtunkhwa, Pakistan
[3] Department of Computer Science, Edge Hill University, Lancashire L39 4QP, UK
amr.ahmed@edgehill.ac.uk
[4] School of Computer Science, Faculty of Science, University of Nottingham,
Nottingham NG8 1BB, UK

Abstract. Methods for post-hoc interpretability are essential for understanding neural network results. Recent years have seen the emergence of numerous post-hoc techniques, but their application to certain tasks, such as histopathological image classification for breast cancer, can produce varied and unpredictable outcomes. Frameworks for quantitative assessment are essential for evaluating each method's effectiveness. The implementation of post-hoc interpretability methodologies is however hampered by the shortcomings of current frameworks, particularly in high-risk industries. In this study, the performance levels of several common post-hoc interpretability methods are systematically evaluated and compared in the context of histopathological image classification for breast cancer. The study is based on six post-hoc interpretability methods, 3 datasets, and 3 deep neural network models, compared via a RemOve And Retrain (ROAR) approach. The results show that Shapley value sampling obtains the best overall performance in the context of the chosen breast cancer histopathological image datasets.

Keywords: Breast cancer · Histopathological images · ROAR · Post-hoc interpretability

1 Introduction

Worldwide, breast cancer affects a lot of women, and the key to better outcomes is early detection [1]. The gold standard for detecting breast cancer is histopathological examination, however due to its complexity and subjectivity, it can be difficult [2]. Deep learning provides automated classification, but given that it is

J. Ren et al. (Eds.): BICS 2023, LNAI 14374, pp. 95–104, 2024.
https://doi.org/10.1007/978-981-97-1417-9_9

a black box, confidence issues are raised [3,4]. Post-hoc interpretability methods provide insights into model decisions, improving transparency, confidence, and error correction [5].

This study evaluates post-hoc interpretability methods for the classification of breast cancer histopathological images. Our objective is to close the gap between the high accuracy of Deep Neural Networks (DNNs) and the requirements for clinical interpretability. Contrary to intrinsic interpretability approaches, which demand interpretable models from scratch, post-hoc methods enable us to understand DNNs after training without changing their architecture. In this study, we employed DeepLIFT (Deep Learning Important FeaTures) [6], Shapley values (SHAP) [7], DeepLIFTShap [8], GradShap [8], Integrated Gradients (IG) [9], and KernelShap [8], six well-known post-hoc interpretability approaches. These methods were selected due to their wide acceptability, effectiveness, and practicability. These methods are widely used for post-hoc interpretability of deep networks. For instance, in [10] DeepLIFT was used to improve the post interpretability of a deep learning model for forecasting energy demands. Similarly in another study, the researchers in [11] utilised integrated gradients for explaining a graph neural network based model for predicting hidden links in supply chains. Many researchers have compared post-hoc methods for specific tasks in many fields. Researchers in [12] have evaluated and compared thirteen post-hoc interpretability methods for the MIMIC-IV dataset. In their study, they have applied it on 5 different deep models and presented quantitative evaluations of interpretability methods on the MIMIC-IV dataset for mortality prediction. Similarly in another study, researchers in [13] presented quantitative comparisons of seven (7) post-hoc interpretability methods on time series classification.

Our analysis is based on a large collection of breast cancer subtypes and tissue samples from several histopathological imaging datasets. We ran experiments on 3 different datasets i.e. BreakHis [14], BACH2018 [15] and a closely related Kaggle-Breast Cancer Histopathology (KBCH) dataset [16,17]. Considerable efforts has been put into these datasets by multiple researchers, and as a result relatively high classification accuracies have been obtained [2].

In this article we have presented a quantitative analysis of the aforementioned six post-hoc interpretability methods applied to breast cancer histopathological images on three different networks i.e. DHA-Net (deep high-order attention neural network) and two student networks of the DHA network. Details regarding the networks are presented in Sect. 2 of this paper. We have used the ROAR (RemOve And Retrain) approach [18] for evaluating the accuracy of the interpretability methods. ROAR is a common evaluation approach that is used to assess the estimated accuracy of interpretability techniques that gauge the significance of input features in deep neural networks. We have utilised the Area Under the Precision-Recall Curve (AUPRC) and the Area Under the Receiver Operating characteristic Curve (AUROC) metrics for presenting our results. The results are shown in Tables 2, 3 and 4. The experiments have shown that, in most cases, Shapely value sampling performed better than other methods. The main contribution of this article lies in providing for the first time, and to the best of our knowledge, a systematic evaluation and comparison of different post-hoc interpretability methods for breast cancer histopathological images.

The remainder of this paper is organized as follows: Sect. 2 describes the models used in this study. Section 3 presents the datasets and experimental setup used in this study. Section 4 presents the evaluation results and analysis. Finally, Sect. 5 concludes the paper and outlines future research directions.

2 The Proposed Work

In this section, we will first discuss the post-hoc interpretability methods, ROAR benchmark and the architectures we used in our study.

2.1 Post-hoc Interpretability Methods

The main goal of this study is to thoroughly evaluate the performance of the six well-known post-hoc interpretability approaches, DeepLIFT (Deep Learning Important FeaTures), Shapley values (SHAP), DeepLIFTShap, GradShap, Integrated Gradients (IG), and KernelShap, for the extraction of meaningful and clinically relevant features from deep learning models for breast cancer histopathological image classification.

DeepLIFT (Deep Learning Important FeaTures). By comparing neuron activations to a reference input, DeepLIFT [6], a post-hoc interpretability technique, assesses the contributions of input features to the predictions made by the model. In order to identify critical regions and pixels in histopathological images that influence the decision-making process, it provides feature importance ratings.

Shapley Values (SHAP). SHAP [7], a principled approach to link a model's prediction to input attributes, is based on cooperative game theory. It takes into account every potential feature combination, calculating how much each feature contributes to predictions, and grading each feature's relevance. This method provides a thorough explanation of feature significance in the classification of breast cancer histopathology imaging data.

DeepLIFTShap. DeepLIFTShap [8] provides thorough knowledge of feature relevance and significance at both the local and global levels by combining DeepLIFT with Shapley values. It improves the interpretability of deep learning models in breast cancer histopathology image classification by combining the global interpretability of Shapley values with the local feature importance of DeepLIFT.

Integrated Gradients (IG). From a reference baseline to the input image, IG [9] estimates the average gradients of the model's output with respect to input features along a straight path. Each characteristic is given a significance value, indicating how important it is to the model's prediction. This approach offers a simple way to understand deep learning models and spot important traits in the classification of breast cancer histopathological images.

GradShap. Shapley values and gradient-based saliency maps are used by Grad-Shap [8] to assign relevance scores to each feature. GradShap offers a single framework for analysing deep learning models, taking into account the significance of both local and global characteristics, by combining gradients using Shapley values. This technique aids in locating important regions and noteworthy characteristics in histopathology pictures, affecting the predictions of the model.

KernelShap. Using sampling and kernel density estimation, KernelShap [8] is a Shapley value technique version that approximates feature importance scores. With the use of this scalable technology, deep learning models are analysed to determine the importance of particular features in categorising breast cancer histopathology images.

This study thoroughly assesses post-hoc interpretability methodologies to see how well they extract clinically significant and meaningful characteristics from deep learning models. It aids in the selection of appropriate interpretability strategies to boost adoption and trust in automated breast cancer classification systems by highlighting the advantages and disadvantages of each method.

2.2   ROAR (RemOve and Retrain)

As a benchmark for assessing the approximate accuracy of interpretability techniques that gauge the significance of input features in DNNs, ROAR (RemOve And Retrain) [18] is used. The process entails retraining the model after removing the most crucial elements determined by each interpretability method. The capacity of an estimator to recognise inputs as crucial—those whose removal leads to the greatest loss in model performance relative to other estimators—determines how accurate it is.

While working with ROAR with any post-hoc estimator e, the following steps are followed:

1. Apply the post-hoc estimator e on the model M, and then rank the features in descending order of importance.
2. Remove the top $t\%$ most important features (specific pixels in our case) from the dataset and generate a new dataset. In the new dataset, replace the features that were removed with the per-channel mean from the original dataset. The average value for each channel in an image is referred to as the per-channel mean. In case of histopathological images, the average value of the red, green, and blue channels, for instance, would make up the per-channel mean.
3. Retrain the new generated dataset and re-evaluate it.
4. Compare the accuracy of the new model and the original model.

Table 1. Image distribution in BreakHis dataset according to Magnification Levels and Types of Breast Cancer

Type	Sub-Category	40X	100X	200X	400X	No. of Patients		Total Images	
Benign	Adenosis	114	113	111	106	4	24	444	2480
	Fibroadenoma	253	260	264	237	10		1014	
	Phyllodes Tumor	109	121	108	115	3		453	
	Tubular Adenoma	149	150	140	130	7		569	
Malignant	Ductal Carcinoma	864	903	896	788	38	58	3451	5429
	Lobular Carcinoma	156	170	163	137	5		626	
	Mucinous Carcinoma	205	222	196	169	9		792	
	Papillary Carcinoma	145	142	135	138	6		560	
Total		1995	2081	2013	1820	82		7909	

2.3 Deep Models

In this study we applied the aforementioned 6 post-hoc interpretability methods and the ROAR benchmark on three models i.e. DHA-Net (Deep High order Attention Network) and two corresponding student networks.

The DHA-Net consists of a ResNet-18 transfer learning model implemented and used to classify breast cancer histopathological images by combining Effective Attention Module (EAM) and Second-Order covariance pooling (MPN-COV). The MPN-COV is only injected into the last layer of the model before the last fully connected layer of the network while the EAM is connected in the last layer of each residual block of the model.

The two student networks are obtained by knowledge distillation for the same DHA-Net. They are lightweight (with a much smaller number of parameters) and perform almost the same as the teacher network [2]. The ghost module and inverted MobileNetV2 Residual blocks make up the student networks. The difference between the two students is based on the number of inverted residual blocks (IRBs) used, where Student Network-1 uses 8 IRBs and Student Network-2 uses 7 IRBs.

3 Experiments

In this section, we will first discuss the datasets and then the experimental settings and the evaluation criteria adopted.

3.1 Datasets

In this study, we evaluated six different post-hoc interpretability methods on BreakHis, KBCH, and BACH datasets. We selected these datasets because they allow us to evaluate on a wider range of data, including those that are binary

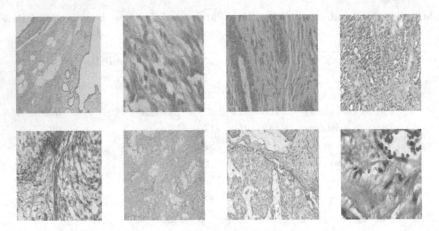

Fig. 1. Sample Images from the BreakHis dataset. The top four images are benign and the bottom four are malignant. Top (left to right): adenosis 40X, fibroadenoma 400X, phyllodes_tumor 200X and tubular_adenoma 100X. Bottom (left to right): ductal_carcinoma 200X, lobular_carcinoma 40X, mucinous_carcinoma 100X and papillary_carcinoma 400X

class, 4 class, 8 class, Magnification Dependent (MD), and Magnification Independent (MI) classification.

A total of 7909 images (2480 benign and 5429 malignant) from 82 patients make up the 2016-built BreakHis dataset. This dataset offers the chance to work at several magnification levels, including 40X, 100X, 200X, and 400X. Table 1 displays the BreakHis dataset's image distribution. Few sample images from BreakHis datasets are shown in Fig. 1. The KBCH dataset is a binary class dataset with a single magnification factor. There are 162 whole slide images in all. 277,524 images with a size of 50×50 were taken from these images (198,738 IDC negative and 78,786 IDC positive). The BACH dataset was also utilised with 400 H&E-stained images from 4 classes—normal tissue, benign abnormalities, malignant in-situ carcinoma, and malignant invasive carcinoma.

3.2   Data Pre-processing

If required, we have re-sized the input image as per the requirement of our teacher network. Image resizing may affect the quality of images, hence we have normalized our images by using Global Contrast Normalization (GCN). Smaller datasets like BreakHis and BACH more often lead to bad generalization when applied to DNNs. To solve this issue, we have increased the number of images on both BreakHis and BACH datasets by adopting data augmentation. We have applied three transformations i.e. horizontal flip, vertical flip and 45° rotations. For applying the ROAR benchmark method on all the datasets and models, we have used t = [0,10, ... , 100].

Table 2. Area Under the Precision-Recall Curve (AUPRC) of interpretability methods for each model and each classification performance metric evaluated using ROAR on BreakHis Magnification Dependent (MD) images. SN-I and SN-II stands for Student Network-1 and Student-Network-2 respectively.

		40X 2 class	100X 2 class	200X 2 class	400X 2 class	40X 8 class	100X 8 class	200X 8 class	400X 8 class
DHA Net	DeepLift	0.525	0.581	0.584	0.520	0.593	0.532	0.531	0.576
	DeepLiftShap	0.468	0.514	0.470	0.474	0.469	0.463	0.480	0.520
	GradShap	**0.405**	**0.412**	0.462	0.447	0.438	**0.417**	0.435	0.423
	IG	0.543	0.494	0.547	0.550	0.541	0.586	0.585	0.549
	KernelShap	0.471	0.512	**0.391**	**0.396**	0.582	0.419	0.417	0.403
	Shapley	0.418	0.448	0.461	0.414	**0.350**	0.498	**0.391**	**0.369**
SN-I	DeepLift	0.579	0.606	0.533	0.569	0.639	0.599	0.605	0.596
	DeepLiftShap	0.465	0.467	0.471	0.522	0.501	0.514	0.511	0.513
	GradShap	**0.428**	0.465	0.473	0.433	0.440	0.459	0.447	0.471
	IG	0.570	0.499	**0.466**	0.545	0.538	0.532	0.557	0.459
	KernelShap	0.500	0.516	0.499	**0.361**	0.455	0.545	**0.393**	0.473
	Shapley	0.474	**0.384**	0.480	0.387	**0.356**	**0.408**	0.433	**0.433**
SN-II	DeepLift	0.575	0.587	0.560	0.587	0.540	0.563	0.546	0.563
	DeepLiftShap	0.470	0.553	0.507	0.515	0.486	0.489	0.530	0.491
	GradShap	0.461	**0.425**	0.443	0.500	0.510	0.465	0.491	0.467
	IG	0.533	0.478	0.478	0.549	0.536	0.511	0.571	0.522
	KernelShap	0.437	0.509	**0.433**	**0.411**	0.456	0.452	**0.434**	0.416
	Shapley	**0.434**	0.461	0.461	0.462	**0.416**	**0.447**	0.464	**0.363**

4 Results and Discussion

This section discusses the results obtained after applying the six post-hoc interpretability methods. We have presented the results in Tables 2, 3 and 4.

Table 2 presents the performance of each post-hoc interpretability method on BreakHis Magnification Dependent (MD) images w.r.t. AUPRC. The lowest value in each column represents the best performing method for that dataset because it indicates a more rapid prediction rate and better feature importance interpretation. So after analysing all the results for BreakHis MD images, we can see that mainly Shapley value sampling is performing better than other methods in 11 out of 24 instances. KernalShap also shows promising performance on this dataset as it performed better than others in 7 instances. For BreakHis MD classification, GradShap and Integrated Gradient performed better than others for 5 and 1 instances respectively.

Table 3 presents the performance of each post-hoc interpretability method on BreakHis Magnification Dependent (MD) images w.r.t. AUROC. Shapley sampling performed better than other methods for 15 out of 24 instances. On the remaining 9 instances KernalShap obtained the best performance.

Table 4 shows the results for BreakHis Magnification Independent (MI), BACH and KBCH datasets. This table shows the results for AUPRC and AUROC combined for all these datasets. For AUPRC, we can see from the table that Shapley sampling is the best approach for 5 instances while KernelShap and GradShap obtained the best performance for 4 and 3 instances respectively. Finally, for AUROC, Shapley and KernelShap obtained the best performance for 6 instances.

Table 3. Area Under the Receiver Operating characteristic Curve (AUROC) of interpretability methods for each model and each classification performance metric evaluated using ROAR on BreakHis Magnification Dependent (MD) images. SN-I and SN-II stands for Student Network-1 and Student-Network-2 respectively.

		40X 2 class	100X 2 class	200X 2 class	400X 2 class	40X 8 class	100X 8 class	200X 8 class	400X 8 class
DHA Net	DeepLift	0.891	0.898	0.877	0.905	0.900	0.901	0.896	0.880
	DeepLiftShap	0.904	0.920	0.900	0.915	0.870	0.883	0.871	0.898
	GradShap	0.885	0.855	0.823	0.832	0.893	0.831	0.856	0.836
	IG	0.866	0.854	0.839	0.856	0.873	0.878	0.862	0.828
	KernelShap	0.759	0.766	**0.743**	**0.758**	**0.720**	0.765	0.760	0.754
	Shapley	**0.734**	**0.731**	0.773	0.766	0.770	**0.758**	**0.729**	**0.700**
SN-I	DeepLift	0.891	0.872	0.873	0.910	0.887	0.896	0.912	0.884
	DeepLiftShap	0.898	0.928	0.943	0.940	0.893	0.942	0.894	0.911
	GradShap	0.864	0.859	0.880	0.822	0.870	0.837	0.863	0.864
	IG	0.854	0.874	0.824	0.874	0.847	0.865	0.877	0.863
	KernelShap	**0.720**	0.769	**0.733**	0.759	0.723	0.742	0.771	**0.741**
	Shapley	0.778	**0.763**	0.785	**0.704**	**0.705**	**0.697**	**0.761**	0.772
SN-II	DeepLift	0.910	0.905	0.895	0.892	0.912	0.931	0.877	0.893
	DeepLiftShap	0.909	0.950	0.901	0.905	0.901	0.900	0.941	0.905
	GradShap	0.841	0.857	0.847	0.834	0.883	0.867	0.872	0.894
	IG	0.852	0.851	0.888	0.886	0.895	0.848	0.858	0.833
	KernelShap	0.775	**0.740**	**0.759**	0.795	**0.738**	0.786	0.766	0.749
	Shapley	**0.708**	0.789	0.789	**0.715**	0.759	**0.769**	**0.715**	**0.693**

Table 4. AUPRC and AUROC values for each model and each classification performance metric evaluated using ROAR on BreakHis (BH) MI, BACH and KBCH Images. SN-I and SN-II stands for Student Network-1 and Student-Network-2 respectively.

		AUC-PRC				AUC-ROC			
		BH-MI 2 Class	BH-MI 8 Class	KBCH 2 class	BACH 4 class	BH-MI 2 Class	BH-MI 8 Class	KBCH 2 class	BACH 4 class
DHA Net	DeepLift	0.572	0.568	0.562	0.560	0.874	0.884	0.871	0.916
	DeepLiftShap	0.508	0.520	0.510	0.465	0.895	0.877	0.896	0.891
	GradShap	**0.407**	0.425	**0.462**	0.415	0.856	0.826	0.895	0.891
	IG	0.486	0.450	0.594	0.513	0.820	0.891	0.827	0.852
	KernelShap	0.599	0.453	0.576	0.517	**0.720**	**0.758**	0.752	**0.750**
	Shapley	0.478	**0.382**	0.467	**0.386**	0.737	0.770	**0.710**	0.759
SN-I	DeepLift	0.524	0.52	0.558	0.560	0.896	0.930	0.895	0.900
	DeepLiftShap	0.536	0.526	0.499	0.502	0.883	0.899	0.893	0.921
	GradShap	0.443	0.489	0.445	0.444	0.870	0.849	0.826	0.856
	IG	0.508	0.495	0.537	0.525	0.859	0.820	0.831	0.858
	KernelShap	0.490	**0.427**	0.528	**0.405**	**0.745**	0.753	0.752	0.740
	Shapley	**0.389**	0.491	**0.426**	0.419	0.783	**0.719**	**0.704**	**0.718**
SN-II	DeepLift	0.576	0.620	0.602	0.594	0.949	0.921	0.921	0.921
	DeepLiftShap	0.529	0.533	0.541	0.483	0.919	0.906	0.916	0.886
	GradShap	**0.426**	0.449	0.501	0.466	0.832	0.888	0.861	0.841
	IG	0.462	0.559	0.526	0.537	0.862	0.856	0.876	0.847
	KernelShap	0.454	**0.435**	**0.390**	0.430	0.749	**0.757**	**0.744**	0.770
	Shapley	0.467	0.496	0.433	**0.387**	**0.703**	0.772	0.780	**0.701**

Overall, the results above indicate that Shapley value sampling demonstrated the best overall performance across all of the datasets.

5   Conclusion and Future Work

We investigated the significance of post-hoc interpretability methods in the context of breast cancer histopathology image classification. We assessed six methods on three deep neural network models using three datasets. According to our research, Shapley value sampling works better than other techniques at interpreting breast cancer histopathology images. This emphasises Shapley value sampling as a trustworthy and efficient method for comprehending neural network decisions in the classification of breast cancer. Examining alternative post-hoc interpretability techniques to determine their advantages and disadvantages in this application is one of the future research directions. It would also be beneficial to look at how well the discovered interpretability techniques work with other cancer types and medical imaging domains. In conclusion, post-hoc interpretability techniques have great promise for enhancing the clinical utility, openness, and consistency of deep learning models in identifying cancer.

Acknowledgment. This work was made possible by the financial support from the University of Malakand, Higher Education Commission (Government of Pakistan), and the University of Nottingham Malaysia Campus. Iman Yi Liao also expresses gratitude to the Malaysian Ministry of Education for the grant received under the Fundamental Research Grant Scheme (No. FRGS/1/2014/ ICT07/UNIM/02/1).

References

1. Ginsburg, O., et al.: Breast cancer early detection: a phased approach to implementation. Cancer **126**, 2379–2393 (2020)
2. Waqas, M., Maul, T., Liao, I.Y., Ahmed, A.: Lightweight deep network for the classification of breast cancer histopathological images. In: 2022 15th International Congress on Image and Signal Processing, BioMedical Engineering and Informatics (CISP-BMEI), pp. 1–6. IEEE (2022)
3. Petch, J., Di, S., Nelson, W.: Opening the black box: the promise and limitations of explainable machine learning in cardiology. Can. J. Cardiol. **38**(2), 204–213 (2022)
4. Li, J., Lopez, S.A.: A look inside the black box of machine learning photodynamics simulations. Acc. Chem. Res. **55**(14), 1972–1984 (2022)
5. Madsen, A., Reddy, S., Chandar, S.: Post-hoc interpretability for neural NLP: a survey. ACM Comput. Surv. **55**(8), 1–42 (2022)
6. Li, J., Zhang, C., Zhou, J.T., Fu, H., Xia, S., Hu, Q.: Deep-lift: deep label-specific feature learning for image annotation. IEEE Trans. Cybern. **52**(8), 7732–7741 (2021)
7. Messalas, A., Kanellopoulos, Y., Makris, C.: Model-agnostic interpretability with Shapley values. In: 2019 10th International Conference on Information, Intelligence, Systems and Applications (IISA), pp. 1–7. IEEE (2019)
8. Lundberg, S.M., Lee, S.-I.: A unified approach to interpreting model predictions. In: Advances in Neural Information Processing Systems, vol. 30 (2017)
9. Sundararajan, M., Taly, A., Yan, Q.: Axiomatic attribution for deep networks. In: International Conference on Machine Learning, pp. 3319–3328. PMLR (2017)
10. Shajalal, Md., Boden, A., Stevens, G.: Towards user-centered explainable energy demand forecasting systems. In: Proceedings of the Thirteenth ACM International Conference on Future Energy Systems, pp. 446–447 (2022)
11. Kosasih, E.E., Brintrup, A.: A machine learning approach for predicting hidden links in supply chain with graph neural networks. Int. J. Prod. Res. **60**(17), 5380–5393 (2022)

12. Meng, C., Trinh, L., Nan, X., Enouen, J., Liu, Y.: Interpretability and fairness evaluation of deep learning models on MIMIC-IV dataset. Sci. Rep. **12**(1), 7166 (2022)
13. Turbé, H., Bjelogrlic, M., Lovis, C., Mengaldo, G.: Evaluation of post-hoc interpretability methods in time-series classification. Nat. Mach. Intell. **5**(3), 250–260 (2023)
14. Spanhol, F.A., Oliveira, L.S., Petitjean, C., Heutte, L.: A dataset for breast cancer histopathological image classification. IEEE Trans. Biomed. Eng. **63**(7), 1455–1462 (2015)
15. Aresta, G., et al.: Grand challenge on breast cancer histology images. Bach. Med. Image Anal. **56**, 122–139 (2019)
16. Janowczyk, A., Madabhushi, A.: Deep learning for digital pathology image analysis: a comprehensive tutorial with selected use cases. J. Pathol. Informa. **7** (2016)
17. Cruz-Roa, A., et al.: Automatic detection of invasive ductal carcinoma in whole slide images with convolutional neural networks. In: Medical Imaging 2014: Digital Pathology, vol. 9041, p. 904103. SPIE (2014)
18. Hooker, S., Erhan, D., Kindermans, P.-J., Kim, B.: Evaluating feature importance estimates (2018)

Red Blood Cell Detection Using Improved Mask R-CNN

Hongfang Pan, Han Su, Jin Chen$^{(\boxtimes)}$, and Ying Tong

Tianjin Key Laboratory of Wireless Mobile Communications and Power Transmission, Tianjin Normal University, Tianjin 300387, China
cjwoods@163.com

Abstract. Automatic segmentation of microscopy images is an important task in medical image processing and analysis. Red blood cell detection is an important example of this task. Manual detection is not only labor-intensive, but also prone to misdirection and omission. In order to enhance the speed and accuracy, an improved mask regional convolution neural network (Mask R-CNN) is proposed in this paper. The algorithm utilizes Split-Attention Networks (ResNeSt) as a feature extraction network. ResNeSt combines channel attention with multi-path representation, and feature extraction is performed in combination with Feature Pyramid Network (FPN). The experimental results show that the improved Mask R-CNN has an average precision increase of 2.55%, and improves the efficiency of red blood cell detection.

Keywords: Deep Learning · Red blood cell detection · Mask R-CNN · Split-attention network

1 Introduction

Blood is the unique fluid tissue in the human body, which can be subdivided into peripheral blood and bone marrow blood. The cellular component of peripheral blood consists of three types of blood cells: red blood cells (RBCs), white blood cells (WBCs) and platelets, whereas the fluid component, known as plasma, primarily transports blood cells, substances essential to sustain human life, and metabolic wastes produced by the body. In the body's blood circulation system, different cells play different roles. RBCs, which represent about 93% to 96% of the total number of peripheral blood cells, transport substances such as oxygen and carbon dioxide in the circulatory system [1]. Red blood cell testing is critical because many blood disorders in human bodies are related to RBCs, including polycythemia, iron deficiency anaemia, sickle cell disease and pure red cell aplasia. In addition to this, the detection of RBCs provides strong support for subsequent studies, such as the classification, counting and tracking of red blood cells. As a result, the red blood cell test is fundamental and vital. In this paper, microscopic cell images of peripheral blood smears are used to examine the red blood cells in the images.

© The Author(s), under exclusive license to Springer Nature Singapore Pte Ltd. 2024
J. Ren et al. (Eds.): BICS 2023, LNAI 14374, pp. 105–112, 2024.
https://doi.org/10.1007/978-981-97-1417-9_10

In recent years, many cell image segmentation algorithms have been proposed by domestic and foreign researchers. These algorithms are mainly divided into two categories: traditional algorithms and deep learning algorithms. In the traditional cell image segmentation, Wang et al. [2] proposed an algorithm combining graph theory and skeleton distance histogram to segment blurred and adherent cell images. Although the algorithm is simple and easy to implement, it has a high error rate for cells with irregular shapes and uneven sizes, and the segmentation results are not satisfactory. Miao et al. [3] proposed a marker-controlled watershed algorithm to simultaneously segment red and white blood cells in blood smear images. This algorithm simplifies the operation and saves computational time, but the medium image quality can negatively affect the accuracy of the segmentation algorithm. Some traditional segmentation algorithms cannot show good robustness due to noise, image quality and other issues, while deep learning algorithms use a large amount of sample data to learn cellular image features, and these algorithms have shown higher segmentation accuracy than traditional image segmentation algorithms [4]. Marek Kowal et al. [5] proposed a method that combined convolution neural network (CNN) classifier and seeded watershed algorithm to segment cell images. The algorithm first preprocessed the images using the CNN classifier, and then conducted semantic segmentation of preprocessed images using the seeded watershed algorithm. Reena M. Roy et al. [6] proposed a framework combining DeepLabv3+ and Residual Network (ResNet) as a semantic feature extraction network segmentation technology for segmenting leukocytes of microscopic blood images. The algorithm demonstrated good segmentation accuracy for red blood cell images and whole blood smear images.

In summary, the traditional cell image segmentation algorithms have limitations in terms of accuracy and robustness due to image noise and quality issues. Deep learning algorithms have shown significant improvements in cell image segmentation accuracy compared to their traditional counterparts, due to their ability to extract features from large amounts of sample data. The deep learning-based target detection networks can be divided into two-stage and one-stage models, where Mask R-CNN is a popular two-stage model with significant advantages for small object detection. In this paper, we use ResNeSt in the backbone of Mask R-CNN, which combines channel attention with a multi-path representation for enhanced feature extraction. The improved Mask R-CNN was validated by comparative experiments on a microscopic blood smear data set, which showed a significant improvement in detection and can be well applied to the red blood cell detection task.

2   A Deep Learning Model for Red Blood Cells Detection

When a network connects a large number of simple calculations together, intelligent behavior can be achieved. Deep learning is the process of automatically combining simple features into complex features, and then using complex features to solve some problems [7]. The progress of deep learning provides more direct methods for solving object detection problems. CNN and networks built on them have created possibilities for solving object detection tasks, and R-CNN are the core of this method. R-CNN series of object detection algorithms utilize feature maps to make it a powerful tool for object detection.

In 2017, Kaiming He et al. proposed a further improvement on R-CNN called Mask R-CNN [8]. The basic idea of Mask R-CNN is to extend Faster R-CNN for pixel level semantic segmentation. It adds an additional mask branch to predict object masks, parallel to existing branches used for bounding box (bbox) recognition. And It replaces the RoI (Region of Interest) Pooling with RoI Align to construct accurate instance segmentation masks. The structure of Mask R-CNN can be divided into the several parts, including a backbone network, a region proposal network (RPN), RoI Align, a bbox branch, and a mask branch.

2.1 Backbone Network

ResNet is the backbone network of Mask R-CNN, which provides feature maps with high semantic and spatial information. However, since the feature map has a fixed resolution, it cannot handle objects of varying sizes effectively. To solve this problem, FPN is integrated into the network. FPN constructs a feature pyramid with features of different scales, and thus, it can handle objects of different sizes effectively.

FPN is usually composed of two parts: a bottom-up pathway and a top-down pathway. The bottom-up pathway is used to extract features at all scales from the input image using ResNet [9]. Each scale's feature maps are then used to construct the feature pyramid. The top-down pathway is used to propagate the high-level semantic information to lower levels of the pyramid, where object details are finer. To construct the top-down pathway, FPN employs lateral connections, which connect the high-level feature maps with the low-level feature maps. Each lateral connection performs a 1×1 convolution to reduce the number of channels of the high-level feature maps to match the number of channels of the corresponding low-level feature maps. Then, the resulting features are upsampled by a factor of two using bi-linear interpolation. After that, the resulting features are added element-wise to the corresponding low-level feature maps. The result of this operation is a feature map that incorporates both high-level semantic information and fine details.

In summary, ResNet and FPN are combined in the Mask R-CNN architecture to provide a feature pyramid that is suited for object detection and segmentation tasks. The ResNet backbone provides high-level semantic features, while the FPN provides features at different scales, which enables the model to detect objects of different sizes effectively.

2.2 Regional Proposal Network

Regional Proposal Network (RPN) is a lightweight neural network that scans images through sliding windows to find the region where the target is located. It consists of three main stages: convolutional layers for feature extraction, binary classification for calculating the fraction indicating the probability of objects in each anchor point, and boundary box regression for width and height of each anchor point.

2.3   Region of Interest Align

Faster R-CNN employs a RoI pooling layer to extract fixed-length feature vectors for each RoI, which can then be classified and regressed to obtain object instances. In Mask R-CNN, the RoI pooling layer is replaced with a RoI align layer. RoI Align preserves the spatial information on the feature map, which largely solves the error caused by the two quantification of the feature map in the RoI Pooling layer, and solves the problem of regional mismatch of the image object. The bi-linear interpolation [10] is used to calculate the exact values of the input features at four regular sampling positions in each RoI, and aggregates the results to obtain pixel-level detection segmentation.

2.4   Output Branches

The Mask R-CNN model comprises three parallel branches, where the class and box branches function similarly to those of Faster R-CNN, where each RoI is classified and regressed separately. The newly added mask branch uses fully convolutional network (FCN) to recognize semantic mask, generating binary masks for each RoI. Then, bi-linear interpolation is used to enlarge the feature image to the original image size.

The training and prediction process of the Mask R-CNN model is different. During training, all three branches (class, box, and mask) are trained simultaneously. When predicting, first, the class and box branches are run to obtain the results, and subsequently, these results are passed to the mask branch to generate the train masks for each RoI.

3   Methods

3.1   Data Preprocessing

The data set format used in this paper is a COCO type data set. The COCO data set, also Known as Microsoft Common Objects in Context, which was funded by Microsoft in 2014 and has been used as the data set for the competition. The COCO data set has three annotation types: Object Instances, Object Keypoints, and Image Captions, all of which use JSON files to store information [11]. The data set used in this paper consists of light microscope image of blood smears. The original data set contains 360 clear images of the blood smear and its .xml annotation file, which contains a large number of red cells and a small number of white cells. According to the requirements of the COCO data set, 60 images were selected randomly as the validation set and the rest as the training set. Then the data set is converted into the Object Instance format, which is used for image recognition detection.

3.2   Backbone Network

Microscopic blood images often contain irregular red blood cells with a high degree of overlapping cell adhesion, making it difficult to detect and segment cells accurately. Traditional cell detection methods are difficult to segment the adherent cells accurately. In this paper, based on the Mask R-CNN, ResNeSt is proposes as its feature extraction network. Figure 1 depicts an overview of a ResNeSt block.

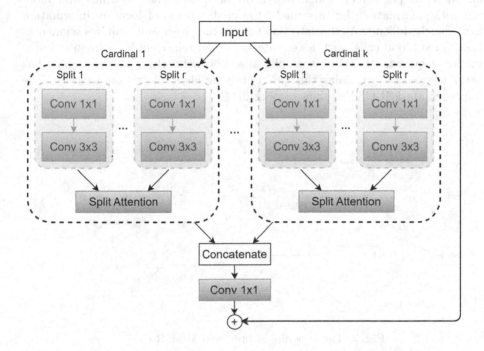

Fig. 1. Network structure of the ResNeSt block

Zhang et al. [12] proposed ResNeSt in 2020, which proposed a Split-Attention Network based on ResNeXt, SE-Net and SK-Net for replacing the four stage's conv in ResNet. The network structure block diagram of ResNeSt is shown in Fig. 1. ResNeSt extends channel orientation concerns to feature mapping group representations and uses a unified CNN algorithm for modularity and acceleration. The network structure of its core module combines the ideas of multi-path mechanisms, packet convolution, channel attention mechanisms and feature mapping attention mechanisms. Multi-path representation has shown success in GoogleNet [13], in which each network block consists of different convolutional kernels. ResNeXt [14] adopts group convolution in the ResNet bottle block, which converts the multi-path structure into a unified operation. SE-Net [15] introduces a channel attention mechanism by adaptively recalibrating the channel feature responses. Recently, In SK-Net [16], an attention mechanism for feature mapping was proposed and applied to two branches of the network.

The ResNeSt architecture leverages channel attention with multi path representations into a single unified split attention block. The network generally improves the representation of the learned features with stronger feature extraction capabilities, which can optimise the segmentation of small targets, increase the depth of the model, and improved the feature extraction performance. The Split-attention networks are ResNeSt-50, ResNeSt-101, ResNeSt-200, etc. As we all know, deeper layers of neural network lead to higher accuracy and more semantic information, but also make the confirmation of location information increasingly difficult. As the detection task in the paper only requires segmentation of red blood cells, with fewer categories and relatively less semantic information required, ResNeSt-50, a split attention network, is chosen, which has fewer layers, saving training time and obtaining higher detection accuracy. The improved Mask R-CNN is shown in Fig. 2.

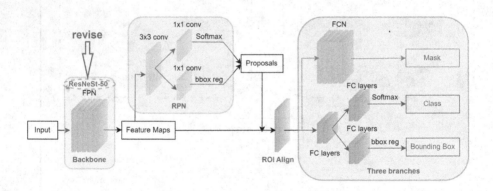

Fig. 2. The structure of Improved Mask R-CNN

4 Results

The processor used for this experiment is Intel Core i7 11800H, the GPU is NVIDIA GeForce RTX 3050, the deep learning framework is PyTorch, and the development environment is PyTorch 1.12.1, Python 3.9.12, and CUDA 12.1.

In order to study the detection performance of the improved algorithm on the blood cell data set, it is compared with the advanced detection algorithm Mask R-CNN, and the results are shown in Table 1.

As can be seen from Table 1, AP50:95 of the improved detection algorithm is 2.55% higher than the Mask R-CNN, and AP50 is 5.27% higher. Although AP50 is slightly decreased, the overall performance of the model has been improved. Some of the image test results of the red blood cell detection algorithm are shown

Table 1. Comparison of average precision (%).

Algorithm	AP50:95	AP75	AP50
Mask R-CNN	66.76	83.63	94.73
Improved Mask R-CNN	69.31	88.90	94.26

in Fig. 3. The algorithm not only realizes the detection of independent cells in the image, but also has a good detection effect on the group of cells with high adhesion and overlap. The white blood cells in the image are not mistaken for red blood cells, indicating that the network can fully extract various types of feature information in the image, such as size, texture and color information. These feature networks distinguish red blood cells from other tissues.

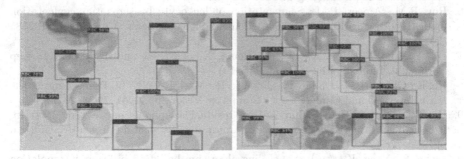

Fig. 3. Selected image detection results

5   Conclusions

In the work described in this paper, a detection algorithm based on Mask R-CNN for red blood cell detection is proposed. ResNeSt is used in the algorithm. ResNeSt uses multiple convolutional kernel branches in the same layer to extract features separately, allowing the network to extract more diverse features and introducing split attention blocks. The attention module of a channel is used to assign different weights to channels to extract effective features, enabling efficient extraction of feature information. After testing and improvement, the improved Mask R-CNN based method showed higher detection accuracy and robustness compared to previous detection algorithms, achieving good results for fuzzy and adherent cells. The proposed algorithm contributes to the progress in object detection and segmentation in medical image analysis, which is important in medical research and practice.

References

1. Zhang, Z.: Introduction to Basic Medicine. China Science and Technology Press (2012)
2. Wang, W., Tian, L., Wang, Y.: Improved graph-theoretic minimum spanning tree and skeleton distance histogram-based segmentation of cell images. Opt. Precis. Eng. **21**(9), 2464–2472 (2013)
3. Miao, H., Xiao, C.: Simultaneous segmentation of leukocyte and erythrocyte in microscopic images using a marker-controlled watershed algorithm. Comput. Math. Methods Med. (2018)
4. Taghanaki, S.A., Abhishek, K., Cohen, J.P., Cohen-Adad, J., Hamarneh, G.: Deep semantic segmentation of natural and medical images: a review. Artif. Intell. Rev. **54**, 137–178 (2021)
5. Kowal, M., Żejmo, M., Skobel, M., Korbicz, J., Monczak, R.: Cell nuclei segmentation in cytological images using convolutional neural network and seeded watershed algorithm. J. Digit. Imaging **33**, 231–242 (2020)
6. Roy, R.M., Ameer, P.M.: Segmentation of leukocyte by semantic segmentation model: a deep learning approach. Biomed. Signal Process. Control **65**, 102385 (2021)
7. Hu, Y., Luo, D., Hua, K., Lu, H., Zhang, X.: A review and discussion on deep learning. J. Intell. Syst. **14**(1), 1–19 (2019)
8. He, K., Gkioxari, G., Dollár, P., Girshick, R.: Mask R-CNN. In: Proceedings of the IEEE International Conference on Computer Vision, pp. 2961–2969 (2017)
9. He, K., Zhang, X., Ren, S., Sun, J.: Deep residual learning for image recognition. In: Proceedings of the IEEE Conference on Computer Vision and Pattern Recognition, pp. 770–778 (2016)
10. Wang, S., Yang, K.: An image scaling algorithm based on bi-linear interpolation with VC. Autom. Technol. Appl. **07**, 41–45 (2008)
11. Yang, X., Li, C., Chen, S., Li, H., Yin, G.: Text-image cross-modal retrieval based on transformer. Comput. Sci. **50**(04), 141–148 (2023)
12. Zhang, H., et al.: ResNeSt: split-attention networks. In: Proceedings of the IEEE/CVF Conference on Computer Vision and Pattern Recognition, pp. 2736–2746 (2022)
13. Szegedy, C., et al.: Going deeper with convolutions. In: Proceedings of the IEEE Conference on Computer Vision and Pattern Recognition, pp. 1–9 (2015)
14. Xie, S., Girshick, R., Dollár, P., Tu, Z., He, K.: Aggregated residual transformations for deep neural networks. In: Proceedings of the IEEE Conference on Computer Vision and Pattern Recognition, pp. 1492–1500 (2017)
15. Hu, J., Shen, L., Sun, G.: Squeeze-and-excitation networks. In: Proceedings of the IEEE Conference on Computer Vision and Pattern Recognition, pp. 7132–7141 (2018)
16. Li, X., Wang, W., Hu, X., Yang, J.: Selective kernel networks. In: Proceedings of the IEEE/CVF Conference on Computer Vision and Pattern Recognition, pp. 510–519 (2019)

Underwater Object Detection for Smooth and Autonomous Operations of Naval Missions: A Pilot Dataset

Yijun Yan[1], Yinhe Li[1], Hanhe Lin[2], Md Mostafa Kamal Sarker[3], Jinchang Ren[1]([✉]), and John McCall[1]

[1] National Subsea Centre, Robert Gordon University, Aberdeen, UK
{y.yan2,y.li24,j.ren,j.mccall}@rgu.ac.uk
[2] School of Science and Engineering, University of Dundee, Dundee, UK
hlin001@dundee.ac.uk
[3] Institute of Biomedical Engineering, University of Oxford, Oxford, UK
md.sarker@eng.ox.ac.uk

Abstract. Underwater object detection is essential for ensuring autonomous naval operations. However, this task is challenging due to the complexities of underwater environments that often degrade image quality, thereby hampering the performance of detection and classification systems. On the other hand, the absence of a readily available dataset complicates the development and evaluation of underwater object detection approaches, particularly for deep learning approaches. To address this bottleneck, we have created a new dataset, called National Subsea Centre Underwater Images (NSCUI). It is comprised of 243 images, divided into three subsets that are captured in bright, low-light, and dark environments, respectively. To validate the utility of this dataset, we implemented three popular deep learning models in our experiments. We believe that the annotated NSCUI will significantly advance the development of underwater object detection through the application of deep learning techniques.

Keywords: Underwater object detection · Image enhancement · Deep learning

1 Introduction

Obstacles detection and collision avoidance are the key challenges for smooth and autonomous operations of naval missions, where sonar and optics sensors are widely used [1]. Due to the complexity of the underwater environments, the image quality of both sonar and optics system can be severely degraded, resulting in inconsistent performance of target detection and classification.

Sonar has been popularly applied in many underwater inspection tasks [2], featuring long-range detection. However, its working condition is not only affected by internal factors (e.g. latency, narrow bandwidth and self-noise), but also external factors related to ocean environment (e.g. spreading loss, multipath effect, reverberation, ocean noise, target reflection characteristics and radiation noise [3]). Due to poor performance in shallow depth and noisy data [4], it makes it difficult to detect small but important objects such as nets. Another downside is the heavy cost, which has constrained its wide deployment. In addition, military sonar systems may severely affect the lives of marine mammals, leading to deafness and death of dolphins, whales and sea turtles [5].

Optic systems, an economically viable alternative, are celebrated for their substantial bandwidth, contributing rich structure, shape, and texture features, which are particularly beneficial for shallow underwater inspection [6]. Despite their advantages, light transmission in water encounters severe attenuation attributed to both internal factors, such as absorption and scattering, and external factors like turbidity and suspended particles, which collectively result in a relatively low detection accuracy.

Deep learning (DL) based object detection methods offer a promising solution to address these limitations. DL models have demonstrated success in various image modalities, including video surveillance [7], aerial image [8], hyperspectral images [9], etc. Recently, the application of DL-based methods has extended to underwater object detection. Examples of applications include the usage of R-CNN [10] and its extended versions [11] for detecting and recognizing aquatic organisms, for which datasets such as UTDAC2020 and DUO have been used for modelling. Similarly, Mask R-CNN has been utilized for deep-sea litter detection [12], with the Trashcan dataset used for modelling purposes. Additionally, the YOLO series models have been employed for marine animal detection [13, 14], where the datasets UODD [13] and Blackish [14] have been conducted for modelling. However, due to lack of sufficient data and labels, the feasibility of using deep learning models for underwater obstacle recognition remains largely underexplored and presents significant opportunities for further research.

The aim of this study is to construct a new underwater image dataset and investigate the potential of deep learning techniques in underwater obstacle recognition. By integrating image enhancement techniques with deep learning-based object detection, we aspire to advance obstacle recognition under varying lighting conditions. This enhancement is expected to contribute significantly to the autonomous operation of underwater vehicles, offering improved navigation and operational efficiency of naval missions.

2   Dataset Creation

In this experiment, we captured images for simulated 5 categories of underwater targets (i.e., container, props, keel, wreck and net) in different light conditions. The water tank we used has the size of 60 cm × 60 cm × 150 cm and maximum volume of 500 L. The imaging device we used is a scotopic camera, Sony 4K video camera with full 35 mm frame Exmor CMOS sensor. It operates in the maximum 4240 × 2842 (12M) pixels for still image and 3810 × 2160 pixels for video recording with an ISO sensitivity ranges of 50 − 409600, which allows it to capture the high resolution image data in both bright and low light environments. Figure 1. (a) shows the position of imaging system, Fig. 1. (b–c) shows the experimental setting for data acquisition in bright environment, and low light/dark environment. Notably, in the low light environment, we attached black vinyl sheets on the walls of the water tank and used a black sheet to cover half of the top to reduce the incoming light. In the dark environment, the top of water tank was totally covered by the black sheet, resulting a totally black scene. However, the existing object detection models failed to perform in such kind of scene. Instead, a laser projector consisting of a laser diode with 520 nm and 50 mV output power, and a diffractive optical element with 51 × 51 dot matrix, was employed to provide consistent light source and point-cloud data for 3D measurement.

Examples of captured images in different light condition are shown in Fig. 2. Due to the uncertainty of water dynamics, it is impossible to capture the same object with the same position in different light conditions. Thus, we created three subsets separately.

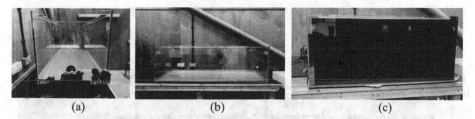

    (a)   (b)   (c)

Fig. 1. (a) Imaging system, (b) experimental setting for bright environment, and (c) dark/low light environment.

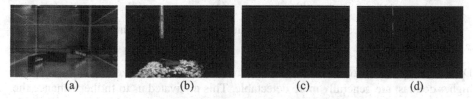

    (a)   (b)   (c)   (d)

Fig. 2. Image captured in (a) bright environment, (b) low light environment, (c) dark environment and (d)dark environment with laser light source.

where the images were captured in bright, low light and dark conditions, and the number of images in each subset is 95, 74, and 74, respectively.

After that, we made the annotation using an open-source annotation tool LabelMe. The ground-truth images annotation format is then converted to MS-COCO format. Figure 3 shows the image data containing single category and its corresponding annotation. To further simulate the complexity in the real world, image data for mixed categories of targets that were randomly distributed in the water were also captured (Fig. 4), and the ground-truth data were also carefully annotated.

3 Objective Object Detection Assessment

Deep learning-based object models are designed to replicate the mechanisms involved in visual perception, aiming to improve object recognition performance and achieve more brain-like processing capabilities. In our study, we evaluated the object detection performance using three popular deep learning methods: Mask R-CNN, Faster R-CNN, and YOLO-X. Since underwater environments often have complex lighting conditions and lower image quality, these factors can negatively impact object detection accuracy.

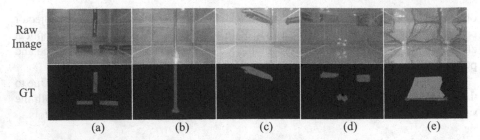

Fig. 3. Single category of data acquisition and annotation. (a) container, (b) props, (c) keel, (d) wreck, and (e) net.

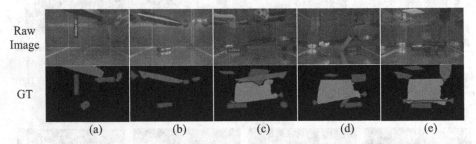

Fig. 4. Mixed categories of data acquisition and annotated ground-truth(GT).

Drawing inspiration from human visual perception, we recognized that objects with higher contrast are generally more detectable. This motivated us to further enhance the performance of these models by integrating image enhancement techniques. Specifically, we fed the enhanced images into the object detection models to evaluate their effectiveness in improving object detection accuracy. According to our previous work [15], three image enhancement methods, including two best performed traditional methods (CLAHE [16], Fusion-based [17]) and one deep learning method (WaterNet [18]), are adopted in this study. For objective evaluation, we have calculated mAP50, mAP75 and mAP to analyse the robustness of models under different Intersection over Union (IoU) thresholds, where IoU threshold is set as 0.5, 0.75 and 0.5:0.05:0.95 for mAP50, mAP75 and mAP, respectively.

4   Results and Analysis

To assess the object detection accuracy of different models under different light condition, three experiments were carried out in this paper. In the first experiment, we utilized a dataset of images captured under bright conditions. This dataset was divided into training and testing sets, consisting of 68 and 27 images, respectively. For the second and third experiments, we focused on different lighting conditions. In the second experiment, we used a dataset of images captured under a specific low-light condition. This dataset was divided into training and testing sets, with 49 images allocated for training and 25 images for testing. Similarly, in the third experiment, we considered a specific dark condition. The dataset of images captured under this condition was also divided into training and testing sets, with 49 images for training and 25 images for testing.

4.1 Object Detection in Bright Environment

Several key findings can be drawn from the results presented in Table 1. YOLOX consistently outperforms Faster-RCNN and Mask-RCNN in terms of mAP50 score, regardless of whether image enhancement is applied. The combination of YOLOX with WaterNet achieves the highest mAP50 score of 0.961. Additionally, Fig. 5 reveals that YOLOX exhibits superior object recognition capabilities, producing tidier and more precise bounding boxes compared to Faster-RCNN and Mask-RCNN.

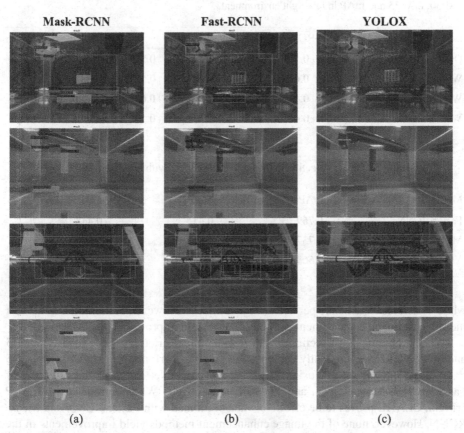

| Mask-RCNN | Fast-RCNN | YOLOX |

(a)   (b)   (c)

Fig. 5. Comparison of three deep learning methods for underwater object detection on the original images

Furthermore, the impact of image enhancement on object detection performance depends on the integration strategy and the specific selection of enhancement and detection methods. Specifically, CLAHE enhances the detection accuracy of Faster-RCNN,

Table 1. Evaluation of deep learning methods with or without image enhancement in terms of mAP50, mAP75 and mAP in bright enviroment.

mAP50/mAP75/mAP	Faster-RCNN			Mask-RCNN			YOLOX		
Original images	0.899	0.780	0.688	**0.901**	0.778	0.688	0.938	0.915	**0.844**
With CLAHE	**0.926**	0.782	**0.695**	0.866	**0.803**	**0.710**	0.945	0.846	0.809
With Fusion	0.903	**0.798**	0.623	0.881	0.789	0.679	0.955	0.889	0.832
With WaterNet	0.883	0.763	0.652	0.884	0.791	0.692	**0.961**	**0.921**	0.807

Table 2. Evaluation of deep learning methods with or without image enhancement in terms of mAP50, mAP75 and mAP in low light environment.

mAP50/mAP75/mAP	Faster-RCNN			Mask-RCNN			YOLOX		
Original images	0.882	0.591	0.56	0.786	0.511	0.519	0.936	0.757	**0.730**
With CLAHE	0.862	**0.660**	0.555	0.864	**0.744**	0.597	0.398	0.237	0.245
With Fusion	**0.897**	0.491	**0.560**	**0.886**	0.725	**0.622**	0.930	**0.795**	0.729
With WaterNet	0.860	0.453	0.504	0.872	0.583	0.576	**0.948**	0.774	0.724

Table 3. Object detection in the dark environment with laser light source

Methods	mAP50	mAP75	mAP
Faster-RCNN	**0.769**	0.415	0.448
Mask-RCNN	0.737	0.458	**0.474**
YOLOX	0.563	**0.488**	0.468

while WaterNet benefits the performance of YOLOX. However, it should be noted that the selected image enhancement methods do not yield improvements in Mask-RCNN.

When evaluating the detection accuracy with more restrictive criteria such as mAP75 and mAP, YOLOX consistently demonstrates superior performance. Under the mAP75 criteria, the integration of Fusion, CLAHE, and WaterNet enhances the precision of Faster-RCNN, Mask-RCNN, and YOLOX, respectively. When considering the mAP criteria, CLAHE proves to be effective when combined with Faster-RCNN and Mask-RCNN. However, none of the image enhancement methods yield improvements in the detection accuracy of YOLOX.

4.2   Object Detection in Low-Light Environment

In general, object detection performance tends to be inferior in low light conditions compared to bright conditions. The reduced contrast between objects of interest and the background in low light conditions poses challenges for object detection methods to accurately recognize objects. However, YOLOX consistently outperforms other object detection methods when applied to both original images and images enhanced using Fusion and WaterNet methods (Table 2).

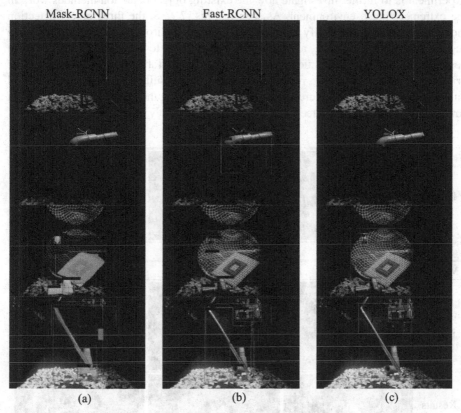

Fig. 6. Visualised object detection results generated by (a) Mask-RCNN, (b) Fast-RCNN, and (c) YOLOX.

An example of applying three DL-based object detection methods on original images is given in Fig. 6. As seen Faster-RCNN and Mask-RCNN either mis-detect the objects or make redundant detections, while YOLOX can precisely recognize the objects. It is surprised that CLAHE dramatically reduces the detection accuracy of YOLOX, and the possible reason is that CLAHE is based on histogram equalization which brings the distortion to the low light image and mislead the object detection methods to make decision. Also, similar to the experimental results in Sect. 4.1, image enhancement can more or less improve the object detection accuracy in the low light condition, though it depends on the selection of image enhancement methods and the combination strategies of image enhancement and object detection methods.

4.3 Object Detection in Dark Environment

Additionally, we extended this experiment by applying the object detection methods to the image acquired in a totally dark environment. The motivation of this extended experiment is to further investigate how the existing object detection methods work in the extremely harsh environment. As seen in Figs. 7 & 8, the three object detection methods are unable to identify the object in the dark environment. However, with the support of laser light source, all three methods are able to make the better detection despite the fact that the detection accuracy (as shown in Table 3) is worse than that of prior experiments. This main reason is that there are insufficient color attributes and only shape and potential texture attributes when capturing the image under a laser light source. Consequently, the detection accuracy will inevitably decline.

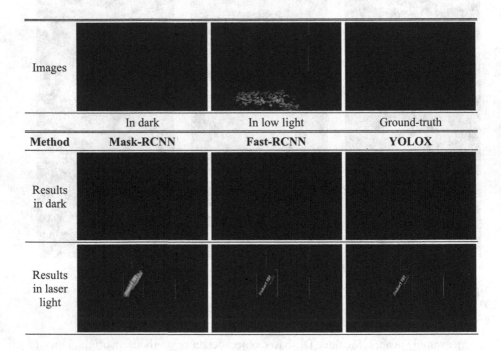

Fig. 7. Object detection in dark condition with laser as light source - Example 1.

To overcome this issue, depth information should be obtained from laser-based triangulation system and then integrated with image data for improved detection performance. Due to the page limitation, we didn't include this work in this paper. However, it will be our future focus.

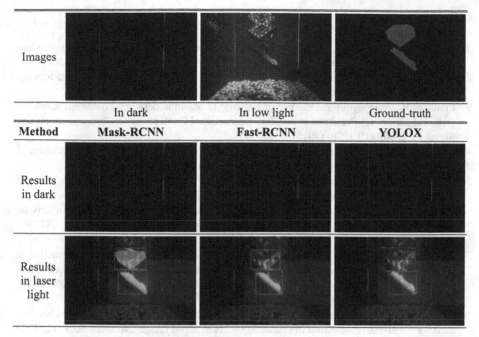

	In dark	In low light	Ground-truth
Method	**Mask-RCNN**	**Fast-RCNN**	**YOLOX**

Fig. 8. Object detection in dark condition with laser as light source - Example 2.

5 Conclusion

In this project, a scotopic imaging system was built up to generate unique underwater object detection dataset consisting of 243 images with five obstacle-alike objects. Comprehensive assessment on this dataset has been carried out by three step-by-step experiments. From the experimental results, some interesting findings can be concluded as follows.

- Current experimental setting can support data acquisition in different experimental settings (bright, low-light and dark);
- Image enhancement can somehow enhance visibility of underwater images and prediction accuracy of the objects in various light conditions, but still has big room to improve;
- Existing deep-learning models can produce good object detection results in the bright and low light underwater environment;
- There is a big challenge for object detection models to make right predication in the dark environment;
- Laser can be useful for object detection as an additional light source in the dark environment and potentially provide supplementary information of depth.

Acknowledgement. This work was partially funded by the Office of Naval Research.

References

1. Braginsky, B., Guterman, H.: Obstacle avoidance approaches for autonomous underwater vehicle: Simulation and experimental results. IEEE J. Oceanic Eng. **41**, 882–892 (2016)
2. Neves, G., Ruiz, M., Fontinele, J., Oliveira, L.: Rotated object detection with forward-looking sonar in underwater applications. Expert Syst. Appl. **140**, 112870 (2020)
3. Pranitha, B., Anjaneyulu, L.: Review of research trends in underwater communications—a technical survey. In: Presented at the 2016 International Conference on Communication and Signal Processing (ICCSP) (2016)
4. Ostashev, V.E.: Sound propagation and scattering in media with random inhomogeneities of sound speed, density and medium velocity. Waves Random Media **4**, 403 (1994)
5. Life, M.: Mitigation of underwater anthropogenic noise and marine mammals: the 'death of a thousand'cuts and/or mundane adjustment? Mar. Pollut. Bull. **102**, 1–3 (2016)
6. Dairi, A., Harrou, F., Senouci, M., Sun, Y.: Unsupervised obstacle detection in driving environments using deep-learning-based stereovision. Robot. Auton. Syst. **100**, 287–301 (2018)
7. Yan, Y., Zhao, H., Kao, F.-J., Vargas, V.M., Zhao, S., Ren, J.: Deep background subtraction of thermal and visible imagery for pedestrian detection in videos. In: Presented at the Advances in Brain Inspired Cognitive Systems: 9th International Conference, BICS 2018, Xi'an, China, July 7–8, 2018, Proceedings 9 (2018)
8. Fang, Z., Ren, J., Sun, H., Marshall, S., Han, J., Zhao, H.: SAFDet: A semi-anchor-free detector for effective detection of oriented objects in aerial images. Remote Sens. **12**, 3225 (2020)
9. Li, Y., et al.: CBANet: an end-to-end cross band 2-D attention network for hyperspectral change detection in remote sensing (2023)
10. Song, P., Li, P., Dai, L., Wang, T., Chen, Z.: Boosting R-CNN: reweighting R-CNN samples by RPN's error for underwater object detection. Neurocomputing **530**, 150–164 (2023)
11. Liu, C., et al.: A dataset and benchmark of underwater object detection for robot picking. In: Presented at the - 2021 IEEE International Conference on Multimedia & Expo Workshops (ICMEW) (2021). https://doi.org/10.1109/ICMEW53276.2021.9455997
12. Hong, J., Fulton, M., Sattar, J.: TrashCan: a semantically-segmented dataset towards visual detection of marine debris (2020)
13. Jiang, L., et al.: Underwater species detection using channel sharpening attention. In: Presented at the Proceedings of the 29th ACM International Conference on Multimedia (2021)
14. Pedersen, M., Bruslund Haurum, J., Gade, R., Moeslund, T.B.: Detection of marine animals in a new underwater dataset with varying visibility. In: Presented at the Proceedings of the IEEE/CVF Conference on Computer Vision and Pattern Recognition Workshops (2019)
15. Lin, H., Men, H., Yan, Y., Ren, J., Saupe, D.: Crowdsourced quality assessment of enhanced underwater images - a pilot study. In: Presented at the - 2022 14th International Conference on Quality of Multimedia Experience (QoMEX) (2022). https://doi.org/10.1109/QoMEX5 5416.2022.9900904
16. Reza, A.M.: Realization of the contrast limited adaptive histogram equalization (CLAHE) for real-time image enhancement. 38, 35–44 (2004)
17. Saleem, A., Beghdadi, A., Boashash, B.: Image fusion-based contrast enhancement **2012**, 1–17 (2012)
18. Li, C., et al.: An underwater image enhancement benchmark dataset and beyond. IEEE Trans. Image Process. **29**, 4376–4389 (2019)

RSF-SSD: An Improved SSD Algorithm Based on Multi-level Feature Enhancement

Weiqiang Yang, Shan Zeng$^{(\boxtimes)}$, Chaoxian Liu, Hao Li, and Zhen Kang

Wuhan Polytechnic University, Wuhan, China
zengshan1981@whpu.edu.cn

Abstract. SSD (Single Shot Multi Box Detector) is one of the commonly used object detection algorithms known for its fast detection speed and high accuracy. However, SSD's performance in detecting objects of different scales is suboptimal. This paper proposes the RSF-SSD network based on multi-level feature enhancement. By improving the backbone network of SSD, skip-connections and channel attention mechanism are introduced into VGG16. This operation enhances the ability of the backbone network to extract detailed features. In the feature fusion module, an improved FPN (Feature Pyramid Networks) + PAN (Path Aggregation Network) module is introduced (referred to as FAN) to achieve comprehensive fusion of deep semantic information and shallow localization information. In testing on the PASCAL VOC07 + 12 dataset, the proposed network achieves a 15.3%mAP improvement in detection accuracy compared to traditional SSD. The experiments demonstrate that the RSF-SSD model effectively enhances the capability of extracting detailed features, and incorporates more semantic information in the shallow layers, leading to improved detection performance for objects of different sizes.

Keywords: object detection · multi-level · feature fusion · RSF-SSD

1 Introduction

Object detection is an important branch of image processing and computer vision, which aims to recognize objects in images through computer vision techniques, reducing the need for manual labor. It has been widely applied in various fields such as car navigation [1], intelligent surveillance [2, 3], remote sensing [4], and aerospace [5], providing significant value for scientific research and practical applications.

Currently, deep learning-based object detection algorithms can be broadly categorized into two types. The first type is two-stage object detection algorithms based on the Region Proposal strategy, such as SPPNet (Spatial Pyramid Pooling Network) [6] and Fast R-CNN (Fast Regions with CNN features) [7]. The second type is one-stage object detection algorithms based on regression frameworks, with representative algorithms including the YOLO (You Only Look Once) [8] series and the Single Shot Multi Box Detector [9] series. Two-stage algorithms refer to a step-by-step and progressive detection process. For example, FPN [10] proposed in 2017 combines rich semantic

© The Author(s), under exclusive license to Springer Nature Singapore Pte Ltd. 2024
J. Ren et al. (Eds.): BICS 2023, LNAI 14374, pp. 123–132, 2024.
https://doi.org/10.1007/978-981-97-1417-9_12

information from higher layers with precise object localization information from lower layers, and integrates them into various feature levels to jointly detect objects of different sizes. This algorithm greatly improves the accuracy of object detection, especially for small objects. For one-stage algorithms, they directly predict bounding boxes and class probabilities of objects through a convolution neural network applied to the input image. For instance, YOLO proposed in 2015, as the first one-stage object detection algorithm based on deep learning frameworks, divides the original image into multiple regions and predicts bounding boxes and class probabilities for each region.

Among the object detection algorithms, SSD is an end-to-end deep learning-based algorithm that achieves a good balance between detection speed and accuracy. SSD performs dense and uniform sampling at various positions on the image, while leveraging the anchor box concept from Faster R-CNN [11], enabling it to maintain high detection accuracy even at lower input image resolutions. The end-to-end design of SSD makes it easy to train and achieves a good balance between detection speed and accuracy. SSD utilizes a feature pyramid structure for object detection, where lower-level feature maps with smaller receptive fields are responsible for predicting small objects, and higher-level feature maps with larger receptive fields are responsible for predicting large objects. However, this structure suffers from a unidirectional and loosely connected relationship between the semantic information in the higher-level feature maps and the localization information in the lower-level feature maps. Additionally, SSD mainly uses an improved VGG16 as its backbone network for feature extraction. While VGG16 is characterized by its simplicity, depth, small convolution and pooling kernels, which makes it suitable for larger datasets and reduces the number of parameters and computation for the same receptive field. It also results in the loss of fine-grained information during the multi-level convolution process, leading to poor performance in detecting small objects.

To address these issues, some researchers have proposed improvements in feature fusion, such as pooling shallow-level features into deep-level features, deconvolutional fusion of deep-level features into shallow-level features, and concatenation of features from different layers and scales[12]. Although these adjustments have achieved good results in terms of detection accuracy, they significantly sacrifice detection speed. Other researchers have focused on improving the backbone network, replacing the VGG16 backbone in SSD with ResNet-101 [13] or using an improved DenseNet [14] module instead of VGG-16. While these improvements have increased the detection accuracy of SSD, ResNet-101 and DenseNet are more complex than the original VGG16 backbone, resulting in no advantage in terms of detection speed.

Although the aforementioned improvements have addressed some issues in SSD, there are still unresolved problems and new directions to explore: (1) reducing model computation while achieving comprehensive feature fusion [15], (2) reducing model complexity while enhancing the detail extraction capability of the backbone network. In this paper, we draw inspiration from the FPN and PAN [16] modules for feature fusion, and the ResNet and DenseNet networks for the backbone network, while introducing the channel attention mechanism [17]. The main contributions of this paper are as follows: (1) Introduce skip-connection mechanisms and channel attention mechanisms to each convolution layer of the VGG16 backbone, proposing the RSNet network structure. Compared to the traditional VGG16, the RSNet network structure maximally preserves the detailed information of features, significantly improving the expressive power of effective features and accelerating model convergence. (2) The feature fusion mechanism is introduced for the features extracted from the backbone network, and the FAN network structure is proposed as its feature fusion module, and the features processed by this module can be fully integrated with the low-level location features and high-level semantic features relative to the features that have not been processed by this module.

2 RSF-SSD

2.1 Overall Structure of RSF-SSD

This paper builds upon the basic framework of SSD to address the issues of significant loss of detailed information and insufficient feature extraction capabilities in the multilayer convolution process of the SSD algorithm. Firstly, residual layers are added to the VGG-16 network to enhance the feature extraction capabilities and avoid network degradation issues. Next, channel attention mechanism is incorporated into the residual layers to improve channel expression, significantly enhancing the network feature representation capabilities. Finally, an improved FPN + PAN structure (FAN) is introduced into the original SSD network pyramid structure, facilitating comprehensive fusion between low-level localization features and high-level semantic features. The overall architecture of RSF-SSD is as follows (see Fig. 1. The structure of RSF-SSD).

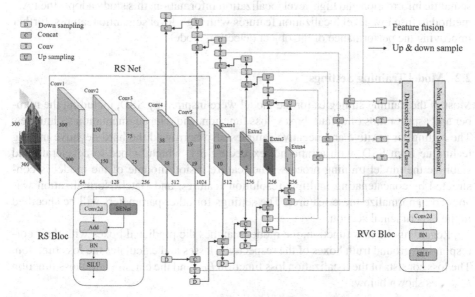

Fig. 1. The structure of RSF-SSD

Specifically, in this study, RSNet was used for feature extraction. This network has the option to include a channel attention mechanism. RSNet consists of 16 blocks, divided into five convolutional layers: 2 blocks, 3 blocks, 3 blocks, 4 blocks, and 4 blocks. Each layer's first block is an RVG block, while the rest are RS blocks. The input image size is set to 300 × 300. After passing through RSNet, the features from the last layer and the fourth layer of RSNet are selected for output, resulting in two feature maps of size 38 × 38 × 512 and 19 × 19 × 1024. Then, these features go through four extra modules, yielding a total of six feature maps with sizes 38 × 38 × 512, 19 × 19 × 1024, 10 × 10 × 512, 5 × 5 × 256, 3 × 3 × 256, and 1 × 1 × 256. In the next step, the first layer of features is up sampled through deconvolution and fused with the corresponding features from the upper layer by summation. This process is repeated in a top-down manner until the fusion reaches the topmost layer.

In the second step, the features are down sampled in descending order of their sizes and fused with the features from the lower layer. In the third step, the fused features from the previous two steps are added together based on their corresponding feature map sizes. A convolution module is then used to smooth the features before inputting them into the detector and classifier for network prediction. Finally, a non-maximum suppression algorithm is applied to obtain the optimal results from six detection classifiers, which are then used as the final output of the SSD network. Therefore, inspired by ResNet and DenseNet, this study introduces partial residual connections in the backbone network based on VGG16 and constructs RSNet with channel attention mechanism (SENet) to address the issues of network degradation and feature disappearance. Furthermore, conventional FPN structures enhance the entire feature pyramid by transmitting high-level semantic features top-down. However, while semantic information is enhanced, the localization information is not transmitted. To address the insufficient low-level semantic information and high-level localization information, this study adopts the FAN method to fuse low-level localization features with high-level semantic features, further improving the performance of the object detection model.

2.2 Model Training Settings

Most of the training strategies for the model were inspired by SSD, including the number and aspect ratios of default boxes, loss function, and data augmentation techniques. The setting for positive and negative samples also borrowed the hard negative mining technique from SSD, with a ratio not exceeding 3:1, which helped to accelerate and stabilize the model training process. The feature fusion module of the model is constructed by concatenating multiple resolution features, and Batch Normalization was applied to normalize these features. The settings for other parameters will be specified in the experimental section.

Regarding the loss function, the model matches the predicted boxes with their corresponding ground truth boxes of the respective classes and calculates the loss function. The loss consists of the localization loss function L_{loc} and the classification loss function L_{conf}, as shown below:

$$L(x, c, l, g) = \frac{1}{N}(L_{conf}(x, c) + \alpha L_{loc}(x, l, g)) \tag{1}$$

Here, N represents the number of matched default boxes. If N = 0, the loss function value is 0. The purpose of the localization loss function L_{loc} is to regress the offset of the center point and the width and height of the default box:

$$L_{loc}(x, l, g) = \sum_{i \in Pos}^{N} \sum_{m \in \{cx,cy,\omega,h\}} x_{ij}^k smooth_{L1}(l_i^m - \hat{g}_j^m) \qquad (2)$$

In the above equation, g represents the ground truth box, $\hat{g}$ represents the predicted box from the network. Additionally, the classification loss function is defined as:

$$L_{conf}(x, c) = -\sum_{i \in Pos}^{N} x_{ij}^p \log(\hat{c}_i^p) - \sum_{i \in Neg} \log(\hat{c}_i^0) \; where \; \hat{c}_i^p = \frac{\exp(c_i^p)}{\sum_p \exp(c_i^p)} \qquad (3)$$

3 Experiment and Analysis

To fully demonstrate the impact of the improved RSNet module and FAN module on the SSD model, this paper conducted model validation using the VOC07 + 12 dataset. The VOC07 + 12 dataset includes the training and validation sets of VOC2007 and VOC2012, with the test set of VOC2007 used as the validation set. It consists of a total of 20 object categories. The training set contains 16,551 annotated images with 40,058 annotated objects, while the validation set contains 4,952 annotated images with 12,032 annotated objects.

The experimental platform was configured with an Intel® Core™ i9–10900 CPU and an NVIDIA GeForce RTX 3090 GPU. The operating system used was Windows 10 Professional Workstation Edition, and the deep learning framework employed was PyTorch 1.8. The model was evaluated on the PASCAL VOC2007, PASCAL VOC2012, and RSOD [18, 19] datasets. These datasets consist of 20, 20, and 4 target classes, respectively. The model was trained from scratch using the corresponding datasets, with the Adam optimization algorithm as the optimizer. The maximum learning rate (Init_lr) was set to 0.0006, the minimum learning rate (Min_lr) was set to 0.000006, and the learning rate decayed gradually as the training epoch increased. The batch size was set to 32, and training was stopped after 150 epochs. The evaluation metrics primarily used for the model included mean Average Precision (mAP), Frames Per Second (FPS), Precision, and Recall.

3.1 Analysis of Test Results

Five sets of images with different scales and complex backgrounds were selected from the PASCAL dataset for instance verification (see Fig. 2. Comparison of test results of improved model). By analyzing the detection results of the original algorithm and the improved model on these 5 sets of images, it is evident that the improved model outperforms the original algorithm in detecting dense small objects and overlapping objects.

The experimental results in Table 1. Comparison of experimental results with main-stream models on VOC07 + 12 dataset demonstrate that the improved models, compared

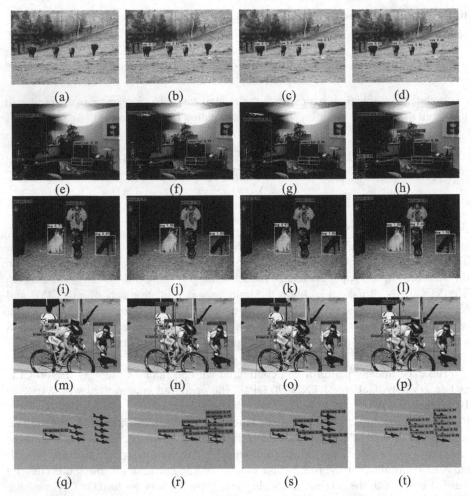

Fig. 2. Comparison of test results of improved model.The results obtained from the original SSD algorithm are shown in figures (a), (e), (i), (m), and (q). The detection results from the SSD + RS model are shown in figures (b), (f), (j), (n), and (r). The detection results of the SSD + FAN model are displayed in figures (c), (g), (k), (o), and (s). Finally, the detection results from the SSD + RSF model are shown in figures (d), (h), (l), (p), and (t).

to the improved models without feature fusion, showed a performance improvement of 1.4% in mAP. Directly incorporating the feature fusion module into SSD resulted in a performance improvement of 7.7% in mAP and a speed increase of 28 fps. It can be observed that the added feature fusion module had a significant effect, but its improvement on RS-SSD was not as rapid as on SSD. The reason behind this can be attributed to the distinct differences in feature channel processing between RSNet and the feature fusion module. RSNet excels at enhancing the transmission of neighboring convolution information, while the feature fusion module excels at the interaction of information across different-scale feature layers.

Furthermore, by replacing VGGNet16, the backbone network of SSD, with RSNet, the detection performance of RS-SSD improved by 13.9% in mAP (78.5% mAP VS 64.6% mAP). Moreover, the detection speed increased by 10 fps (23 fps compared to 33 fps). Since both models were trained from scratch after initialization, the improved RSNet exhibited strong feature extraction capabilities compared to VGG16, making it more suitable for model inference.

Table 1. Comparison of experimental results with mainstream models on VOC07 + 12 dataset

Algorithm	Backbone	mAP	Input	FPS
Faster RCNN	VGG16	73.9	600 × 600	7
YOLO	VGG16	63.5	448 × 448	46
YOLO9000	Darknet19	69.0	288 × 288	91
YOLOv4	VGG16	70.1	416 × 416	65
YOLOV5	CSPdarknet	72.1	640 × 640	197
YOLOV8	Darknet	77.2	640 × 640	84
DSSD	Resnet101	72.7	320 × 320	33
SSD300	VGG16	64.6	300 × 300	23
SSD512	VGG16	71.0	512 × 512	15
SSD-FAN	VGG16	72.3	300 × 300	51
SSD-RS	RSNet	78.5	300 × 300	33
SSD-RSF	**RSNet**	**79.9**	**300 × 300**	**31**

3.2 Analysis of Ablation Experiments

This section presents the results of controlled ablation experiments conducted on the PASCAL VOC07 + 12 dataset to study the individual modules of the RSF-SSD model and analyze their experimental outcomes. The specific experimental results are shown in Table 2. Experimental results comparison of different modules To avoid issues such as overfitting due to insufficient data and enhance the robustness of the network, various data augmentation techniques were employed during training, including random grayscale transformations, random flipping, random cropping, and random rotation. These operations were applied to the training data to increase its diversity and volume.

From the VGG & RS module, it can be observed that regardless of the dataset size, the RS backbone network outperforms the traditional VGG16, with a maximum improvement of 13.9% in mAP. When comparing the detection speed (FPS), it can be seen that the improvement in this module has significantly increased the speed of object detection. In summary, the improved module not only achieves more accurate object detection but also enables faster detection speed.

On the other hand, from the feature fusion module, it can be observed that the addition of this module results in an increase in parameter count of approximately 40.5M.

Table 2. Experimental results comparison of different modules on the PASCAL VOC07 + 12 dataset

Method	data	Input	VGG	RS	FAN	mAP	Par	FPS
SSD	07 + 12	300	✔	×	×	64.6	26.2 M	23
SSD	07 + 12	300	✔	×	✔	72.3	66.7 M	51
SSD	07 + 12	300	×	✔	×	78.5	54.0 M	33
SSD	07 + 12	300	×	✔	✔	79.9	94.5 M	31

However, it leads to improvements in both detection speed and model performance (mAP). The detection speed increases by approximately 28 fps, while the performance improves by around 7.7% in mAP. Overall, the inclusion of the feature fusion module plays a significant role in enhancing the performance of the algorithm.

3.3 Validation on Other Datasets

Experimental results on the VOC07 + 12 public dataset demonstrate that the proposed model in this paper performs well on small objects and objects with special relationships. However, to further validate the generalization performance of this approach, we selected the RSOD remote sensing dataset for further generalization verification. The RSOD dataset consists of 976 images and a total of 6,950 objects, including four categories: Oil tank, Aircraft, Overpass, and Playground. In this study, we used the metrics of Recall

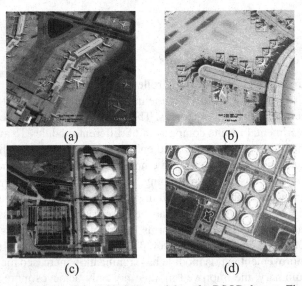

(a)   (b)

(c)   (d)

Fig. 3. Partial detection results of the RSF-SSD model on the RSOD dataset. Figures (a) and (b) depict smaller and densely clustered aircraft targets, while figures (c) and (d) represent smaller and closely packed oil tank targets.

and Precision to quantitatively evaluate each category. Some validation results are shown below (see Fig. 3. Partial detection results of the RSF-SSD model on the RSOD dataset).

By combining the results of various models in Table 3. Comparison of detection results of the RSF-SSD model on the RSOD dataset, it can be observed that the improved modules perform well in terms of detection performance on the RSOD remote sensing dataset. Moreover, remote sensing datasets typically have high image resolutions, with relatively small objects to be detected and a higher proportion of small objects. Therefore, this highlights the superiority of the improved model in detecting small objects from another perspective.

Table 3. Comparison of detection results of the RSF-SSD model on the RSOD dataset

	Oil tank		Aircraft		Overpass		Playground	
Met	P	R	P	R	P	R	P	R
M1	0.9364	0.9349	0.9296	0.9152	0.7651	0.8636	0.6770	0.8456
M2	0.9392	0.9177	0.8336	0.8863	0.5260	0.7652	0.8933	0.9404
M3	0.8000	0.9852	0.4758	0.8113	0.6481	0.9722	0.8108	1.0000
M4	0.9729	0.8516	0.9640	0.2599	0.7576	0.6944	0.9062	0.9667
M5	0.9796	0.8546	0.8789	0.8546	0.8611	0.8611	0.9677	1.0000

Note: M1 represents the AlexNet method [19], M2 represents the GoogleNet method [19], M3 represents the Faster R-CNN method, M4 represents the SSD method, and M5 represents the RSF-SSD method

4 Conclusion

RSF-SSD is an object detection algorithm based on VGG16, FPN, PAN, channel attention mechanism, and residual modules. The model is trained from scratch, which enables it to achieve excellent detection performance using a smaller training dataset compared to other pre-trained models. RSF-SSD not only achieves better results on the PASCAL VOC07 + 12 dataset but also exhibits fast detection speed. Moreover, it demonstrates impressive detection performance on small-sized objects and objects with specific relationships, as validated on the RSOD dataset. In the future, further deep optimization will be conducted to focus on reducing parameter size and improving detection speed under the premise of guaranteeing the detection accuracy.

References

1. Ess, A., Leibe, B., Van Gool, L.: Object detection and tracking for autonomous navigation in dynamic environments. Inter. J. Robot. Res. **29**, 1707–1725 (2010)
2. Feng, D., et al.: Deep multi-modal object detection and semantic segmentation for autonomous driving: datasets, methods, and challenges. IEEE Trans. Intell. Transp. Syst. **22**, 1341–1360 (2021)

3. Mhalla, A., Chateau, T., Gazzah, S., Amara, N.E.B.: An embedded computer-vision system for multi-object detection in traffic surveillance. IEEE Trans. Intell. Transp. Syst. **20**, 4006–4018 (2019)
4. Li, Y., et al.: CBANet: an end-to-end cross-band 2-D attention network for hyperspectral change detection in remote sensing. IEEE Trans. Geosci. Remote Sens. **61**, 1–11 (2023)
5. Han, J., Ding, J., Li, J., Xia, G.-S.: Align deep features for oriented object detection. IEEE Trans. Geosci. Remote Sens. **60**, 1–11 (2022)
6. He, K., Zhang, X., Ren, S., Sun, J.: Spatial pyramid pooling in deep convolutional networks for visual recognition. IEEE Trans. Pattern Anal. Mach. Intell. **37**, 1904–1916 (2015)
7. Girshick, R.: Fast R-CNN. In: 2015 IEEE International Conference on Computer Vision (ICCV), pp. 1440–1448 (2015)
8. Redmon, J., Divvala, S., Girshick, R., Farhadi, A.: You only look once: unified, real-time object detection. In: Presented at the Proceedings of the IEEE Conference on Computer Vision and Pattern Recognition (2016)
9. Liu, W., et al.: SSD: single shot MultiBox detector. In: Leibe, B., Matas, J., Sebe, N., Welling, M. (eds.) Computer Vision – ECCV 2016, pp. 21–37. Springer International Publishing, Cham (2016)
10. Lin, T.-Y., Dollár, P., Girshick, R., He, K., Hariharan, B., Belongie, S.: Feature pyramid networks for object detection. In: 2017 IEEE Conference on Computer Vision and Pattern Recognition (CVPR), pp. 936–944 (2017)
11. Ren, S., He, K., Girshick, R., Sun, J.: Faster R-CNN: towards real-time object detection with region proposal networks. IEEE Trans. Pattern Anal. Mach. Intell. **39**, 1137–1149 (2017)
12. Zhai, S., Shang, D., Wang, S., Dong, S.: DF-SSD: an improved SSD object detection algorithm based on DenseNet and feature fusion. IEEE Access **8**, 24344–24357 (2020)
13. He, K., Zhang, X., Ren, S., Sun, J.: Deep residual learning for image recognition. In: 2016 IEEE Conference on Computer Vision and Pattern Recognition (CVPR), pp. 770–778 (2016)
14. Huang, G., Liu, Z., Van Der Maaten, L., Weinberger, K.Q.: Densely connected convolutional networks. In: 2017 IEEE Conference on Computer Vision and Pattern Recognition (CVPR), pp. 2261–2269 (2017)
15. Luo, F., Zhou, T., Liu, J., Guo, T., Gong, X., Ren, J.: Multiscale Diff-changed feature fusion network for hyperspectral image change detection. IEEE Trans. Geosci. Remote Sens. **61**, 1–13 (2023)
16. Liu, S., Qi, L., Qin, H., Shi, J., Jia, J.: Path aggregation network for instance segmentation. In: 2018 IEEE/CVF Conference on Computer Vision and Pattern Recognition, pp. 8759–8768 (2018)
17. Hu, J., Shen, L., Sun, G.: Squeeze-and-excitation networks. In: 2018 IEEE/CVF Conference on Computer Vision and Pattern Recognition, pp. 7132–7141 (2018)
18. Long, Y., Gong, Y., Xiao, Z., Liu, Q.: Accurate object localization in remote sensing images based on convolutional neural networks. IEEE Trans. Geosci. Remote Sens. **55**, 2486–2498 (2017)
19. Xiao, Z., Liu, Q., Tang, G., Zhai, X.: Elliptic fourier transformation-based histograms of oriented gradients for rotationally invariant object detection in remote-sensing images. Int. J. Remote Sens. **36**, 618–644 (2015)

GaitMG: A Multi-grained Feature Aggregate Network for Gait Recognition

Jiwei Wan[1], Huimin Zhao[1], Rui Li[1,2(✉)], Rongjun Chen[1], Tuanjie Wei[1], and Yongqi Ren[1]

[1] School of Computer Science, Guangdong Polytechnic Normal University, Guangzhou 510665, China
ruili_gpnu@163.com
[2] College of Fine Arts, Guangdong Polytechnic Normal University, Guangzhou 510665, China

Abstract. In order to make better use of the temporal information of gait data and improve the accuracy of gait recognition. In this paper, we propose a multi-grained feature aggregate network(GaitMG), which contains two important modules: Multi-Grained Feature Aggregator(MGFA) and Spatio-Temporal Feature Fusion Module(STFFM). The MGFA: a novel applying of convolution, can tackle the problem of poor representation ability of single grained temporal features. STFFM can fuse the multi-grained temporal features obtained by MGFA, get a discriminative representation. On CASIA-B, our method can achieve rank-1 accuracy of 98.0% under normal walking condition, 94.1% under bag-carrying condition, 85.3% under coat-wearing condition.

Keywords: Gait Recognition · Multi-Grained · Spatio-Temporal Feature

1 Introduction

Gait recognition, as an emerging biometric technology, which can recognize the identity of a person by extracting body shape, head shape and continuous movements characteristics from the walking pattern of human body. Gait recognition aims to identify a person at a distance, so compared with other commonly used biometric technologies like face recognition, fingerprint recognition, iris recognition; gait recognition is non-invasive, does not affect the normal activities of the identified pedestrians, and does not need the cooperation of the identified object. Based on the above characteristics that the application scenarios of gait recognition are broad: forensic identification, criminal investigation, social security and stability. In order to make better use of inherent temporal information contained in video frames, we assume that the amount of information contained in video frames with different temporal grained is different, and these multi-grained temporal information plays an important role in gait recognition. Thus we propose a Multi-Grained Feature Aggregate Network for gait recognition called GaitMG. GaitMG is made up with two simple but effective module: Multi-Grained Feature Aggregator(MGFA) and Spatio-Temporal Feature Fusion Module(STFFM).

J. Ren et al. (Eds.): BICS 2023, LNAI 14374, pp. 133–142, 2024.
https://doi.org/10.1007/978-981-97-1417-9_13

The main contributions of this article can be summarized as the following three aspects:

In MGFA, we propose a novel, simple and effective temporal information aggregator, which can collect temporal information of different grained so as to obtain features with stronger discrimination ability.

In STFFM, STFFM can further extract the temporal information of different grained, and fuse fine-grained spatial features and multi-grained temporal features enables the final representation to have stronger generalization ability.

Extensive experiments conducted on the popular public datasets CASIA-B show that our method achieve state-of-the-art performance.

2   Related Work

2.1   Gait Recognition

At present, gait recognition methods based on deep learning can be roughly divided into two categories: Model-Based [5,12,16,23] and Appearance-Based [1,3,6,8,9,11,17,18,20,21]. Model-based gait recognition methods model gait structure using a set of parameters and a set of logic to obtain discriminative gait features. [22] propose a gait lateral network structure, which can directly obtain discriminative and compact features from the gait contour sequence. The compressed blocks contained in the network can significantly reduce the number of gait features without affecting accuracy. GaitPart [3] believe that different parts of the body have unique temporal and spatial characteristics, and proposed a model based on local characteristics to divide the gait sequence contour into different parts to obtain fine-grained spatial features. GaitSlice [11] propose SED module to enhance the subtle information between different gait parts and connect the adjacent body parts to enhance the spatial features. In terms of temporal features, frame attention mechanism is used to determine key frames.

2.2   Temporal Operation

The biggest difference between gait recognition and other biometric recognition technologies is the temporal information contained in gait frames. Therefore, how to operate temporal information become the key factor in achieving excellent results in gait recognition. Current gait recognition methods can be grouped into four main categories according to the way they operate on temporal information: 1D convolution based [2,3,7,9,11,13,21,25], 3D convolution based [14,20], LSTM-based [24] and set-based [1]. 1D convolution based methods extract temporal features of gait frames by using 1D convolution on temporal dimension. The way of 3D convolution based methods use 3D convolution directly to extract spatio-temporal features from low-level features. LSTM based method add continuous gait features through the long-term and short-term memory module to obtain discriminative features that contain temporal information. GaitSet [1] regard gait sequence as an unordered set to extract temporal feature through a pooling layer in a global-local structure.

3   Proposed Method

In this section, we first introduce the pipeline of GaitMG, and then we give a detailed information of each module in GaitMG, including Multi-Grained Feature Aggregator(MGFA) and Spatio-Temporal Feature Fusion Module(STFFM).

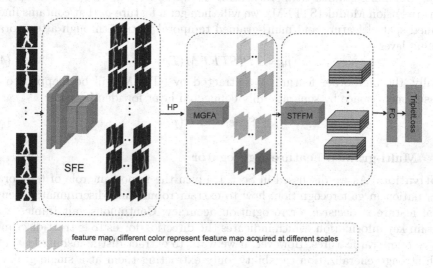

Fig. 1. The overall structure of GaitMG. The first lyaer is Shallow Feature Extractor(SFE), consists of 4 Conv2d layers. HP denotes Horizontal Pooling layer, MGFA represents Multi-Grained Feature Aggregator, STFFM denotes Spatio-Temporal Feature Fusion Module, FC represents Fully Connected layer.

3.1   Overview

As shown in Fig. 1, the input of GaitMG is a sequence of gait silhouette that contains N * T frames, where N denotes the quantity of objects in each batch, and T denotes gait frames that every object has. So the raw input data could be represented as X_{ij} $(i \in 1, 2, ..., T. \ j \in 1, 2, ..., N)$. For the convenience of the following illustration, we only select gait frames of one subject to describe. The representation of input is X_i. And then the raw data of input $X_i (i \in 1, 2, ..., T)$ will be sent to Shallow Feature Extractor(SFE). Through 4 convolution layers in SFE, the frame-level feature S_i could be formulated as:

$$S_i = SFE(X_i) \tag{1}$$

Then the frame-level feature S_i will pass through an intermediate transition layer, Horizontal Pooling(HP) module. In this part, we choose the maximum function like [1], more representative feature F_i would be got:

$$F_i = HP(S_i) \tag{2}$$

By inputting F_i into the MGFA module we designed, the temporal information of different grained is gathered, and the representation ability of the obtained feature T_i is enhanced, the process can be formulated as:

$$T_i = MGFA(F_i) \tag{3}$$

Then, the multi-grained temporal feature T_i would be sent to Spatio-Temporal Feature Fusion Module(STFFM), we will then get a feature p_i that contains fine-grained spatial feature and multi-grained temporal feature through a temporal pooling layer :

$$p_i = TP(STFFM(T_i)) \tag{4}$$

Finally, the high-order features p_i extracted by STFFM will be mapped to a measurement space by separate fully connected layer for identification:

$$f_i = SeperateFC(p_i) \tag{5}$$

3.2 Multi-grained Feature Aggregator

Motivation. As we discussed in Sect. 1, given the important role of temporal information in gait recognition, how to extract robust and discriminative temporal feature is decisive for recognition accuracy. Continuous gait frames will contain key information, which indicates the direction for us to extract discriminative temporal features. However, if we want to extract distinguishing features with strong generalization capability, only extracting them at a single grained will lose key information. Therefore, we propose the concept of multi-grained temporal information aggregation, which makes the obtained temporal features more robust by operating the temporal information at different grained.

Operation. Convolution, as a simple and effective feature extraction method, has been widely used as a feature extractor. Convolution kernel, as a crucial part of convolution operation, the size of convolution kernel will affect the effect of feature extraction. Convolution kernels of different sizes have different receptive fields, and the amount of information it can get is different, which opens the way for us to extract temporal features of different grained. Therefore, in the multi-grained temporal information aggregation layer, we aggregate temporal features of different grained by weighted summation of temporal information of different grained based on the convolution operation with convolution kernels of different sizes. (In addition, in order to maintain the boundary information of the feature to the greatest extent and ensure feature dimension does not change, here we need to fill the corresponding blank elements in the border) (Fig. 2).

3.3 Spatio-Temporal Feature Fusion Module

Motivation. Through the operations in previous section, we extracted features from different temporal grained, and obtained multi-grained temporal features through fusion and aggregation. However, the ability of this feature to distinguish subtle changes is limited, and further operations are needed to obtain fine grained features to enhance the feature's ability of distinguishing and representing.

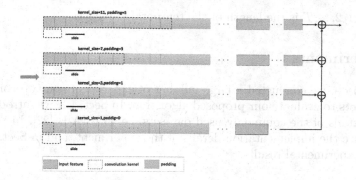

Fig. 2. The detailed operation of MGFA.

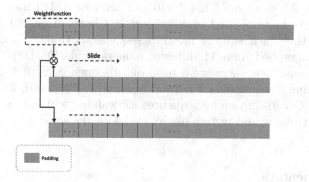

Fig. 3. Schematic diagram of re-weighting each feature point.

Operation. Due to the wide use of attention mechanism and its effect gain, we use attention mechanism to obtain fine-grained information here. For the obtained features with multi-grained temporal information aggregation, we re-weight them by the following formula:

$$WeightFunction(F_i) = Sigmoid(Conv1d(BN(Conv1d(F_i)))) \qquad (6)$$

Actually, since the multi-grained feature extractor does not change the input dimension, the input dimension of this part is the same as the previous section. For this part, our weight is still obtained through 1D convolution. For the vector corresponding to each channel, we operate them as shown in Fig. 3. First, according to the operation rules corresponding to the weight function, we perform corresponding operations on several feature points to obtain the weight that should be given to the corresponding points, then directly carry out point-wise multiplication between the feature point and its corresponding weight, and obtain the value of each feature point after it was re-weighted. The features after re-weighting can be expressed by the following formula:

$$ReWeight(F_i) = WeightFunction(F_i) * F_i \qquad (7)$$

Where $i \in 1, 2, ..., T$ is the length of the temporal sequences.

4 Experiments

In this section, we conducted a large number of experiments on two commonly used datasets to evaluate our proposed algorithm. In Sect. 4.1, we introduce the relevant details of the commonly used datasets, i.e. CASIA-B [22]. In Sect. 4.2, we introduce the implementation details of the experiment, and in Sect. 4.3, we show our experimental results.

4.1 Datasets

The CASIA-B [22] consists of 124 * 110 gait sequences. 124 means that there are 124 subjects in total, and each subject is labeled from 001 to 124. 110, which means that each subject has 110 sequences. Among them, each subject was photographed from 11 different camera angles($0°, 18°, 36°, ..., 180°$.). Under each perspective, there are a total of 10 sequences with 3 different conditions, including 6 sequences for normal walking(NM:#01-06), 2 sequences for backpacking(BG:#01-02), and 2 sequences for walking with a coat on(CL:#01-02). In model training and testing phase, we adopt the same dataset settings as [19] and [1].

4.2 Implementation

Parameter Settings: All our experiments were carried out on a computer containing 2 NVIDIA 3090. Each frame is resized to the size of 64 * 44. Adam [10] optimizer is the optimization method we used to train GaitMG. The learning rate is set to 1e-4 and momentum is set to 0.2. We use the same sampling method as [3] to obtain our input sequence. During the process of training, the loss function used is batch all+ separate triplet loss [4], which margin was set to 0.2. The format of batch size used in this experiment is (P, K) where P refers to the number of objects to be identified, and K refers to the number of sample sequences contained in each object. Specifically, in the experiment of CASIA-B, this paper sets batch size to (8, 12), the number of iterations of training is 120 k.

4.3 Comparison with State-of-the-Art Methods

Table 1 reports the comparison between our proposed method GaitMG and state-of-the-art method. The comparison use the rank-1 accuracy on CASIA-B, LT, seetings, three walking conditions: NM/BG/CL and 11 camera viewing angles. From Table 1, we can see that our method exceed the accuracy of GaitGL in both NM and CL, especially under CL condition, increased by 1.7%. Although under the condition of BG, the accuracy of our method is slightly lower than that of GaitGL(We think it may be that the backpack makes the gait contour sequence change greatly, which reduces the extraction accuracy of multi-grained

Table 1. Average rank-1 accuracy(%) on CASIA-B under three different experimental settings, excluding identical-view cases.

Gallery NM #1-4			$0° - 180°$											mean
	Probe		0°	18°	36°	54°	72°	90°	108°	126°	144°	162°	180°	
LT (74)	NM (#5-6)	GaitSet[1]	90.8	97.9	99.4	96.9	93.6	91.7	95.0	97.8	98.9	96.8	85.8	95.0
		GaitPart[3]	94.1	98.6	99.3	98.5	94.0	92.3	95.9	98.4	99.2	97.8	90.4	96.2
		GaitSlice[11]	95.5	99.2	**99.6**	99.0	94.4	92.5	95.0	98.1	**99.7**	98.3	92.9	96.7
		GaitGL[15]	96.0	98.3	99.0	97.9	96.9	95.4	97.0	98.9	99.3	98.8	94.0	97.5
		GaitMG(ours)	**96.3**	**99.0**	99.4	**98.8**	**97.2**	**95.5**	**97.9**	**99.0**	99.3	**98.9**	**96.0**	**98.0**
	BG (#1-2)	GaitSet[1]	83.8	91.2	91.8	88.8	83.3	81.0	84.1	90.0	92.2	94.4	79.0	87.2
		GaitPart[3]	89.1	94.8	96.7	95.1	88.3	84.9	89.0	93.5	96.1	93.8	85.8	91.5
		GaitSlice[11]	90.2	96.4	96.1	94.9	89.3	85.0	90.9	94.5	96.3	95.0	88.1	92.4
		GaitGL[15]	**92.6**	96.6	96.8	**95.5**	**93.5**	**89.3**	92.2	**96.5**	**98.2**	**96.9**	**91.5**	**94.5**
		GaitMG(ours)	92.0	**97.7**	**97.3**	95.3	92.5	88.5	**92.3**	96.0	97.0	95.4	91.2	94.1
	CL (#1-2)	GaitSet[1]	61.4	75.4	80.7	77.3	72.1	70.1	71.5	73.5	73.5	68.4	50.0	70.4
		GaitPart[3]	70.7	85.5	86.9	83.3	77.1	72.5	76.9	82.2	83.8	80.2	66.5	78.7
		GaitSlice[11]	75.6	87.0	88.9	86.5	80.5	77.5	79.1	84.0	84.8	83.6	70.1	81.6
		GaitGL[15]	76.6	90.0	90.3	87.1	**84.5**	79.0	84.1	87.0	87.3	84.4	69.5	83.6
		GaitMG(ours)	**79.4**	**90.4**	**93.7**	**89.6**	83.3	**81.5**	**84.3**	**87.6**	**86.9**	**85.6**	**76.0**	**85.3**

temporal information). The accuracy improvement under NM/CL conditions is 0.5% and 1.7% respectively, the average rank-1 accuracy improvement under three conditions(NM/BG/CL) is 0.6

The bar chart(Fig. 4) vividly shows the effect comparison of GaitMG and various state-of-the-art methods. Under the condition of three datasets, the recognition accuracy of our method is higher than GaitGL under normal walking(NM) condition, and slightly lower than GaitGL under backpacking(BG) condition, but significantly higher than GaitGL under coat wearing condition(CL).

4.4   Effectiveness of Each Module

Table 2. Study of the effectiveness of modules in GaitMG on CASIA-B in terms of averaged rank-1 accuracy(%).

Module	Rank-1 Accuracy			
	NM	BG	CL	Mean
Baseline	96.0	91.2	80.5	89.2
Baseline + MGFA	96.8	92.8	82.5	90.7
Baseline + STFFM	97.2	93.0	80.9	90.4
Baseline + MGFA + STFFM(GaitMG)	**98.0**	**94.1**	**85.3**	**92.5**

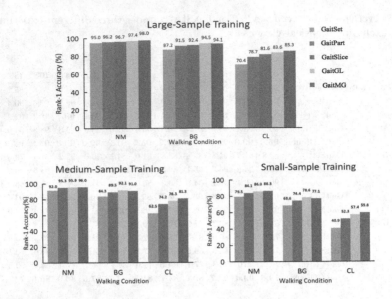

Fig. 4. Comparison of average recognition accuracy of our method GaitMG and various SOTA methods under three dataset settings(including three walking conditions).

From Table 2, we can see the effect gain of two modules MGFA and STFFM in the proposed model on the accuracy of gait recognition. Among them, our baseline is composed of a convolution layer with a four layer convolutional neural network and the MCM layer in GaitPart. It can be seen that the average recognition accuracy under 3 different walking conditions is improved 1.2% after replacing MCM with STFFM; After adding MGFA to Baseline, the recognition accuracy increased by 0.8% and 2.0% respectively under BG and CL conditions. When use MGFA and STFFM at the same time, our network structure was formed. Compared with the baseline, the average accuracy improved by 3.3%.

5   Conclusion

We propose a Multi-Grained Convolutional Neural Network structure(GaitMG) for gait recognition. The core of GaitMG is Multi-Grained Feature Aggregator(MGFA) and Spatio-Temporal Feature Fusion Module(STFFM). MGFA aggregate multi-grained temporal features through convolution with different sizes of convolution kernels, and STFFM further completes the screening of high-quality discriminative spatial-temporal information. Experiments on commonly used gait datasets CASIA-B show the superiority of our network.

Acknowledgement. This work was supported by the National Natural Science Foundation of China (62072122), Key Construction Discipline Scientific Research Capacity Improvement Project of Guangdong Province (No.2021ZDJS025), Postgraduate

Education Innovation Plan Project of Guangdong Province (2020SFKC054). The Special Projects in Key Fields of Ordinary Universities of Guangdong Province (2021ZDZX1087) and Guangzhou Science and Technology Plan Project (2023B03 J1327).

Data Availability Statement and Compliance with Ethical Standards. The data that supports the findings of this study are available on request from the corresponding author. The data are not publicly available due to privacy or ethical restrictions. The authors declare that they have no conflict of interest. This article does not contain any studies with human participants performed by any of the authors.

References

1. Chao, H., He, Y., Zhang, J., Feng, J.: GaitSet: regarding gait as a set for cross-view gait recognition. In: Proceedings of the AAAI Conference on Artificial Intelligence, vol. 33, pp. 8126–8133 (2019)
2. Chen, R., et al.: Rapid detection of multi-QR codes based on multistage stepwise discrimination and a compressed MobileNet. IEEE Internet Things J. (2023)
3. Fan, C., et al.: GaitPart: temporal part-based model for gait recognition. In: Proceedings of the IEEE/CVF Conference on Computer Vision and Pattern Recognition, pp. 14225–14233 (2020)
4. Hermans, A., Beyer, L., Leibe, B.: In defense of the triplet loss for person re-identification. arXiv preprint: arXiv:1703.07737 (2017)
5. Hong, C., Yu, J., Tao, D., Wang, M.: Image-based three-dimensional human pose recovery by multiview locality-sensitive sparse retrieval. IEEE Trans. Industr. Electron. **62**(6), 3742–3751 (2014)
6. Hou, S., Cao, C., Liu, X., Huang, Y.: Gait lateral network: learning discriminative and compact representations for gait recognition. In: Vedaldi, A., Bischof, H., Brox, T., Frahm, J.M. (eds.) Computer Vision - ECCV 2020. Lecture Notes in Computer Science(), vol. 12354, pp. 382–398. Springer, Cham (2020). https://doi.org/10.1007/978-3-030-58545-7_22
7. Hou, Z., Li, F., Wang, S., Dai, N., Ma, S., Fan, J.: Video object segmentation based on temporal frame context information fusion and feature enhancement. Appl. Intell. **53**(6), 6496–6510 (2023)
8. Hu, M., Wang, Y., Zhang, Z., Little, J.J., Huang, D.: View-invariant discriminative projection for multi-view gait-based human identification. IEEE Trans. Inf. Forensics Secur. **8**(12), 2034–2045 (2013)
9. Huang, X., et al.: Context-sensitive temporal feature learning for gait recognition. In: Proceedings of the IEEE/CVF International Conference on Computer Vision, pp. 12909–12918 (2021)
10. Kingma, D.P., Ba, J.: Adam: a method for stochastic optimization. arXiv preprint: arXiv:1412.6980 (2014)
11. Li, H., et al.: GaitSlice: a gait recognition model based on spatio-temporal slice features. Pattern Recogn. **124**, 108453 (2022)
12. Li, X., Makihara, Y., Xu, C., Yagi, Y., Yu, S., Ren, M.: End-to-end model-based gait recognition. In: Proceedings of the Asian Conference on Computer Vision (2020)

13. Li, Y., et al.: CBANet: an end-to-end cross band 2-D attention network for hyperspectral change detection in remote sensing. IEEE Trans. Geosci. Remote Sens. (2023)
14. Lin, B., Zhang, S., Bao, F.: Gait recognition with multiple-temporal-scale 3D convolutional neural network. In: Proceedings of the 28th ACM International Conference on Multimedia, pp. 3054–3062 (2020)
15. Lin, B., Zhang, S., Yu, X.: Gait recognition via effective global-local feature representation and local temporal aggregation. In: Proceedings of the IEEE/CVF International Conference on Computer Vision, pp. 14648–14656 (2021)
16. Liu, X., You, Z., He, Y., Bi, S., Wang, J.: Symmetry-driven hyper feature GCN for skeleton-based gait recognition. Pattern Recogn. **125**, 108520 (2022)
17. Ma, P., et al.: Multiscale superpixelwise prophet model for noise-robust feature extraction in hyperspectral images. IEEE Trans. Geosci. Remote Sens. **61**, 1–12 (2023)
18. Qin, H., Chen, Z., Guo, Q., Wu, Q.J., Lu, M.: RPNet: gait recognition with relationships between each body-parts. IEEE Trans. Circuits Syst. Video Technol. **32**(5), 2990–3000 (2021)
19. Takemura, N., Makihara, Y., Muramatsu, D., Echigo, T., Yagi, Y.: Multi-view large population gait dataset and its performance evaluation for cross-view gait recognition. IPSJ Trans. Comput. Vis. Appl. **10**, 1–14 (2018)
20. Wolf, T., Babaee, M., Rigoll, G.: Multi-view gait recognition using 3D convolutional neural networks. In: 2016 IEEE International Conference on Image Processing (ICIP), pp. 4165–4169. IEEE (2016)
21. Wu, H., Tian, J., Fu, Y., Li, B., Li, X.: Condition-aware comparison scheme for gait recognition. IEEE Trans. Image Process. **30**, 2734–2744 (2020)
22. Yu, S., Tan, D., Tan, T.: A framework for evaluating the effect of view angle, clothing and carrying condition on gait recognition. In: 18th International Conference on Pattern Recognition (ICPR'06), vol. 4, pp. 441–444. IEEE (2006)
23. Zeng, W., Wang, C., Li, Y.: Model-based human gait recognition via deterministic learning. Cogn. Comput. **6**, 218–229 (2014)
24. Zhang, Z., et al.: Gait recognition via disentangled representation learning. In: Proceedings of the IEEE/CVF Conference on Computer Vision and Pattern Recognition, pp. 4710–4719 (2019)
25. Zhao, H., et al.: SC2Net: a novel segmentation-based classification network for detection of COVID-19 in chest x-ray images. IEEE J. Biomed. Health Inform. **26**(8), 4032–4043 (2022)

Saliency Detection on Graph Manifold Ranking via Multi-scale Segmentation

Yuxin Yao, Yucheng Jin, Zhengmei Xu, and Huiling Wang[✉]

School of Computer and Information, Fuyang Normal University, Fuyang 342001, Anhui, China
wanghuilingfy@foxmail.com

Abstract. Saliency detection is an essential task in the field of computer vision. Its role is to identify significant or prominent regions from an image, weigh the visual information, and then better understand the visual scene. The existing saliency detection methods based on graph manifold ranking usually achieve good results in salient object detection tasks. However, they do not consider the influence of different segmentation scales on the saliency detection results. To solve this problem, based on the original saliency detection method based on graph manifold ranking, this paper performs multi-scale segmentation. It constructs a multi-scale fusion saliency detection algorithm. The experimental results show that the proposed method achieves better than seven classical saliency detection algorithms on four data sets.

Keywords: Saliency detection · Multi-scale fusion · Superpixel segmentation · Manifold ranking

1 Introduction

Image significance through the characteristics of the human visual system, intelligent algorithm simulation by predicting human vision staring point and eye movement, and extracting the image in the area, can be widely used in target recognition [1], image segmentation [2], video summarization [3], and many other areas, is critical in the field of computer vision image analysis technology.

Since Itti [4] first proposed a computational model to obtain image saliency based on multi-scale feature contrast, many researchers have begun to work on the detection of saliency detection, and there are many models. Existing models can be roughly divided into top-down task-driven and bottom-up data-driven models.

The latest top-down saliency detection model is based on the research of deep learning networks, but it requires time-consuming training. Most of the bottom-up data-driven models use global and local features such as image color, orientation, and contrast to detect image saliency. However, the local contrast method is easy to highlight the contour of the salient object, and the global detection algorithm could be better when the foreground and background features are similar.

Perazzi et al. [5] proposed a contrast-based saliency filter that measures saliency by the uniqueness and spatial distribution of regions on an image. Wei et al. [6] pay more

© The Author(s), under exclusive license to Springer Nature Singapore Pte Ltd. 2024
J. Ren et al. (Eds.): BICS 2023, LNAI 14374, pp. 143–153, 2024.
https://doi.org/10.1007/978-981-97-1417-9_14

attention to the background and construct salient object detection based on two background priors: boundary and connectivity. Zhang et al. [7] construct a Bayesian-based top-down model by integrating top-down and bottom-up information, where saliency is computed locally. Jiang et al. [8] proposed a learning-based approach that treats saliency detection as a regression problem and builds a model based on integrating many descriptions extracted from training samples with truth labels. Achanta et al. [9] compute global saliency based on the color contrast of each pixel with the whole image. Li et al. [10] proposed two saliency measures through dense and sparse reconstruction errors and incorporated them into an efficient Bayesian framework. Margolin et al. [11] measure the pattern saliency of patches using PCA and integrate it with color saliency and high-level cues for salient object detection.

In recent years, the algorithm that combines prior knowledge with graph models also has a good detection effect. Jiang et al. [12] Perform superpixel segmentation of the image and use an absorbing Markov chain for saliency detection. Yang et al. [13] proposed a saliency detection method based on graph manifold ranking (MR Algorithm), which considers the global context information in the image. By constructing the graph manifold in the image, the saliency calculation is extended to the global scope to better capture the salient regions in the image.

However, the traditional saliency detection method based on graph manifold ranking has some things that could be improved. The graph manifold ranking method needs to calculate the similarity matrix between pixels, which has high computational complexity. Due to the small number of edge pixels, it can be challenging to establish the similarity relationship between pixels. When the image has noise or complex scenes, the similarity calculation between pixels may have problems, resulting in inconsistent saliency detection results and other issues.

As shown in Fig. 1, when the MR Algorithm is used for saliency image detection, the original Fig. 3 has a better effect using the superpixel block 290. When the superpixel value is 200, the original image 1 has the most accurate detection. The original Fig. 2 has higher accuracy when the superpixel block is 230. Therefore, due to the difference in the image itself, the effect of the saliency map obtained by different superpixel segmentation is quite different.

To solve the above problems, the algorithm in this paper integrates the multi-scale segmentation theory based on the MR Algorithm, changes the original single-layer model into five layers, and constructs a multi-scale fusion image saliency detection model. Through multi-scale superpixel segmentation value, adaptive parameters are used for different images, and multi-scale segmentation is added so that images can be processed more finely, saliency information is obtained at different levels, and the model has better saliency detection ability. Experiments show that the proposed method achieves promising results on four data sets.

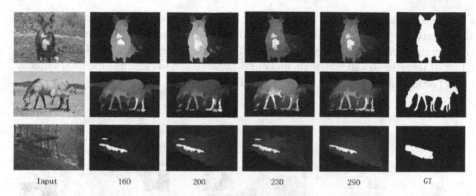

| Input | 160 | 200 | 230 | 290 | GT |

Fig. 1. Saliency maps segmented by different superpixels

2 Theory Related to Sorting on Graph Manifolds

The classical graph-based popular ranking saliency method process is based on constructing a graph model. Define a graph model $G = (V, E)$, where V is the set of all nodes and E is the edge between nodes.

The graph-based manifold ranking problem can be simply described as follows: some nodes in the graph are defined as queries, and the remaining nodes are ranked according to their relevance to the query objects. Specifically, given a set of data $X = \{x_1, x_2, \ldots, x_n\} \in R^m$, n is the number of these data, and m is the feature dimension, where some data are labeled query objects. We define a ranking function $f: X \rightarrow Rm$, assigning a ranking score to each data relative to the query object.

Think off as a vector $f = [f_1, f_2, \ldots, f_n]^T$. Also, define the indicator vector $y = [y_1, y_2, \ldots, y_n]^T$, where $y_i = 1$ if x_i is the query object and $y_i = 0$ otherwise. This data set corresponds to the graph model $G = (V,E)$, whose edge weights are represented by the incidence matrix $W = [w_{ij}]_{m*n}$. If the corresponding degree matrix of G is $D = \text{diag} \{d_{11}, d_{12}, \ldots, d_{mn}\}$, then the ranking score for a given query object can be obtained by solving the optimization problem shown in Eq. (1):

$$f^* = \arg_f \min \frac{1}{2} \left(\sum_{i,j=1}^{n} w_{ij} \| \frac{f_i}{\sqrt{d_{ii}}} - \frac{f_j}{\sqrt{d_{jj}}} \|^2 + \mu \sum_{i=1}^{n} \| f_i - y_i \|^2 \right) \quad (1)$$

where $d_{ii} = \sum_j w_{ij}$ for the ith element, and the parameter μ is used to balance the smoothness constraint of the first term and the fitting of the second term constraints.

3 Our Method

Aiming at the characteristic that the image background usually presents local or global appearance connections with each of its four boundaries. In contrast, the foreground presents coherence and consistency of appearance, and we calculate pixel saliency based on the ranking of superpixels. Firstly, a closed-loop graph was constructed for each image, where each node was a superpixel. Then, saliency detection is modeled as a manifold ranking problem, and a multi-scale segmentation image saliency detection algorithm is proposed. The main contents of the algorithm include multi-scale superpixel segmentation, calculation of background saliency map based on the multi-scale, calculation of foreground saliency map, and multi-scale fusion to calculate the final saliency map. The algorithm framework is shown in Fig. 2:

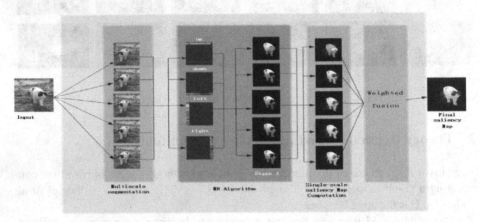

Fig. 2. Framework diagram of the proposed algorithm

3.1   Multiscale Segmentation

The pixel-based saliency detection algorithm does not consider the spatial organization relationship between image pixels, and the operation efficiency is low. The superpixel segmentation algorithm can improve the efficiency of image post-result processing and make the feature extraction of middle-level image information provide support. In this paper, the SLIC(Simple Linear Iterative Clustering) method is used to segment the image into several superpixels, which means that the pixels that are adjacent to each other and have similar colors are clustered into small local regions by considering the similarity of color and distance between pixels.

Based on the research of a single-layer popular sorting algorithm model, for different types of images, different superpixel values need to be selected according to the image, considering the integrity of the detection results and the detailed information. In this paper, the results of various segmentation scales are experimentally analyzed, and the statistical results are shown in Fig. 3:

Taking the SOD dataset as an example, in the SOD dataset, the model is built from five scales respectively. The superpixel value of the single-layer model is 200, the superpixel value of the two-layer model is 130 and 230, the three superpixel values of the three-layer model are 240, 280, 160, and 200, the four-layer model is 240, 280, 160, 260 and 200, and the five-layer model is 240, 280, 160, 260 and 200. The experiment is carried out using mean distribution, and the Precision value, Recall value, and F-measure value of the five scale experiments are compared by histogram. The experimental results show that the five-layer model has the best practical effect. Therefore, the superpixel scale of this paper is chosen as five layers.

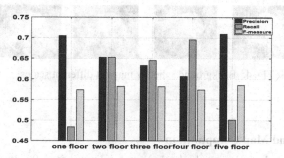

Fig. 3. Analysis histogram of detection results at different scales

3.2 Saliency Map Computation

3.2.1 Background Image Computation

The superpixels at the four boundaries of the image are selected as background seed points in turn, and the ranking score of the nodes in the graph is calculated. The background saliency value of each superpixel block based on the upper boundary is shown in Eq. (2):

$$S_t(i) = 1 - \overline{f^*}(i) \tag{2}$$

where $\overline{f^*}(i)$ is the ranking score of node i normalized to the range [0,1]. Similarly, according to formula (2), the saliency $S_d(i)$, $S_l(i)$ and $S_r(i)$ of the query object at the bottom, left and right boundaries are calculated, respectively, and the final saliency map based on the background is obtained according to Eq. (3). The detection results of five scales are shown in Fig. 4, and it can be seen that the detection effects of different scales are different.

$$S_{bg}(i) = S_t(i) \times S_b(i) \times S_l(i) \times S_r(i) \tag{3}$$

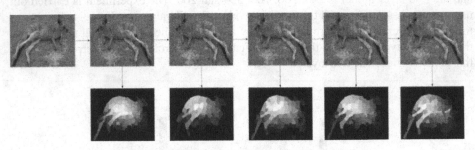

Fig. 4. Background saliency maps at different scales

3.2.2 Foreground Map Computation

. Based on the four boundaries, the background saliency map S_{bg} basis is calculated, the adaptive threshold segmentation method is used for binarization processing, and the super image value block with a high saliency value is selected to obtain the query node based on the foreground. Calculate the ranking scores of the nodes in the graph concerning these query objects according to Eq. (4):

$$S_{fq}(i) = \overline{f^*}(i) \tag{4}$$

According to the calculated foreground-based saliency value of node i, combined with the region-based calculation results, the pixel-level saliency is set as the Gaussian linear weighting of the saliency value of the superpixel block it is located in. The saliency value of the surrounding two neighboring regions and the saliency value of the image is used to map from the superpixel level to the pixel level to obtain the final saliency map, as shown in Fig. 5.

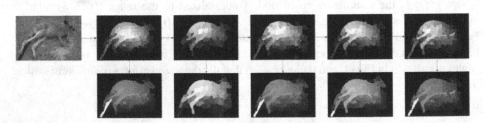

Fig. 5. Foreground saliency maps at different scales

3.3 Multi-scale Saliency Map Fusion

In the multi-level model, the single-level model's superpixel values with better effects are selected for each layer. However, the superpixel values with better effects also have advantages and disadvantages. If they are averaged, the effect is not optimal. In this paper, we build two five-layer models on the SOD dataset. Group A uses the average weighted statistical results; group B considers the detection effects of different scales, sets the ratio as 0.1:0.1:0.6:0.1, and calculates the precision, Recall, and F-measure values. The experimental results show that Group B's effect is better than Group A's. Therefore, Eq. (5) shows the proposed algorithm's multi-scale fusion.

$$f(i) = \sum_{i=1}^{5} f(i)a(i)i \in [1,5], i \in N, \sum_{i=1}^{5} a(i) = 1 \tag{5}$$

where a(i) represents the proportion of each scale, and in different data sets, different data may correspond to different proportions, the adaptive selection is carried out for weighted fusion. Through multi-scale fusion, the final saliency map of the proposed algorithm is obtained, and the fused saliency map can obtain a more complete and accurate saliency map

4 Experimental Results and Analysis

On the four data sets compares with BS [14], CA [15], FT [16], SR [17], SF [18], IM [19] and other methods, it finally produced effective improvement. The following compares the FM graph and PR graph effect between other methods and our method on four data sets. The results show that the effect of the proposed method is better than other methods.

4.1 Quantitative Analysis

4.1.1 PR (Precision-Recall) Diagram

Precision-Recall (PR) plot stands for precision-recall plot and is a standard method used to evaluate the performance of binary classification models. Precision measures the fraction of examples predicted as positive that positive, also known as accuracy. Recall is the fraction of instances that were correctly predicted to be positive out of all positive instances. Recall measures the proportion of all accurate positive class samples that are correctly predicted as positive.

True positives represent the number of samples that are genuinely positive class, and false negatives represent the number of samples that are actually positive class but are incorrectly predicted as negative class. By calculating recall, you can see how good the model is at predicting genuinely positive samples.

Often, neither Precision nor Recall can comprehensively evaluate the quality of a saliency map. The F-measure is estimated as a weighted harmonic mean of them with non-negative weigh β^2:

$$F_\beta = \frac{(1+\beta^2)\text{Precision} \times \text{Recall}}{\beta^2 \text{Precision} \times \text{Recall}} \tag{6}$$

where we set $\beta^2=0.3$ in our experiments. As shown in Fig. 6, the proposed method is compared with the PR maps btained by the experimental results of other methods on the four data sets, respectively. The proposed method has improved the effect compared with other methods.

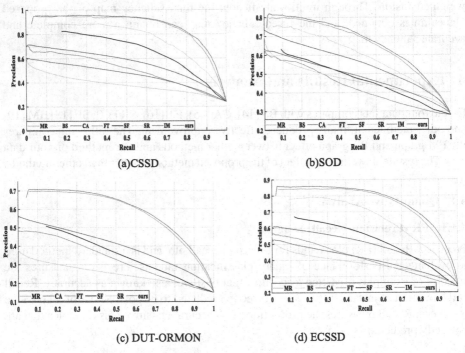

(a)CSSD   (b)SOD

(c) DUT-ORMON   (d) ECSSD

Fig. 6. PR curves of the four data sets

4.1.2 F-measure

. In F-Measure plots, the abscissa usually represents the prediction threshold of a classification model, and the ordinate represents the value of the F-Measure metric. F-Measure is a metric that considers both precision and recall to evaluate the performance of classification models (Fig. 7).

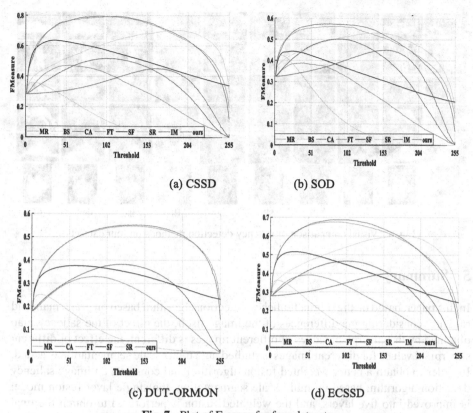

(a) CSSD   (b) SOD

(c) DUT-ORMON   (d) ECSSD

Fig. 7. Plot of F-score for four datasets

A higher value of F-Measure indicates that the model achieves a better balance between precision and recall.As shown in Fig. 8, the F-figure obtained by the experimental results of the proposed method and MR, BS, CA, FT, SR, SF, IM, and other methods on four data sets is compared. The proposed method has a better effect than other methods.

4.2 Qualitative Analysis

As shown in Fig. 8, compared with the other seven saliency detection methods, the proposed method has a more obvious distinction between foreground and background. For example, the dog in the second image, the saliency map generated by the proposed method, is closer to the truth map in detail. In the flower in the third image, the saliency map of the proposed method is more precise, and the contour is more distinct.

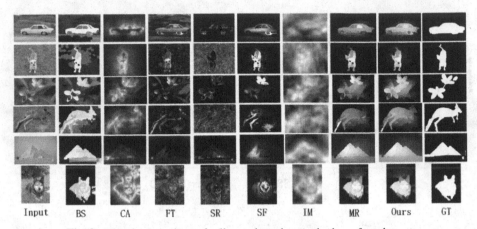

| Input | BS | CA | FT | SR | SF | IM | MR | Ours | GT |

Fig. 8. Visual comparison of saliency detection methods on four datasets

5 Summary

In this paper, based on the original saliency detection algorithm based on graph manifold ranking, considering the differences of the image itself, the effect of the saliency map obtained by the superpixel values of different images is different, the effect of different superpixel values for different images is studied, and multi-scale segmentation is added. In order to obtain a better-weighted fusion algorithm and construct an image saliency detection algorithm based on multi-scale segmentation, the single-layer fusion model is improved into five layers, and the weighted fusion is performed to obtain the final saliency map. The experimental results on four data sets show that compared with a variety of classical saliency detection algorithms, the proposed method can obtain better detection results, and the completeness of the detected saliency map must be higher.

Funding. This study was funded by Anhui Provincial Education Department Key Project (2022 AH051324).

References

1. Ren, Z., et al.: Region-based saliency detection and its application in object recognition. IEEE Trans. Circ. Syst. Video Technol. **24**(5) (2014)
2. Yo, U., et al.: Salient object detection with importance degree. IEEE Access **2020**, 99 (2020)
3. Zhang, M., et al.: Dynamic context-sensitive filtering network for video salient object detection. In: International Conference on Computer Vision, vol. 2021 (2021)
4. Itti, L., Koch, C., Niebur, E.: A model of saliency-based visual attention for rapid scene analysis. IEEE Trans. Pattern Anal. Mach. Intell. **20**(11), 1254–1259 (1998)
5. Perazzi, F., Kr¨ahenb¨uhl, P., Pritch, Y., Hornung, A.: Saliency filters: contrast based filtering for salient region detection. In: 2012 Proceedings of the IEEE International Conference on Computer Vision and Pattern Recognition, pp. 733–740. IEEE (2012)
6. Wei, Y., Wen, F., Zhu, W., Sun, J.: Geodesic saliency using background priors. In: Fitzgibbon, A., Lazebnik, S., Perona, P., Sato, Y., Schmid, C. (eds.) ECCV 2012. LNCS, vol. 7574, pp. 29–42. Springer, Heidelberg (2012). https://doi.org/10.1007/978-3-642-33712-3_3

7. Zhang, L., Tong, M.H., Marks, T.K., Shan, H., Cottrell, G.W.: Sun: A Bayesian framework for saliency using natural statistics. J. Vision **8**(7) (2008)
8. Jiang, H., Wang, J., Yuan, Z., Wu, Y., Zheng, N., Li, S.: Salient object detection: a discriminative regional feature integration approach. In: 2013 Proceedings of the IEEE International Conference on Computer Vision and Pattern Recognition, IEEE (2013)
9. Achanta, R., Estrada, F., Wils, P., Süsstrunk, S.: Salient region detection and segmentation. Comput. Vis. Syst. **2008**, 66–75 (2008)
10. Liu, R., Cao, J., Lin, Z., Shan, S.: Adaptive partial differential equation learning for visual saliency detection. In: 2014 Proceedings of the IEEE International Conference on Computer Vision and Pattern Recognition (2014)
11. Margolin, R., Tal, A., Zelnik-Manor, L.: What makes a patch distinc. In: 2013 Proceedings of the IEEE International Conference on Computer Vision and Pattern Recognition (2013)
12. Jiang, B., et al.: Saliency detection via absorbing Markov chain. In: Proceedings of the IEEE International Conference on Computer Vision, pp. 1665–1672 (2013)
13. Yang, C., Zhang, L., Lu, H., Ruan, X., Yang, M.: Saliency detection via graph-based manifold ranking. In: 2013 Proceedings of the IEEE Conference on Computer Vision and Pattern Recognition, pp. 3166–3173 (2013)
14. Xie, Y., Lu, H., Yang, M.H.: Bayesian saliency via low and mid level cues. IEEE Trans. Image Process. Publ. IEEE Sign. Process. Soc. **22**(5), 1689–1698 (2013)
15. Achanta, R., et al.: Frequency-tuned salient region detection. In: 2009 Proceedings of the IEEE International Conference on Computer Vision and Pattern Recognition, Miami, pp. 1597–1604 (2009)
16. Goferman, S., Zelnik-Manor, L., Tal, A.: Context-aware saliency detection. IEEE Trans. Pattern Anal. Mach. Intell. **34**(10), 1915–1926 (2012)
17. Hou, X., Zhang, L.: Saliency detection: a spectral residual approach. In: 2007 Proceedings of the IEEE International Conference on Computer Vision and Pattern Recognition, pp. 1–8. IEEE (2007)
18. Perazzi, F., Kraehenbuehl, P., Pritch, Y.A.: Saliency filters: contrast based filtering for salient region detection. In: 2012 IEEE Conference on Computer Vision and Pattern Recognition, pp. 733–740. IEEE Press, Providence, USA (2012)
19. Murry, N., et al.: Saliency estimation using a non-parametric low-level vision mode. In: 2011 Proceedings of the IEEE International Conference on Computer Vision and Pattern Recognition, pp. 433–440. IEEE Press, Providence (2011)

Research on Improved Algorithm of Significance Object Detection Based on ATSA Model

Yucheng Jin, Yuxin Yao, Huiling Wang$^{(\boxtimes)}$, and Yingying Feng$^{(\boxtimes)}$

School of Computer and Information, Fuyang Normal University, Fuyang 342001, Anhui, China
wanghuilingfy@foxmail.com, 517491038@qq.com

Abstract. Saliency detection refers to accurately positioning and extracting significant objects or regions in the image. Most effective object detection methods are based on RGB-D and adopt the dual-flow architecture with RGB and depth symmetry. At the same time, the asymmetric dual-flow architecture can also effectively extract rich global context information. The existing ATSA model uses asymmetric dual-flow architecture to locate significant objects accurately. However, this model's initial learning rate needs to be improved, and choosing a suitable learning rate takes work. Therefore, in order to improve the overall performance of the model, the RangerQH optimizer algorithm was introduced to enable the model to adjust the learning rate during the training process dynamically, and the cross-entropy loss function of the original model was replaced with a mixed loss function composed of Focal loss and Dice loss. The results show that E-measure, S-measure, F-measure, and MAE improve the seven existing public RGB-D datasets.

Keywords: Saliency detection · RGB-D · Asymmetric dual-flow architecture · RangerQH optimizer algorithm · Mixed loss function

1 Introduction

Salient object detection is designed to automatically detect the most prominent objects in an image or video. Itti et al. [1] first proposed a computational model for calculating visual salience. With the progress of the research, subsequent researchers began to explore the use of RGB features-based methods for salient object detection. With the development of machine learning, deep learning, and other methods, more and more RGB-based methods have been proposed.

However, using only RGB images can be affected by lighting conditions, making distinguishing between objects with similar colors or textures complex, especially when the object's appearance is similar to the background; salient object detection is susceptible to interference. Therefore, the effective object detection method based on RGB-D began attracting people's attention. Compared with the traditional RGB saliency detection method, salient object detection based on RGB-D can capture the visual and geometric features of objects in the scene more comprehensively by combining the information of RGB image and depth image. In recent years, with the continuous progress of technology

J. Ren et al. (Eds.): BICS 2023, LNAI 14374, pp. 154–165, 2024.
https://doi.org/10.1007/978-981-97-1417-9_15

and in-depth research, many methods and technologies based on RGB-D have emerged for research in this field.

Existing RGB-D saliency detection methods can be roughly divided into two categories. One is the traditional method. Ren et al. [2] proposed a method that utilizes global prior knowledge to improve the performance of RGB-D saliency detection. This method fuses RGB and depth information under prior global constraints and deduces by graph cut algorithm to achieve more accurate and robust significant detection results. Desingh et al. [3] emphasized the importance of using depth information in significant detection and proposed a depth-based method to improve the detection performance of the significant region. Zhang et al. [4] proposed a depth-enhanced significance method, which uses depth difference and depth gradient as weights to generate the final depth-enhanced significance map by weighted fusion of the initial significance map.

The other method is based on deep learning. Huang et al. [5] proposed a novel method to improve the effect of RGB-D significant object detection by combining cross-modal and single-modal features, providing valuable ideas and methods for research and application in related fields. Ji et al. [6] propose a new learning framework that innovatively integrates deep and salience learning into high-level feature learning processes in a mutually beneficial manner. Ren et al. [7] proposed a new accurate RGB-D significant detection framework considering the two modes' global position and local detail complementarity. By designing complementary interaction modules, useful representations are selectively selected from RGB and depth data, and cross-modal features are efficiently integrated. Zhao [8] et al. proposed an RGB-D effective object detection method using self-supervised representation learning. The method trains the network through self-generated tasks and self-adversarial training strategies to learn discriminant and representational features from RGB-D data.

However, most RGB-D-based salient object detection uses a symmetric dual-flow architecture, which processes the RGB image and the depth image through two symmetric streams and fuses the RGB and depth information. However, this approach can lead to mismatches in feature extraction and feature fusion and fails to take full advantage of the differences between RGB and depth information.

For this reason, Zhang et al. [9] designed an asymmetric dual-flow network architecture model ATSA in 2020 and cleverly fused RGB and depth information through the deep attention mechanism to detect significance accurately. As the initial learning rate set by the model is too small, it takes much time to extract features during the training process effectively. This paper introduces the RangerQH optimizer algorithm to dynamically adjust the learning rate during the training process to improve the model's performance. On this basis, in order to make the model better handle class unbalance and improve segmentation details and boundary accuracy by paying more attention to boundaries, the initial cross-entropy loss function was replaced by a mixed loss function composed of Focal loss [10], and Dice loss [11].

The results show that in SSD [12], NLPR [13], STEREO [14], NJUD [15], LFSD [16], RGBD135 [17], DUT-RGBD [18], Quantitative comparison of E-measure, S-measure, F-measure, and MAE on seven widely used RGB-D public datasetshave improved the effect.

2 Basic Theory of ATSA Model

ATSA [9] is a salient object detection model based on RGB-D. This architecture comprises a lightweight depth flow and an RGB flow with a flow ladder module. The model framework is shown in Fig. 1:

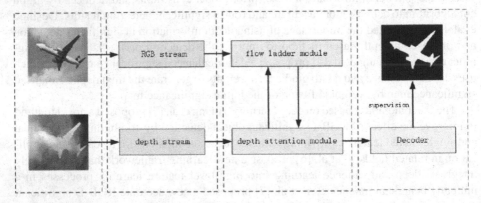

Fig. 1. Frame diagram of ATSA model

The asymmetric structure means that under the premise of treating the two modes equally, the network design is different according to the characteristics of the two modes; that is, the processing of RGB features and depth features is inconsistent.

For the depth stream, the model designs a lightweight architecture by taking the original depth map as input, and then the extracted features are input into the RGB stream through the depth attention mechanism; for the RGB stream, the model designs a flow ladder module. Through evolution to preserve details. Moreover, global position information is received from other vertical and parallel branches to locate significant areas. The realization formula is as follows:

$$B_i L_j = \begin{cases} trans(Conv2)i=1, j=1 \\ trans(Conv(i+1)i=j+1, j \in [1,3] \\ \sum_{n=1}^{j} f(B_n L_{j-1})i \in [1,j], j \in [2,3] \end{cases} \quad (1)$$

$$F_{RGB}^j = \sum_{n=1}^{j+1} f(B_n L_j)j \in [1,2] \quad (2)$$

$$F_{RGB}^3 = cat(f(B_n L_3))n \in [1,4] \quad (3)$$

where B_i and L_j represent the i branch and the j layer, respectively. $f(\cdot)$ Indicates when n > i is up-sampled, when n < i, i − n downsampling. When n is equal to i, $f(\cdot)$ means no operation. $Conv(i)$ Refers to the output feature of the ith Conv block in VGG-19, and $trans(\cdot)$ is operated by the convolution layer to achieve the conversion of the number of channels. $cat(\cdot)$ Denotes concatenating all features together. The final output of the module, F_{RGB}^3, is a concatenating of multi-scale features extracted from the four branches. Features with local and global information move to parallel branches in an evolutionary way.

3 Improved Model Based on ATSA

ATSA [9] model uses a stochastic gradient descent algorithm with a fixed learning rate to select learning rate 1e−10 for training. However, the learning rate is set too small, resulting in the model needing to spend much time training many rounds to extract features effectively. Moreover, the stochastic gradient descent algorithm has slow convergence speed, high sensitivity to the initial learning rate, and difficulty dealing with non-stationary data and noise. Therefore, the Ranger QH optimizer algorithm is introduced to replace the stochastic gradient descent algorithm so that the model can dynamically adjust the learning rate during the training process to better adapt to the characteristics of the data.

Focal Loss loss function can effectively reduce the impact of category imbalance by introducing regulatory factors so that the model pays more attention to samples that are difficult to classify, thus improving the problem of category imbalance. When calculating the similarity index, the Dice Loss function is more suitable for boundary frame matching evaluation. Compared with the cross-entropy Loss function, Dice Loss has a smoother optimization curve, which is conducive to model convergence and stability of the training process. Therefore, this paper proposes to combine Focal loss and Dice loss into a mixed loss function to replace the cross-entropy loss function to improve the overall performance of the ATSA model. The comparison framework of the algorithms in this paper is shown in Fig. 2:

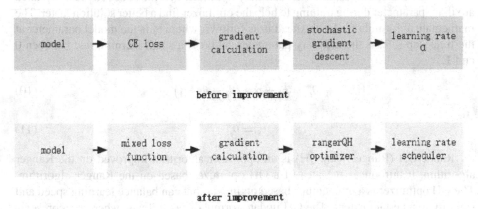

Fig. 2. Comparison diagram of training process before and after improvement

3.1 RangerQH Optimizer Algorithm

The RangerQH optimizer algorithm is an improvement of the Ranger algorithm. The Ranger algorithm can improve the training effect by adapting the learning rate and momentum parameters when training neural networks. It uses the adaptive learning rate mechanism of the RAdam [19] algorithm and the prediction mechanism of the LookAhead [20] algorithm to better deal with the problems in the process of gradient updating.

158   Y. Jin et al.

Among them, RAdam is an improvement to the Adam optimizer, which corrects the performance problems that may exist in the early stages of Adam training. First, the first and second-order momenta of the gradient are calculated. Then, the deviation of first-order momentum is corrected. The update steps of RAdam are as follows: where θ_t is the model parameter at the time step t, $\beta1$ and $\beta2$ are the exponential decay rates of the first and second-order momenta, and ε is a small constant for numerical stability.

$$m_t = \beta1 * m_{\{t-1\}} + (1 - \beta1) * g_t \tag{4}$$

$$v_t = \beta2 * v_{\{t-1\}} + (1 - \beta2) * g_t^2 \tag{5}$$

$$m_{that} = m_t/(1 - \beta1^t) \tag{6}$$

$$r_t = sqrt\left((v_{that}/(1 - \beta2^t)) + \varepsilon\right) \tag{7}$$

$$lr_t = learning_{rate} * sqrt(1 - \beta2^t)/(1 - \beta1^t) \tag{8}$$

$$\theta_t = \theta_{\{t-1\}} - lr_t * r_t * m_{that} \tag{9}$$

The LookAhead algorithm is an auxiliary optimizer that introduces a "fast" updated auxiliary parameter during training to help the optimizer find a better solution faster. The mathematical formula for LookAhead is as follows: where θ_t is the model parameter at the time step t, θ_t^* is the auxiliary parameter, and α is a small learning rate between 0 and 1.

$$\theta_t^* = \theta_t + \alpha * \left(\theta_t - \theta_{\{t-1\}}\right) \tag{10}$$

$$\theta_t = \theta_t^* \tag{11}$$

RangerQH (Ranger with QH) is an optimizer algorithm improved on the Ranger algorithm. It introduces the idea of a QH optimizer based on the Ranger algorithm. The QH optimizer is a momentum-based optimizer that can balance learning speed and stability to a certain extent. The QH update formula is as follows, where p_t represents the update direction of the QH (Quasi-Hyperbolic) optimizer, lr_{qh} is the learning rate of the QH optimizer, and α is the mixed parameter of QH:

$$p_t = p_t - \left(lr_{qh} * gradient\right) \tag{12}$$

$$w_t = (1 - \alpha) * w_t + \alpha * p_t \tag{13}$$

3.2 Design of Mixed Loss Function

In order to improve the overall performance of the ATSA model, this paper proposes a hybrid loss function combining focal loss and dice loss to replace the cross entropy loss function of the original model.

Focal Loss [10] is a function used to solve the class imbalance problem. By reducing the weight of easily classified samples, it pays attention to and balances the influence of complex classified samples to improve the model's performance on complex classified samples. Its design is based on the cross-entropy loss function by introducing a tunable paramete γ and a scaling factor α to adjust the weight of the sample. The loss function is calculated as follows, where a factor that reduces the weight of easily classified samples is a non-negative regulating parameter, and α is a factor that balances positive and negative samples:

$$Focalloss = -\alpha * (1 - p)^{\gamma} * log(p) \qquad (14)$$

Dice Loss [11] is a loss function used in image segmentation tasks that measures the similarity of the predicted result to the actual label based on the Dice coefficient. The goal of Dice Loss is to minimize the difference between the predicted results and the actual label to improve the accuracy of the segmentation model. The Dice coefficient is calculated as follows: Intersection represents the intersection of the Prediction result and the actual label; prediction represents the total number of pixels of the prediction result; Ground Truth represents the total number of pixels of the actual label:

$$Dice = (2 * Intersection)/(Prediction + GroundTruth) \qquad (15)$$

$$DiceLoss = 1 - Dice \qquad (16)$$

In order to give full play to the advantages of the two loss functions and make the model better play its performance on different data sets, the two are combined to form a mixed function, whose formula is as follows:

$$HybridLoss = FocalLoss + DiceLoss \qquad (17)$$

There is an obvious category imbalance problem in salient object detection; the sample number of the significance target and the background is very different. An accurate boundary prediction is significant for locating and segmenting salient targets. By mixing Dice Loss with Focal Loss, the imbalance between salient targets and background can be better handled, the accuracy of a classification and boundary prediction can be optimized, and the accuracy of salient targets detection and positioning can be further improved. At the same time, the limitation of a single loss function can be reduced, and the stability and generalization ability of the model can be improved by integrating different types of loss functions.

4   Experimental Results and Analysis

4.1   Experimental Environment and Setting

Experiments were conducted on seven widely used RGB-D datasets, including STERE, SSD, NLPR, NJUD, LFSD, DUT-RGBD, and RGBD135. They contain 1000, 80, 1000, 1985, 100, 1200, and 135 images, respectively. We trained with 800 images from DUT-RGBD, 1485 from NJUD, and 700 from NLPR and tested with the remaining images from these three and five other datasets.

Regarding implementation details, the input RGB image and depth image are adjusted to 256 × 256. In order to increase the diversity and robustness of data, the pictures in the training set are horizontally flipped, scaled, and randomly clipped to expand the training set. This article uses a single RTX 1080Ti GPU for training and testing. The model's momentum, weight decay, batch size, and initial learning rate were set to 0.9, 0.0005, 2, and 3e-5, respectively.

4.2   Evaluation Index

In order to evaluate various methods comprehensively, four evaluation indexes F_measure, MAE, E_measure, and S_measure, are adopted in this paper.

F_measure is a commonly used evaluation index used to consider the Precision and Recall of classification models comprehensively [21, 22]. It is the harmonic average of accuracy and Recall. The calculation formula is as follows: Precision refers to the precision, and Recall refers to the recall rate:

$$F1 = 2 * (Precision * Recall) / (Precision + Recall) \tag{18}$$

MAE is a regression model evaluation index used to measure the average absolute error between the model's predicted and actual values. The smaller the MAE value, the better, indicating that the average error between the model's prediction and the actual value is more minor. Its calculation formula is as follows, where n is the number of samples and Σ represents the summation operation:

$$MAE = (1/n) * \Sigma |Predicted\ value - True\ value| \tag{19}$$

E-measure is a commonly used significance evaluation index used to measure the performance of the sign detection model in significant object detection tasks. Its calculation formula is as follows:

$$E - measure = (1 + \beta^2) * Pre * Recall / (\beta^2 * Pre + Recall) \qquad (20)$$

S-measure is an index used to evaluate the performance of the sign detection model. The formula is as follows, where S_object indicates the similarity score of the target level, S_region indicates the similarity score of the region level, and α is an equilibrium parameter:

$$S - measure = \alpha * S_{object} + (1 - \alpha) * S_{region} \qquad (21)$$

4.3 Experimental Comparison

4.3.1 Quantitative Analysis

In this paper, the RangerQHoptimizer algorithm was introduced to adjust the learning rate during thetraining process dynamically, increase the initial learning rate from 1e-10 to 3e-5,and replace the loss function with a mixed loss function composed of Focal loss-sand Dice loss. Quantitative comparison of E-measure (E_γ),S-measure (S_λ),F_measure (F_β), andMAE (M) was performedon seven widely used RGB-D datasets with the original model, respectively. Theresults are shown in Table 1, Where A, B, and C are the results provided by theoriginal model, respectively. Using the results of the RangerQH optimizeralgorithm and the results of replacing the loss function on its basis, theoptimal results are shown in red font:

Table 1. Comparison of experimental results

method	STEREO				SSD				NLPR				NJUD	
	E_γ↑	S_λ↑	F_β↑	M↓	E_γ↑	S_λ↑	F_β↑	M↓	E_γ↑	S_λ↑	F_β↑	M↓	E_γ↑	S_λ↑
A	0.924	0.910	0.886	0.039	0.870	0.856	0.827	0.050	0.925	0.900	0.876	0.028	0.928	0.905
B	0.924	0.916	0.887	0.038	0.876	0.861	0.827	0.051	0.923	0.902	0.881	0.032	0.922	0.903
C	0.940	0.914	0.893	0.036	0.920	0.872	0.858	0.043	0.935	0.907	0.890	0.026	0.942	0.911

method	NJUD		LFSD				DUT-RGBD				RGBD135			
	F_β↑	M↓	E_γ↑	S_λ↑	F_β↑	M↓	E_γ↑	S_λ↑	F_β↑	M↓	E_γ↑	S_λ↑	F_β↑	M↓
A	0.893	0.040	0.900	0.891	0.862	0.064	0.933	0.868	0.918	0.043	0.946	0.911	0.885	0.020
B	0.890	0.041	0.902	0.902	0.866	0.068	0.926	0.866	0.916	0.043	0.948	0.907	0.900	0.020
C	0.902	0.038	0.911	0.901	0.873	0.058	0.942	0.868	0.920	0.043	0.962	0.909	0.904	0.020

In the above experiment, the original model converges at round 60, while the improved model converges at round 6. The comparison of the first 30 epochs of loss function curves between the original model and the improved model is shown in Fig. 3:

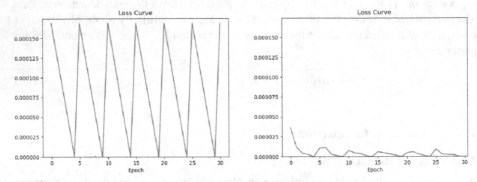

Fig. 3. Loss function curve comparison diagram

In this paper, the improved model is quantitatively compared with other advanced RGB-D significance detection models on seven datasets by E-measure (E_γ), S-measure (S_λ), F_measure (F_β), and MAE (M). The experimental results are shown in Table 2. The two best results are shown in red and yellow:

Table 2. Results comparison chart

Method	STEREO				SSD				NLPR				NJUD	
	$E_\gamma \uparrow$	$S_\lambda \uparrow$	$F_\beta \uparrow$	$M \downarrow$	$E_\gamma \uparrow$	$S_\lambda \uparrow$	$F_\beta \uparrow$	$M \downarrow$	$E_\gamma \uparrow$	$S_\lambda \uparrow$	$F_\beta \uparrow$	$M \downarrow$	$E_\gamma \uparrow$	$S_\lambda \uparrow$
Ours	0.940	0.914	0.893	0.036	0.920	0.872	0.858	0.043	0.935	0.907	0.890	0.026	0.942	0.911
DMRA	0.920	0.886	0.868	0.047	0.892	0.857	0.821	0.058	0.942	0.899	0.855	0.031	0.908	0.886
CPFP	0.897	0.871	0.827	0.054	0.832	0.807	0.725	0.082	0.924	0.888	0.822	0.036	0.906	0.878
TANet	0.911	0.877	0.849	0.060	0.879	0.839	0.767	0.063	0.916	0.886	0.795	0.041	0.893	0.878
PDNet	0.903	0.874	0.833	0.064	0.813	0.802	0.716	0.115	0.876	0.835	0.740	0.064	0.890	0.883

Method	NJUD		LFSD				DUT-RGBD				RGBD135			
	$F_\beta \uparrow$	$M \downarrow$	$E_\gamma \uparrow$	$S_\lambda \uparrow$	$F_\beta \uparrow$	$M \downarrow$	$E_\gamma \uparrow$	$S_\lambda \uparrow$	$F_\beta \uparrow$	$M \downarrow$	$E_\gamma \uparrow$	$S_\lambda \uparrow$	$F_\beta \uparrow$	$M \downarrow$
Ours	0.902	0.038	0.911	0.901	0.873	0.058	0.942	0.868	0.920	0.043	0.962	0.909	0.904	0.020
DMRA	0.872	0.051	0.899	0.847	0.849	0.075	0.927	0.888	0.883	0.048	0.945	0.901	0.857	0.029
CPFP	0.877	0.053	0.867	0.828	0.813	0.088	0.814	0.749	0.736	0.099	0.927	0.874	0.819	0.037
TANet	0.844	0.061	0.845	0.801	0.794	0.111	0.866	0.808	0.779	0.093	0.916	0.858	0.782	0.045
PDNet	0.832	0.062	0.872	0.845	0.824	0.109	0.861	0.799	0.757	0.112	0.915	0.868	0.800	0.050

4.3.2 Qualitative Analysis

As shown in Fig. 4, compared with the original model results, complete salient object generation was achieved in the second and third graphs, and the background and foreground were better distinguished in the fourth and fifth graphs. To sum up, the improved model can better separate salient objects from the background, making them more prominent and noticeable, and the generated salient graph is closer to the truth graph.

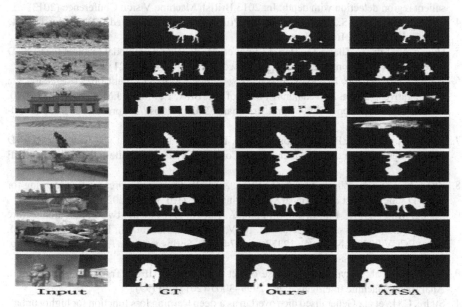

Fig. 4. Comparison between the improved model and the original model

5 Summary

In this paper, the existing RGB-D asymmetric dual-flow architecture model ATSA is improved, and the stochastic gradient descent algorithm is replaced by the RangerQH optimizer algorithm so that the learning rate can be dynamically adjusted to improve the training speed of the original model. In addition, to further improve the overall performance of the original model, the cross-entropy loss function was replaced by a mixed loss function composed of Focal loss and Dice loss. Experiments showed that the training speed was faster, and the significant graph detected was more accurate. E-measure, S-measure, F_measure, and MAE are also improved.

Funding. This study was funded by Anhui Provincial Education Department Key Project (2022 AH051324).

References

1. Itti, L., Koch, C., Niebur, E.: A model of saliency-based visual attention for rapid scene analysis. IEEE Trans. Pattern Anal. Mach. Intell. **20**(11), 1254–1259 (1998)

2. Ren, J., Gong, X., Yu, L., Zhou, W., Yang, M.Y.: Exploiting global priors for RGB-D saliency detection. In: 2015 Computer Vision & Pattern Recognition Workshops, pp. 25–32. IEEE (2015)

3. Desingh, K., Krishna, K.M., Rajan, D., Jawahar, C.V.: Depth really matters: Improving visual salient region detection with depth. In: 2013 British Machine Vision Conference (2013)

4. Zhang, J., Sclaroff, S., Lin, Z., Shen, X., Price, B.: Depth-enhanced saliency for RGB-D images. IEEE Trans. Image Process. (2013)

5. Huang, N., Liu, Y., Zhang, Q., Han, J.: Joint cross-modal and unimodal features for RGB-D salient object detection. IEEE Trans. Multimed. **23**, 2428–2441 (2021)

6. Ji, W., Li, J., Zhang, M., Piao, Y., Lu, H.: Accurate RGB-D salient object detection via collaborative learning. In: Vedaldi, A., Bischof, H., Brox, T., Frahm, J.M. (eds.) ECCV 2020. LNCS, vol. 12363, pp. 52–69. Springer, Cham (2020). https://doi.org/10.1007/978-3-030-58523-5_4

7. Zhang, M., Ren, W., Piao, Y., Rong, Z., Lu, H.: Select, supplement and focus for RGB-D saliency detection. In: 2020 Conference on Computer Vision and Pattern Recognition. IEEE (2020)

8. Zhao, X., Pang, Y., Zhang, L., Lu, H., Ruan, X.: Self-supervised representation learning for RGB-D salient object detection. arXiv preprint arXiv:2101.12482.(2021)

9. Zhang, M., Fei, S.X., Liu, J., Xu, S., Piao, Y., Lu, H.: Asymmetric two-stream architecture for accurate RGB-D saliency detection. In: Vedaldi, A., Bischof, H., Brox, T., Frahm, J.M. (eds.) ECCV 2020. LNCS, vol. 12373, pp. 374–390. Springer, Cham (2020). https://doi.org/10.1007/978-3-030-58604-1_23

10. Lin, T.Y., et al.: Focal loss for dense object detection. In: IEEE Transactions on Pattern Analysis & Machine Intelligence, pp. 2999–3007 (2017). PP(99)

11. Sudre, C.H., et al.: Generalised dice overlap as a deep learning loss function for highly unbalanced segmentations. In: Third International Workshop, DLMIA 2017, and 7th International Workshop, ML-CDS 2017, pp. 240–248 (2017)

12. Zhu, C., Li, G.: A three-pathway psychobiological framework of salient object detection using stereoscopic technology. In: 2017 IEEE International Conference on Computer Vision Workshop IEEE (2017)

13. Peng, H., Li, B., Xiong, W., Hu, W., Ji, R.: RGBD salient object detection: a benchmark and algorithms. In: Fleet, D., Pajdla, T., Schiele, B., Tuytelaars, T. (eds.) ECCV 2014. LNCS, vol. 8691, pp. 92–109. Springer, Cham (2014). https://doi.org/10.1007/978-3-319-10578-9_7

14. Niu, Y., et al.: Leveraging stereopsis for saliency analysis. In: 2012 IEEE Conference on Computer Vision & Pattern Recognition. IEEE (2012)

15. Ju, R., et al.: Depth saliency based on anisotropic center-surround difference. In: 2014 IEEE International Conference on Image Processing (ICIP). IEEE (2015)

16. Li, N., et al.: Saliency detection on light field. In: 2014 IEEE Conference on Computer Vision and Pattern Recognition (2014)

17. Cheng, Y.P., et al.: Depth enhanced saliency detection method. In: 2014 Proceedings of the International Conference on Internet Multimedia Computing and Service. ACM, Xiamen, China (2014)

18. Piao, Y., et al.: Depth-induced multi-scale recurrent attention network for saliency detection. In: 2019 IEEE/CVF International Conference on Computer Vision IEEE, 2020. (2020)

19. Liu, L., et al.: On the variance of the adaptive learning rate and beyond. In: 2020 International Conference on Learning Representations (2020)

20. Zhang, R.M., et al.: Lookahead Optimizer: k steps forward, 1 step back. CoRR 2019 (2019)

21. Sun, G., et al.: Large kernel spectral and spatial attention networks for hyperspectral image classification. IEEE Transactions on Geoscience and Remote Sensing (2023)
22. Zhao, C., Qin, B., Feng, S., Zhu, W., Zhang, L., Ren, J.: An unsupervised domain adaptation method towards multi-level features and decision boundaries for cross-scene hyperspectral image classification. IEEE Trans. Geosci. Remote Sens. **60**, 1–16 (2022)

UAV Cross-Modal Image Registration: Large-Scale Dataset and Transformer-Based Approach

Yun Xiao[1], Fei Liu[4], Yabin Zhu[3], Chenglong Li[1,2(✉)], Futian Wang[4], and Jin Tang[2,4]

[1] School of Artificial Intelligence, Anhui University, Hefei, China
xiaoyun@ahu.edu.cn, lcl1314@foxmail.com
[2] Institute of Artificial Intelligence, Hefei Comprehensive National Science Center, Hefei, China
[3] School of Electronic and Information Engineering, Anhui University, Hefei, China
[4] School of Computer Science and Technology, Anhui University, Hefei, China
{wft,tj}@ahu.edu.cn

Abstract. It is common to equip unmanned aerial vehicle (UAV) with visible-thermal infrared cameras to enable them to operate around the clock under any weather conditions. However, these two cameras often encounter significant non-registration issues. Multimodal methods depend on registered data, whereas current platforms often lack registration. This absence of registration renders the data unusable for these methods. Thus, there is a pressing need for research on UAV cross-modal image registration. At present, a scarcity of datasets has limited the development of this area. For this reason, we construct a dataset for visible infrared image registration (UAV-VIIR), which consists of 5560 image pairs. The dataset has five additional challenges including low-light, low-texture, foggy weather, motion blur, and thermal crossover. Furthermore, the dataset covers more than a dozen diverse and complex UAV scences. As far as our knowledge extends, this dataset ranks among the largest open-source collections available in this field. Additionally, we propose a transformer-based homography estimation network (THENet), which incorporates a cross-enhanced transformer module and effectively enhances the features of different modalities. Extensive experiments are conducted on our proposed dataset to demonstrate the superiority and effectiveness of our approach compared to state-of-the-art methods.

Keywords: Visible-thermal infrared · Cross-modal image registration · UAV dataset · Homography estimation

1 Introduction

Cross-modal image registration is a fundamental task in multi-modal computer vision. It aligns multiple images of the same scene captured at different times,

J. Ren et al. (Eds.): BICS 2023, LNAI 14374, pp. 166–176, 2024.
https://doi.org/10.1007/978-981-97-1417-9_16

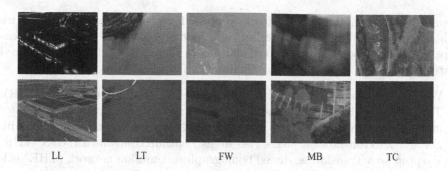

Fig. 1. A brief overview of our dataset. The five columns represent low-light (LL), low-texture (LT), foggy-weather(FW), motion-blur(MB) and thermal crossover(TC) challenges.

from distinct viewpoints, and/or using different sensors [24]. In recent years, multi-rotor unmanned aerial vehicle (UAV) have gained popularity as aerial platforms due to their agility and stability [5,21]. To ensure the reliable execution of specific tasks in challenging environments like low-light, foggy, rainy, or snowy conditions, multi-modal cameras like the thermal infrared one have emerged as the preferred choice (Fig. 1).

In recent years, the task of cross-modal image registration in the visible-thermal infrared has been severely neglected. Davis et al. [6] extract regions of interest in urban scenes by employing images from two modalities, visible-thermal infrared, and offer a dataset OTCBVS consisting of 284 pairs of images related to pedestrians. Morris et al. [13] capture a total of 400 pairs of data using hand-corrected visible-thermal infrared cameras, covering two different scenes. With the rapid development of drone applications, To alleviate the issues concerning radiometric, shape, and texture differences between the visible-thermal infrared modalities, Meng et al. [12] propose a combination of template matching and local max-pooling operation to estimate transformation parameters through the construction of a pyramid similarity map. Additionally, they introduce a dataset for unmanned aerial vehicle (UAV) visible-thermal infrared image registration. The dataset comprises 600 pairs of data, encompassing three main shooting scenes and three sub-scenes, such as wetlands and farmland. However, the aforementioned datasets suffer from issues such as limited data volume, insufficient coverage of diverse challenges, and a lack of representation for various scenes. These limitations make it difficult to train a robust model capable of handling multiple scenarios.

To address this issue, we construct a dataset for visible-thermal infrared image registration (UAV-VIIR), which consists of 5560 image pairs and each one has undergone meticulous annotation to ensure precise labeling.

Moreover, We propose a Transformer-based homography estimation (THENet). The network adopts a coarse-to-fine prediction framework, utilizing multiple subnetworks to progressively predict registration parameters and approach the ground truth. Within our proposed cross-enhancement module, we enhance the structural information of each modality first and then utilize feature information between the two modalities to search for and reinforce shared feature structures. In summary, our main contributions are:

– We publish a drone visible-thermal infrared registration dataset (UAV-VIIR). The dataset at hand encompasses over 5560 meticulously annotated pairs of images, constituting one of the largest fully labeled datasets in this domain. For access to the dataset, please visit https://github.com/ahucslf/UAV-VIIR.
– We propose a Transformer-based homography estimation network (THENet). It enhances the cross-modal registration accuracy by leveraging the cross enhancement of feature representations from two modalities. Our method achieves significant improvements in registration performance on our dataset, surpassing previous state-of-the-art methods and attaining optimal precision.

2   Related Work

2.1   Visible-Thermal Infrared Image Registration

Visible and thermal infrared images have numerous applications [22,23], with image fusion being the most common. However, publicly available datasets consisting of both registered visible and thermal infrared images are scarce. Most datasets [2,4,15] containing visible and thermal infrared images are either already registered, or primarily used in image fusion applications wherein the primary objective is not image registration. Ellmauthaler et al. [9] was initially developed for image fusion but can also serve image registration purposes. It comprises six scenarios, each scenario includes multiple shot sequences. Davis et al. [6] offer a dataset OTCBVS consisting of 284 pairs of images related to pedestrians. With the rapid development of drone applications, Meng et al. [12] propose a dataset for the registration of visible and thermal infrared modalities in drones. The dataset consists of 600 pairs of images, covering three main scenes including wetlands and farmland areas.

Image registration between visible and thermal infrared modalities presents a significant challenge to registration as differences in resolution, radiation, and texture, can affect its accuracy. To address this, Ellmauthaler et al. [9] open source an image fusion dataset and suggest using the parameters of image fusion to represent the quality of image homography estimation. Meng et al. [12] combine template matching with weights, multilevel local max-pooling, and Max index backtracking. However, traditional methods such as SIFT [11], SURF [1], ORB [14] or LIFT [18] rely heavily on the ability to successfully extract salient image features.

With the continuous development of deep learning, methods based on it have achieved satisfactory results [3,7,8,10,16,17,19,20]. In 2016, DeTone et al. [8] bring a deep-learning-based homography solution for single-modal registration tasks. The proposed network processes source and target images to derive the homography matrix using the 4 corner displacement vectors of the source image as input. Then, Zhang et al. [20] further optimize the method by training a mask to focus on the usable image parts, improving the outcome on small baseline data. Recently, Debaqu et al. [7] present a deep homography estimation method for visible and thermal infrared modalities that can only be effective for small baseline displacements, in which weightings of extracted features are determined by a mask prediction layer and anomalous features are removed with the mask as an attention map.

3   UAV-VIIR Dataset

3.1   Various Challenges

We observed the captured dataset and identified five main challenging scenarios, including low-light (LL), motion-blur (MB), low-texture (LT), foggy-weather (FW) and thermal crossover(TC). The first four challenges mentioned pertain to visible, while the last one refers to challenges encountered in thermal infrared imagery. Thermal crossover specifically pertains to scenarios where the environmental temperature during shooting is relatively low and uniform, making it difficult for thermal infrared to capture meaningful textures, resulting in blurry images. The specific quantities of challenges are indicated in Fig. 2, where the regular challenges are not listed. We refer to the image pairs that do not involve these five additional challenges as regular challenges, which are present in every image due to the cross-modal nature of the dataset. For example, there are notable modal differences between visible and thermal infrared, resulting in variations and missing content captured by each modality.

3.2   Rich Scences

The dataset covers more than a dozen diverse and complex UAV scenarios like overpasses, lakes, lawns, skyscrapers, crowded areas, and intersections. And also, the dataset images are selected from diverse angles including both top-down and side views, covering a wide range of applications for UAV. As far as our knowledge extends, this dataset ranks among the largest open-source collections available in UAV visible thermal infrared image registration field. As shown on the right side of Fig. 2, the radius of the circle represents the size of the dataset, the x-axis represents the number of captured scenes, and the y-axis represents the number of challenge types. We have distinct advantages in all three aspects.

3.3   Dataset Detail

We select a total of 4905 image pairs from the entire dataset as the training set, and 655 image pairs as the testing set. Since the data is acquired based on fixed dual cameras, the lack of diversity in registration parameters is a common issue. To train a more robust registration model, we apply random distortions [8] to the annotated images prior to training. By combining these distortion parameters with the actual parameters as ground truth, we enhance the training process.

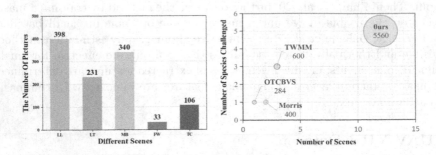

Fig. 2. The relevant information of the dataset. The left figure presents the distribution of various challenges in our dataset, while the right figure depicts a comparative analysis among different datasets, encompassing the number of challenge types, scene diversity, and dataset scale.

We adopt the solution of using the DJI M300 RTK drone equipped with the Zenmuse H20T camera to capture various urban scenes at altitudes below 500 m in 2021. These scenes include parks, overpasses, schools, and more. The H20T camera's thermal imaging sensor type is uncooled vanadium oxide (VOx) microbolometer, with a diagonal field of view (DFOV) of 40.6° and a noise equivalent temperature difference (NETD) of ≤50 mK @ f/1.0.

4   Proposed Algorithm

We propose a transformer-based homography estimation network (THENet), which incorporates a cross-enhanced transformer module and effectively enhances the features of different modalities. The framework diagram is shown in Fig. 3.

THENet is primarily composed of multiple stacked sub-networks, each consisting of our proposed Cross-enhanced Transformer Module (CETM) and a Homography Estimation Net. The last sub-network exclusively predicts the final distortion parameters. The predicted parameters distort the thermal infrared image of this layer as the input for the next layer.

CETM primarily consists of four sub-modules: two Self-Structure Information Enhancement (SSIE) modules and two Cross-Structure Information Enhanceme (CSIE) modules. The SSIE modules enhance the features of the modalities themselves, while the CSIE modules enhance the shared features of the modalities via interaction. At the beginning of the sentence, f_i^{k-1} and $(f_i^{k-1})'$ represent the image features extracted and features enhanced by the CETM.

SSIE enhances the available features of a single modality and assigns different weights adaptively to different positions of the feature maps. Finally, it obtains the features enhanced by SSIE, which can effectively amplify the modality-specific information. To learn location-related feature information,

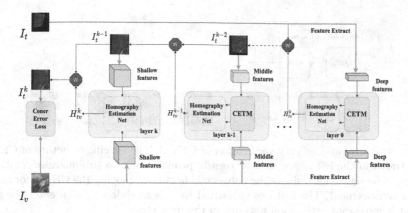

Fig. 3. The framework diagram of our method. We extract different levels of image features as inputs for each sub-network. Through progressive prediction results, it gradually approaches the ground truth. The "w" represents the distortion operation based on the input parameters.

position encoding is incorporated in our approach. Finally, the mechanism of SSIE can be summarized as

$$\mathbf{X}_{self} = \mathbf{X} + \text{MultiHead}\left(\mathbf{X} + \mathbf{P}_x, \mathbf{X} + \mathbf{P}_x, \mathbf{X}\right), \tag{1}$$

where $\mathbf{P}_x \in \mathbb{R}^{d \times N_x}$ is the spatial positional encodings and $\mathbf{X}_{self} \in \mathbb{R}^{d \times N_x}$ is the output of SSIE.

Once the SSIE enhances the feature of a modality, it is received by the CSIE. The CSIE then performs an exchange of features with another modality. It utilizes the feature of the current modality as the query and combines both features, alongside the key and value representations of the other modality, within the CSIE. With the presence of a multi-head attention mechanism, this interaction allows us to discover resemblances between the features of the two modalities and enhance their similarities.

$$\mathbf{X}_{cross} = \widetilde{\mathbf{X}}_{cross} + \text{FFN}\left(\widetilde{\mathbf{X}}_{cross}\right), \tag{2}$$

$$\widetilde{\mathbf{X}}_{cross} = \mathbf{X}_q + \text{MultiHead}\left(\mathbf{X}_q + \mathbf{P}_q, \mathbf{X}_{kv} + \mathbf{P}_{kv}, \mathbf{X}_{kv}\right), \tag{3}$$

where $\mathbf{X}_q \in \mathbb{R}^{d \times N_q}$ is the input of the branch where the module is applied, $\mathbf{P}_q \in \mathbb{R}^{d \times N_q}$ is the spatial positional encoding corresponding to X_q. $\mathbf{X}_{kv} \in \mathbb{R}^{d \times N_{kv}}$ is the input from another branch, and $\mathbf{X}_{kv} \in \mathbb{R}^{d \times N_{kv}}$ is the spatial encoding for the coordinate of X_{kv}. $\mathbf{X}_{kv} \in \mathbb{R}^{d \times N_{kv}}$ is the output of CSIE. According to (3), CSIE calculates the attention map according to multiple scaled products between X_{kv} and X_q, then reweighs X_{kv} according to the attention map, and adds it to X_q to enhance the representation ability of the feature map.

After obtaining the enhanced features of two modalities, we obtain an attention map by correlating them, and then apply a different number of convolution

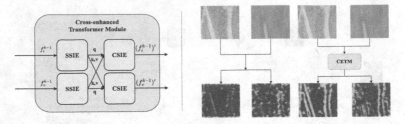

Fig. 4. Structural and validity verification of CETM. The specific structure of CETM is illustrated in the left image, which consists primarily of two sub-modules combined together. The right image shows the enhanced effect of employing the CETM for image feature enhancement. The features outputted by this module will then enter the subsequent homography estimation network in the next step.

blocks to predict eight parameters as [16], representing the eight degrees of freedom of the homography matrix. We represent the position of an entire image by predicting the coordinates of its four corner points. To constrain the gap between predicted values and ground truth, we utilize the L1 loss (Fig. 4).

5   Experimental Results

5.1   Implementation Details

The experiments are carried out by using the NVIDIA Geforce RTX 2080 Ti. To ensure that the distorted images have no black borders when inputted, we resize the images to 256×256. Then, within the image, we randomly perturb four vertices [8] within a radius of 32. Finally, we crop the images to 128×128 for input purposes. Based on the empirical evidence from multiple experiments, we set the learning rate is $1e^{-4}$, the batch size is 32, and the network is trained for 235 epochs.

5.2   Comparison with the State-of-the-Art Methods

Currently, there are few publicly available datasets specifically dedicated to visible-infrared image registration. To demonstrate the effectiveness of our proposed method, we conducted comparative evaluations with existing open-source methods, using our newly introduced dataset.

It can be observed that traditional methods such as SIFT [11], SURF [1], ORB [14] or LIFT [18] perform poorly on our dataset. TWMM [12] is a recently proposed approach, but it does not yield significant improvements either. This could be attributed to the fact that our inputs consist of cropped and compressed images. Both modalities of the images are subject to some level of distortion, deviating from the original proportions. To facilitate fair comparisons between different approaches, standardizing the input is necessary.

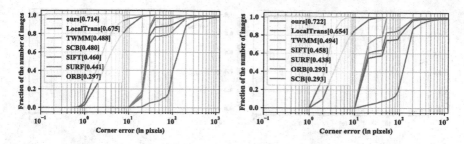

Fig. 5. Comparison of different methods. Different methods are compared on our dataset, with the left image validated on random data and the right image validated on real data.

LocalTrans [16] employs an enhanced version of the transformer for optimization training. Our method is inspired by it, utilizing a progressive model structure and improving modality interaction to enhance image features. As a result, we achieve approximately a 5% improvement in random data and a 10% improvement in real data (Fig. 5).

We conduct comparisons between deep learning methods and non-deep learning methods. By examining the errors between the predicted corner points of the images and the ground truth, we evaluate the quality of individual prediction results, as well as the overall accuracy of the methods using the area under the curve (AUC) metric.

5.3   Ablation Studies

When incorporating the proposed enhancement module, we conduct experiments to evaluate its impact on both randomly distorted and real-world data. The experimental results demonstrated significant improvements in performance with the presence of this module, particularly in the case of randomly distorted data. As shown in Fig. 6, the x-axis represents the coordinate error among the four image corners, while the y-axis represents the percentage of images with prediction results below this error threshold.

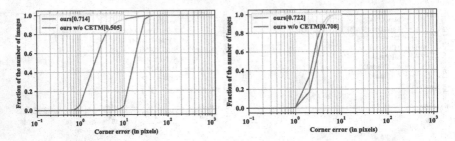

Fig. 6. The results obtained by excluding or including CETN in THENet. The left and right figures represent the results obtained under two different scenarios: random distortion and real-world data, respectively.

6 Conclusion

We present UAV-VIIR, a large-scale dataset for UAV visible-thermal infrared image registration, along with THENet, a transformer-based homography estimation network. Our dataset covers diverse UAV usage scenarios and addresses the primary challenges. It sets a new benchmark for scaled UAV cross-modal image registration. Moreover, our innovative cross-augmentation module demonstrates promising results. Our objective is to offer researchers a valuable platform by providing the dataset, research insights, and contributing to the continuous advancement of related fields.

Acknowledgment. This work was jointly supported by National Natural Science Foundation of China (No. 62006002), the Joint Funds of the National Natural Science Foundation of China (No.U20B2068), Natural Science Foundation of Anhui Province (No. 2208085J18), Natural Science Foundation of Anhui Higher Education Institution (No. 2022AH040014), the University Synergy Innovation Program of Anhui Province (No. GXXT-2022-033) and Anhui Provincial Colleges Science Foundation for Distinguished Young Scholars (No. 2022AH020093).

References

1. Bay, H., Tuytelaars, T., Van Gool, L.: SURF: speeded up robust features. In: Leonardis, A., Bischof, H., Pinz, A. (eds.) ECCV 2006. LNCS, vol. 3951, pp. 404–417. Springer, Heidelberg (2006). https://doi.org/10.1007/11744023_32
2. Bilodeau, G.A., Torabi, A., St-Charles, P.L., Riahi, D.: Thermal-visible registration of human silhouettes: a similarity measure performance evaluation. Infrared Phys. Technol. **64**, 79–86 (2014)
3. Bozcan, I., Kayacan, E.: Au-air: A multi-modal unmanned aerial vehicle dataset for low altitude traffic surveillance. In: 2020 IEEE International Conference on Robotics and Automation (ICRA), pp. 8504–8510. IEEE (2020)
4. Campo, F.B., Ruiz, F.L., Sappa, A.D.: Multimodal stereo vision system: 3D data extraction and algorithm evaluation. IEEE J. Select. Top. Sign. Process. **6**(5), 437–446 (2012)

5. Cao, Z., Huang, Z., Pan, L., Zhang, S., Liu, Z., Fu, C.: Tctrack: temporal contexts for aerial tracking. In: Proceedings of the IEEE/CVF Conference on Computer Vision and Pattern Recognition, pp. 14798–14808 (2022)

6. Davis, J.W., Sharma, V.: Fusion-based background-subtraction using contour saliency. In: 2005 IEEE Computer Society Conference on Computer Vision and Pattern Recognition (CVPR'05)-Workshops, pp. 11–11. IEEE (2005)

7. Debaque, B., et al.: Thermal and visible image registration using deep homography. In: 2022 25th International Conference on Information Fusion (FUSION), pp. 1–8. IEEE (2022)

8. DeTone, D., Malisiewicz, T., Rabinovich, A.: Deep Image Homography Estimation. arXiv:1606.03798 [cs] (2016)

9. Ellmauthaler, A., Pagliari, C.L., da Silva, E.A., Gois, J.N., Neves, S.R.: A visible-light and infrared video database for performance evaluation of video/image fusion methods. Multidimension. Syst. Signal Process. **30**, 119–143 (2019)

10. Hong, M., Lu, Y., Ye, N., Lin, C., Zhao, Q., Liu, S.: Unsupervised homography estimation with coplanarity-aware GAN. In: Proceedings of the IEEE/CVF Conference on Computer Vision and Pattern Recognition, pp. 17663–17672 (2022)

11. Lowe, D.G.: Distinctive image features from scale-invariant keypoints. Int. J. Comput. Vision **60**, 91–110 (2004)

12. Meng, L., et al.: A robust registration method for UAV thermal infrared and visible images taken by dual-cameras. ISPRS J. Photogramm. Remote. Sens. **192**, 189–214 (2022). https://doi.org/10.1016/j.isprsjprs.2022.08.018

13. Morris, N.J., Avidan, S., Matusik, W., Pfister, H.: Statistics of infrared images. In: 2007 IEEE Conference on Computer Vision and Pattern Recognition, pp. 1–7. IEEE (2007)

14. Rublee, E., Rabaud, V., Konolige, K., Bradski, G.: ORB: an efficient alternative to SIFT or SURF. In: 2011 International conference on computer vision, pp. 2564–2571. IEEE (2011)

15. Saponaro, P., Sorensen, S., Rhein, S., Kambhamettu, C.: Improving calibration of thermal stereo cameras using heated calibration board. In: 2015 IEEE International Conference on Image Processing (ICIP), pp. 4718–4722. IEEE (2015)

16. Shao, R., Wu, G., Zhou, Y., Fu, Y., Fang, L., Liu, Y.: LocalTrans: a multiscale local transformer network for cross-resolution homography estimation. In: Proceedings of the IEEE/CVF International Conference on Computer Vision, pp. 14890–14899 (2021)

17. Ye, N., Wang, C., Fan, H., Liu, S.: Motion basis learning for unsupervised deep homography estimation with subspace projection. In: Proceedings of the IEEE/CVF International Conference on Computer Vision, pp. 13117–13125 (2021)

18. Yi, K.M., Trulls, E., Lepetit, V., Fua, P.: LIFT: learned invariant feature transform. In: Leibe, B., Matas, J., Sebe, N., Welling, M. (eds.) ECCV 2016. LNCS, vol. 9910, pp. 467–483. Springer, Cham (2016). https://doi.org/10.1007/978-3-319-46466-4_28

19. Yu, H., et al.: The unmanned aerial vehicle benchmark: object detection, tracking and baseline. Int. J. Comput. Vision **128**, 1141–1159 (2020)

20. Zhang, J., et al.: Content-aware unsupervised deep homography estimation. arXiv:1909.05983 (2020)

21. Zhang, P., Zhao, J., Wang, D., Lu, H., Ruan, X.: Visible-thermal UAV tracking: a large-scale benchmark and new baseline. In: Proceedings of the IEEE/CVF Conference on Computer Vision and Pattern Recognition, pp. 8886–8895 (2022)

22. Zhang, T., Guo, H., Jiao, Q., Zhang, Q., Han, J.: Efficient RGB-T tracking via cross-modality distillation. In: Proceedings of the IEEE/CVF Conference on Computer Vision and Pattern Recognition, pp. 5404–5413 (2023)
23. Zhu, J., Lai, S., Chen, X., Wang, D., Lu, H.: Visual prompt multi-modal tracking. In: Proceedings of the IEEE/CVF Conference on Computer Vision and Pattern Recognition, pp. 9516–9526 (2023)
24. Zitova, B., Flusser, J.: Image registration methods: a survey. Image Vis. Comput. **21**(11), 977–1000 (2003)

Vision and Object Tracking

How Challenging is a Challenge for SLAM? An Answer from Quantitative Visual Evaluation

Xuhui Zhao[1], Zhi Gao[1(✉)], Hao Li[1], Chenyang Li[1], Jingwei Chen[1], and Han Yi[2]

[1] School of Remote Sensing and Information Engineering, Wuhan University, Wuhan 430079, China
{zhaoxuhui,leoli9901,2018302130131,cjw}@whu.edu.cn, gaozhinus@gmail.com
[2] School of Computing, National University of Singapore, Singapore 117417, Singapore
hany24@u.nus.edu

Abstract. SLAM (Simultaneously Localization and Mapping) is the fundamental technology for the application of unmanned intelligent systems, such as underwater exploration with fish robots. But various visual challenges often occur in practical environments, severely threaten the system robustness. Currently, few research explicitly focus on visual challenges for SLAM and analyze them quantitatively, resulting in works with less comprehensiveness and generalization. Many are basically not intelligent enough in the changing real world and sometimes even infeasible for practical deployment due to the lack of accurate visual cognition in the ambient environment, as many animals do. Inspired by visual perception pathways in brains, we try to solve the problem from the view of visual cognition and propose a fully computational reliable evaluation method for general challenges to push the frontier of visual SLAM. It systematically decomposes various challenges into three relevant aspects and evaluates the perception quality with corresponding scores. Extensive experiments on different datasets demonstrate the feasibility and effectiveness of our method by a strong correlation with SLAM performance. Moreover, we automatically obtain detailed insights about challenges from quantitative evaluation, which is also important for targeted solutions. To our best knowledge, no similar works exist at present.

Keywords: SLAM · Robotics · Visual Challenges · Quantitative Evaluation

1 Introduction

In numerous emerging applications of intelligent unmanned systems, many are challenging to visual SLAM (Simultaneously Localization and Mapping) and threaten its robustness, such as low-light and low-texture scenes. Generally, SLAM is a core for intelligent exploration and sustains the autonomy of robots. Therefore, many efforts are devoted to this topic, and some researchers achieve

J. Ren et al. (Eds.): BICS 2023, LNAI 14374, pp. 179–189, 2024.
https://doi.org/10.1007/978-981-97-1417-9_17

compelling performance in specific challenging environments. However, there is still a long way to go for many methods before practical deployment on real robots. One of the key problem is the simple and straightforward solution regardless of condition changes in ambient environments, resulting in redundant computation and ambiguous decisions. This is especially fatal in power and resource-limited platforms, such as UAVs (Unmanned Aerial Vehicles).

With million years of evolution, many creatures have evolved efficient visual cognition and fast reaction to changes in ambient environments for accomplishing various tasks with less energy consumption. For example, the HVS (Human Visual System) contains two basic photoreceptors - cones and rods in retina, which is responsible for bright-light conditions and dim-light conditions, respectively, as explained by Purkinje shift [22]. Unfortunately, current visual SLAMs cannot act as a real animal and react to environment changes due to the lack of reliable visual evaluation, especially in challenging scenes.

Inspired by the visual perception in animal's brains, we believe that the best solution to various challenges is "enhance on demand according to real-time evaluation of ambient environments". Therefore, we systematically propose a novel quantitative evaluation framework for visual challenges, which is the core in SLAM to achieve more intelligent behaviors and decisions. Specifically, we break down various challenges into illumination-related, scene-related, and sensor-related aspects according to visual perception pathways in brains. Finally, we obtain a challenging score from detailed factors in each aspects and judge the challenging state with insightful tips.

2   Related Work

Generally, the quantitative evaluation of visual degradation is related to the SLAM field and the IQA/VQA (Image Quality Assessment or Video Quality Assessment) field, which we will elaborate in the following subsections.

2.1   Degradation Evaluation in SLAM

The quantitative evaluation of visual degradation in the SLAM field is typically in an early stage - existing methods are too simple or insufficient to complex challenges. We briefly categorize these methods into frame-based and pose-based, according to the evaluation is whether on input frames or estimated poses.

Frame-Based Methods. These methods typically focus on the quality of the input image and evaluate it with various metrics. For example, based on the Shannon information, an evaluation method for outdoor airborne dust and smoke discrimination is proposed in [2]. Later, a prototype visual SLAM with primary perception cognition is developed for adverse conditions [3]. Besides the outdoor dust, another method for illumination change is proposed for the robust localization during the rapid change of day and night in ISS (International Space Station) [12]. Typically, these works are fragile to practical applications due to the limited focus on the complex and changing degradations in our real world.

Pose-Based Methods. Compared with Frame-based methods, pose-based methods usually evaluate the visual degradation by checking estimated poses rather than the input frames. For example, the D-optimality criterion is proposed as a metric for the uncertainty of visual odometry by the CERBERUS team in DARPA Subterranean (SubT) Challenge [6]. While the HeRO [17] is also proposed by the CoSTAR team for failure detection in perceptually degraded environments by checking multi-sensor fusion status. Generally, the health status of pose estimation or odometry reflects visual challenges in ambient environments to some extent. But this correlation may be not very solid and bidirectional. For example, we can only know errors arise in pose estimation, but we cannot know what challenge leads to this degradation and give targeted solutions from very little information.

2.2 Quality Evaluation in IQA/VQA

In the field of IQA/VQA, the main goal is usually to evaluate the perceptual quality of images or videos from the view of HVS with various metrics, such as natural scene statistics [15] and neural networks [13]. Many methods are proposed for different purposes. For example, targeted for underwater and night-time scenes, methods with frequency transformation [23] and subjective metrics [21] are proposed, respectively. Despite compelling performance, these methods may still not be feasible for challenge evaluation for the following reasons. First, few of them consider the efficiency while SLAM usually requires real-time performance in resource-limited platforms, such as UAVs. Second, the definition of "image quality" varies from classic IQA and SLAM. Visually pleasing scenes may be harmful to SLAM as Fig. 3) shows. Finally, temporal change between successive frames is typically ignored in IQA but it is very important in SLAM, while VQA evaluates videos with different focuses.

2.3 Brief Summary and Our Idea

Generally, in the SLAM field, the challenge evaluation may be too simple and specific for target scenes, or inexplicitly estimate the health state of odometry, resulting in fragile performance. While in IQA/VQA field, existing methods are not suitable for SLAM due to the different cognition of "challenge" and heavy computation. Therefore, the quantitative evaluation of visual challenges in SLAM is especially needed for more robust performance in degraded scenes, as already demonstrated in LiDAR SLAM [24]. Based on the aforementioned analysis, we propose an evaluation framework with careful design, which will be elaborated in the following sections.

3   The Proposed Method

We propose an evaluation method from the aforementioned three aspects, where the inputs are only images, and the outputs are challenging scores and judgements, as shown in Fig. 1. For efficient scoring, we adopt the three-segmented

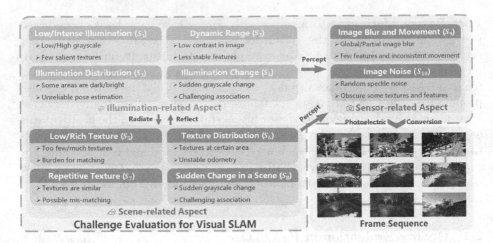

Fig. 1. The proposed quantitative evaluation framework for visual SLAM challenging environments, which involves three aspects and ten corresponding scores.

linear function as the paradigm for quantitative evaluation, as defined in Eq. 1.

$$Score(x;\ \mathbf{C}) = \begin{cases} a_1x + b_1 & t_1 \le x < t_2 \\ a_2x + b_2 & t_2 \le x < t_3 \\ a_3x + b_3 & t_3 \le x \le t_4 \end{cases},\tag{1}$$

where $\mathbf{C} = \{a_1, a_2, a_3, b_1, b_2, b_3, t_1, t_2, t_3, t_4\}$ stands for the coefficient set of scoring functions ($\mathbf{C_i}, i \in [1, 10]$). Without additional explanation, we deal with 8-bit quantified images.

3.1   Illumination-Related Aspect

Good illumination is one of the prerequisites for a robust visual SLAM, we further break down it into four related scores $S_i, i \in [1, 4]$.

Low/Intense Illumination. We believe it is adverse for visual SLAM with too low or intense lights based on the gray world assumption [7]. Therefore, we follow the proposed Eq. 1 and propose the S_1 score as Eq. 2.

$$S_1 = Score(G_{mean};\ \mathbf{C_1}),\tag{2}$$

where $G_{mean} \in [0, 255]$ is the mean grayscale value of the current frame F.

Dynamic Range. Images with narrow dynamic ranges usually lead to challenges in stable feature extraction and matching. Therefore, we evaluate the valid range with the truncation of central 96% in the image histogram to filter out error values. S_2 score is as Eq. 3.

$$S_2 = Score(G_{max} - G_{min};\ \mathbf{C_2}),\tag{3}$$

where $G_{min}, G_{max} \in [0, 255]$ are valid grayscale bounds in current frame F.

Illumination Distribution. Evenly-distributed illumination usually leads to a uniform distribution of visual features, which is beneficial for SLAM [20]. For quantitative evaluation, we apply grids to the input image and calculate the representative grayscale of each grid with Gaussian-weighted averaging. Then, the illumination distribution is obtained with the number histogram of grids, where level n indicates the number of grids with grayscale n. S_3 score with 32 × 32 grids are calculated as Eq. 4.

$$S_3 = Score(\frac{\sum_{j=0}^{255}(N_j - N_{mean})^2}{256}; \ \mathbf{C_3}), \tag{4}$$

where $N_j \in [0, 1024]$ indicates the grid number at grayscale level j and N_{mean} denotes the mean value of $N_j, j \in [0, 255]$.

Illumination Change. Frame-wise data association is crucial for initial pose estimation in visual tracking. But it is fragile to significant illumination variations. Therefore, we propose S_4 by comparing the difference of grayscales between two consecutive frames as Eq. 5.

$$S_4 = Score(abs(G_{mean}^P - G_{mean}); \ \mathbf{C_4}), \tag{5}$$

where $G_{mean}^P \in [0, 255]$ refers to the mean grayscale of previous frame F^P.

3.2   Scene-Related Aspect

The ambient environment and corresponding captured objects play important roles in the imaging process and visual SLAM. We further break down it into four related scores $S_i, i \in [5, 8]$.

Low/Rich Textures. Proper feature sparsity guarantees stable tracking and pose estimation. We think the gradient represents the texture and variations essentially. Therefore, we first conduct the Sobel operator on the input frame and calculate mean values in this gradient map. The S_5 score for texture evaluation is finally obtained with Eq. 6.

$$S_5 = Score(mean(\triangledown^2 F); \ \mathbf{C_5}) \tag{6}$$

Texture Distribution. Evenly distributed textures and features usually lead to more accurate pose estimation with a better approximation of the overall transformation [16]. Similar to S_5, we get the gradient map by the Sobel operator and divide it into 16 × 16 grids. Then, we calculate the gradient sum in each grid and regard it as valid if the corresponding value exceeds the given threshold. Finally, the S_6 score is estimated by the proportion of valid grids N_{val} to all grids $(N_{val} + N_{inv})$, as Eq. 7.

$$S_6 = Score(\frac{N_{val}}{N_{val} + N_{inv}}; \ \mathbf{C_6}) \tag{7}$$

Repetitive Texture. Repetitive textures may bring challenges to matching algorithms and lead to declined performance. Therefore, we adopt the classic GLCM (Grey Level Co-occurrence Matrix) for quantitative evaluation. More specifically, we calculate the homogeneity I_{hom} in GLCM for evaluation [10], where more similar textures usually have bigger homogeneity. Finally, the S_7 score is defined as Eq. 8.

$$S_7 = Score(I_{hom}; \mathbf{C_7}) \tag{8}$$

Sudden Scene Change. The frame-wise matching may fail if there are few overlaps between successive frames, resulting in lost tracking. Therefore, we simply evaluate changes with the similarity of histograms. The S_8 score is obtained with comparison of grayscale level j in $k \in \{Red, Green, Blue\}$ channels, as Eq. 9.

$$S_8 = Score(\sum_{k=0}^{2}\sum_{j=0}^{255} 1 - \frac{abs(N_{jk}^P - N_{jk})}{max(N_{jk}^P, N_{jk})}; \mathbf{C_8}), \tag{9}$$

where N_{jk}^P is the pixel number of grayscale j in channel k within previous frame F^P, while the N_{jk} is the counterpart in the current frame F.

3.3 Sensor-Related Aspect

Finally, the sensor converts photons into electrical signals. It also determines the perception quality to some extent, especially for low-cost cameras. We further break down it into two related scores $S_i, i \in [9, 10]$.

Image Blur and Movement. Both image blur and inconsistent movement bring challenges to reliable matching, especially in low-texture environments. We evaluate the blur by intentional blurring with the Gaussian kernel, where we think the clear images change a lot after blurring while blurred images are not. The blurring is evaluated by the difference of SSIM ($D_{blur} \in [0, 1]$). For movement, we calculate the consistency which is reflected by the proportion $P_{move} \in [0, 1]$ of outliers in RANSAC. Finally, the S_9 score is the combination of D_{blur} and P_{move}, as Eq. 10.

$$S_9 = \frac{1}{2}(Score(D_{blur}; \mathbf{C_{blur}}) + Score(P_{move}; \mathbf{C_{move}})), \tag{10}$$

Image Noise. Image noise is usually ignored due to its subtle appearance in the image. But it indeed influences feature extraction, matching and visual SLAM [14], especially for direct methods that are sensitive to photometric radiance. Similar to S_9, we compare the difference of current frame F and its filtered result F^N. The S_{10} is calculated with the difference of F and F^N, as Eq. 11.

$$S_{10} = Score(mean(abs(F - F^N)); \mathbf{C_{10}}) \tag{11}$$

Fig. 2. Datasets and sequences collected from various sources (downloading, synthesis, simulation) for comprehensive scoring function estimation and performance test.

3.4 The Overall Score

The overall challenging evaluation score S_C is calculated with all aforementioned scores as Eq. 12. The lower score indicates the more challenging states.

$$S_c = \alpha \sum_{i=1}^{4} S_i + \beta \sum_{i=5}^{8} S_i + \gamma \sum_{i=9}^{10} S_i \tag{12}$$

where α, β, and γ are weights for illumination, scene, and sensor aspects, respectively. We think the visual perception is degraded if the condition: $\forall S_i < T_l \vee S_C < T_h$ is satisfied, where T_l and T_h are given thresholds by users.

3.5 Implementation Details

For general coefficient estimation $S_i (i = 1, 2, \cdots, 10)$, we select 1,000 representative images from various sources. For S_C, we set $\alpha = 0.4$, $\beta = 0.4$, and $\gamma = 0.2$. All scores are normalized to $[0, 100]$ and we set $T_h = 70$ and $T_l = 50$. The framework is implemented by Python, with OpenCV and NumPy library.

4 Experiments

4.1 Experiment Preparation

For test datasets, we: (1) download TUM-RGBD [18] and AQUALOC [9], (2) synthesize black frames on EuRoC [4] and KITTI [11], (3) simulate sequences with AirSim software (Unreal Engine) in a European town and Genshin Impact game (Unity Engine) in Liyue harbor, as shown in Fig. 2. Readers may refer to [26] for more details of our Genshin Impact dataset. For evaluation metrics, we adopt PCC (Pearson Correlation Coefficient) [1] for the correlation between evaluation scores and pose estimation accuracy (ATE metric [25]). For compared SLAM, we select DSO (Direct SLAM) [8], ORB-SLAM3 (Indirect SLAM) [5] and DROID-SLAM (Learning-based end-to-end SLAM) [19].

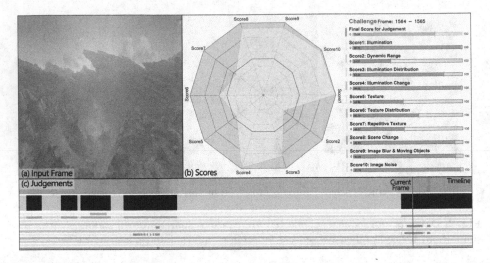

Fig. 3. Evaluation of one frame in our Genshin Impact sequences. (a) is the input frame. (b) are evaluation scores, where the red decagon indicates the threshold T_l. (c) are judgements based on (b), where darker colors indicate more challenging states.

Table 1. The Correlation between Challenge Evaluation and Pose Estimation of Different SLAM on Our EuRoC-Syn Sequence.

SLAM	Challenging Score of EuRoC-Syn Seq.			Frame-wise Estimation Error (Unit: meter)			PCC
	Max	Min	Mean	Max	Min	Mean	
ORB-SLAM [5]	90.158	16.512	81.116	0.153	0.001	0.016	0.944
DSO [8]				0.067	0.006	0.027	0.925
DROID-SLAM [19]				0.117	0.007	0.066	0.795
Mean				0.112	0.005	0.036	0.888

4.2 Results and Analysis

Correlation Validation. We first run compared SLAM (monocular mode) on the EuRoC-Syn dataset to verify the consistency between estimated scores and pose accuracy. Related results are calculated and summarized in Table 1. Generally, ORB-SLAM achieves best tracking performance (ATE of $0.016m$) with a PCC of 0.944. This indicates the high correlation between evaluation score S_C and pose estimation errors in SLAM. Additionally, we achieve the mean PCC of 0.879 in all tests, demonstrating the good generalization to different SLAM algorithms, from direct to indirect, even deep learning ones. Evaluation results on one frame from our Genshin dataset is shown in Fig. 3. Despite the beautiful sunset scenery, it is challenging for SLAM to track features robustly.

Validation of Efficiency. We also test the efficiency of our method on various platforms, from embedded Raspberry Pi 4B+ to powerful ASUS Workstation (Intel i9-9900K CPU). Generally, the module evaluates the TUM-RGBD dataset (640 × 480) stably over 20 FPS on our workstation and 4 FPS even on the Raspberry Pi, which fully meets the real-time demand in SLAM.

5 Conclusions

Inspired by the animal's visual perception of the world, we focus on various challenges for SLAM and propose a quantitative evaluation method for automatic degradation cognition. It is the core for condition-aware visual SLAM and has great potential in many applications. Extensive experiments on multi-source datasets show the high performance of our method. For future work, we will refine the framework and test with more datasets.

Acknowledgements. This research is partially supported by the National Natural Science Foundation of China Major Program (Grant No. 42192580, 42192583), Hubei Province Natural Science Foundation (Grant No. 2021CFA088 and 2020-CFA003), the Science and Technology Major Project (Grant No. 2021AAA010), and Wuhan University - Huawei Geoinformatics Innovation Laboratory. Numerical calculations are supported by Supercomputing Center of Wuhan University.

References

1. Benesty, J., Chen, J., Huang, Y., Cohen, I.: Pearson Correlation Coefficient, pp. 1–4. Springer, Berlin, Heidelberg (2009). https://doi.org/10.1007/978-3-642-00296-0_5

2. Brunner, C., Peynot, T.: Visual metrics for the evaluation of sensor data quality in outdoor perception. In: Proceedings of the 10th Performance Metrics for Intelligent Systems Workshop, pp. 1–8. PerMIS '10, Association for Computing Machinery, New York, NY, USA (2010)

3. Brunner, C., Peynot, T.: Perception quality evaluation with visual and infrared cameras in challenging environmental conditions. In: Khatib, O., Kumar, V., Sukhatme, G. (eds.) Experimental Robotics. Springer Tracts in Advanced Robotics, vol. 79, pp. 711–725. Springer, Berlin, Heidelberg (2014). https://doi.org/10.1007/978-3-642-28572-1_49

4. Burri, M., Nikolic, J., Gohl, P., Schneider, T., Rehder, J., Omari, S., Achtelik, M.W., Siegwart, R.: The euroc micro aerial vehicle datasets. Int. J. Robot. Res. **35**(10), 1157–1163 (2016)

5. Campos, C., Elvira, R., Rodriguez, J.J.G., Montiel, J.M., Tardos, J.D.: ORB-SLAM3: an accurate open-source library for visual, visual-inertial, and multimap slam. IEEE Trans. Robot. **37**(6), 1874–1890 (2021)

6. Carrillo, H., Reid, I., Castellanos, J.A.: On the comparison of uncertainty criteria for active slam. In: IEEE International Conference on Robotics and Automation, pp. 2080–2087 (2012)

7. Cepeda-Negrete, J., Sanchez-Yanez, R.E.: Gray-world assumption on perceptual color spaces. In: Klette, R., Rivera, M., Satoh, S. (eds.) PSIVT 2013. LNCS, vol. 8333, pp. 493–504. Springer, Heidelberg (2014). https://doi.org/10.1007/978-3-642-53842-1_42

8. Engel, J., Koltun, V., Cremers, D.: Direct sparse odometry. IEEE Trans. Pattern Anal. Mach. Intell. **40**(3), 611–625 (2018)

9. Ferrera, M., Creuze, V., Moras, J., Trouvé-Peloux, P.: AQUALOC: an underwater dataset for visual-inertial-pressure localization. Int. J. Robot. Res. **38**(14), 1549–1559 (2019)

10. Gadkari, D.: Image quality analysis using GLCM (2004)

11. Geiger, A., Lenz, P., Stiller, C., Urtasun, R.: Vision meets robotics: the KITTI dataset. Int. J. Robot. Res. **32**(11), 1231–1237 (2013)

12. Kim, P., Coltin, B., Alexandrov, O., Kim, H.J.: Robust visual localization in changing lighting conditions. In: IEEE International Conference on Robotics and Automation, pp. 5447–5452 (2017)

13. Ma, K., Liu, W., Zhang, K., Duanmu, Z., Wang, Z., Zuo, W.: End-to-end blind image quality assessment using deep neural networks. IEEE Trans. Image Process. **27**(3), 1202–1213 (2018)

14. Ma, P., et al.: Multiscale superpixelwise prophet model for noise-robust feature extraction in hyperspectral images. IEEE Trans. Geosci. Remote Sens. **61**, 1–12 (2023)

15. Moorthy, A.K., Bovik, A.C.: Blind image quality assessment: From natural scene statistics to perceptual quality. IEEE Trans. Image Process. **20**(12), 3350–3364 (2011)

16. Mur-Artal, R., Montiel, J.M.M., Tardós, J.D.: Orb-slam: a versatile and accurate monocular slam system. IEEE Trans. Robot. **31**(5), 1147–1163 (2015)

17. Santamaria-Navarro, A., Thakker, R., Fan, D.D., Morrell, B., Agha-mohammadi, A.A.: Towards resilient autonomous navigation of drones. In: Asfour, T., Yoshida, E., Park, J., Christensen, H., Khatib, O. (eds.) Robotics Research. ISRR 2019. LNCS. Springer Proceedings in Advanced Robotics, vol. 20, pp. 922–937. Springer, Cham (2022). https://doi.org/10.1007/978-3-030-95459-8_57

18. Sturm, J., Engelhard, N., Endres, F., Burgard, W., Cremers, D.: A benchmark for the evaluation of RGB-D SLAM systems. In: IEEE/RSJ International Conference on Intelligent Robots and Systems, pp. 573–580 (2012)

19. Teed, Z., Deng, J.: Droid-slam: deep visual slam for monocular, stereo, and rgb-d cameras. Adv. Neural Inf. Process. Syst. **34**, 16558–16569 (2021)

20. Tranzatto, M., et al.: Cerberus: autonomous legged and aerial robotic exploration in the tunnel and urban circuits of the darpa subterranean challenge. arXiv preprint arXiv:2201.07067 (2022)

21. Xiang, T., Yang, Y., Guo, S.: Blind night-time image quality assessment: subjective and objective approaches. IEEE Trans. Multimed. **22**(5), 1259–1272 (2020)

22. Xiong, J., Wang, J., Heidrich, W., Nayar, S.: Seeing in extra darkness using a deep-red flash. In: 2021 IEEE/CVF Conference on Computer Vision and Pattern Recognition (CVPR), pp. 9995–10004 (2021)

23. Yang, N., Zhong, Q., Li, K., Cong, R., Zhao, Y., Kwong, S.: A reference-free underwater image quality assessment metric in frequency domain. Signal Process. Image Commun. **94**, 116218 (2021)

24. Zhang, J., Singh, S.: Enabling aggressive motion estimation at low-drift and accurate mapping in real-time. In: IEEE International Conference on Robotics and Automation, pp. 5051–5058 (2017)

25. Zhang, Z., Scaramuzza, D.: A tutorial on quantitative trajectory evaluation for visual(-inertial) odometry. In: IEEE/RSJ International Conference on Intelligent Robots and Systems, pp. 7244–7251 (2018)
26. Zhao, X.: The genshin impact dataset (GID) for slam (2023). https://github.com/zhaoxuhui/Genshin-Impact-Dataset

Generalized W-Net: Arbitrary-Style Chinese Character Synthesization

Haochuan Jiang[1]([✉]), Guanyu Yang[2], Fei Cheng[3], and Kaizhu Huang[2]

[1] School of Robotics, XJTLU Entrepreneur College (Taicang), Xi'an
Jiaotong-Liverpool University, Suzhou, China
h.jiang@xjtlu.edu.cn
[2] Data Science Research Center, Duke Kunshan University, Suzhou, China
{guanyu.yang,kaizhu.huang}@dukekunshan.edu.cn
[3] Department of Communications and Networking, School of Advanced Technology,
Xi'an Jiaotong-Liverpool University, Suzhou, China
fei.cheng@xjtlu.edu.cn

Abstract. Synthesizing Chinese characters with consistent style using few stylized examples is challenging. Existing models struggle to generate arbitrary style characters with limited examples. In this paper, we propose the **Generalized W-Net**, a novel class of W-shaped architectures that addresses this. By incorporating Adaptive Instance Normalization and introducing multi-content, our approach can synthesize Chinese characters in any desired style, even with limited examples. It handles seen and unseen styles during training and can generate new character contents. Experimental results demonstrate the effectiveness of our approach.

1 Introduction

Alphabet-based languages like English, German, and Arabic horizontally concatenate basic letters to form words. In contrast, oriental Asian languages such as Chinese, Korean, and Japanese use radicals and strokes as the minimum units for block characters. These units can vary horizontally and vertically, resulting in a diverse range of characters, as shown in Fig. 1(a), with over 50 strokes and 10 radicals per character. Handwritten block calligraphies possess not only messaging functions but also artistic and collectible value, as depicted in Fig. 1(b). This property is rarely seen in alphabet-based languages.

Synthesizing a large number of artistic and calligraphic characters in a consistent style, based on a few or even just one stylized example, is a highly challenging task [9]. Previous attempts in few-shot character generation using deep generative neural networks have not proven effective on public Chinese handwritten datasets. While deep models have been successful in synthesizing vector

K. Huang—This research is funded by XJTLU Research Development Funding 20-02-60. Computational resources utilized in this research are provided by the School of Robotics, XJTLU Entrepreneur College (Taicang), and the School of Advanced Technology, Xi'an Jiaotong-Liverpool University.

J. Ren et al. (Eds.): BICS 2023, LNAI 14374, pp. 190–200, 2024.
https://doi.org/10.1007/978-981-97-1417-9_18

(a) Characters with complex structures.

(b) A calligraphy by the Emperor Huizong of the Chinese Song Dynasty, Ji Zhao (中国宋朝徽宗皇帝赵佶), in the 1100s A.D.

Fig. 1. Examples of Chinese characters and calligraphic work.

fonts, they are less suitable for hand-written glyph generation, where image-based approaches are more appropriate.

One successful approach, the *W-Net*, employs a W-shaped architecture that effectively separates content and style inputs, specifically on public Chinese handwritten datasets [10]. It consists of two parallel encoders: the style reference encoder and the content prototype encoder [9,17]. The style reference encoder takes a set of stylized characters with different contents (referred to as *style references*), while the content prototype encoder receives a set of characters with identical content but varying styles (known as *content prototypes*). The extracted features from both encoders are combined using a feature mixer to generate the desired output. This W-shaped deep generative model, depicted in Fig. 2, leverages adversarial training [1,2] to synthesize a character that closely resembles the style of the *style references* while preserving the content from the *content prototypes*[1].

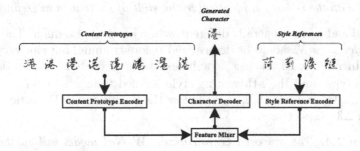

Fig. 2. Structure of W-shaped architectures for the character generation task.

In this paper, the proposed **Generalized W-Net** architecture incorporates residual blocks and dense blocks to enhance model performance. Various

---

[1] The input of the *W-Net* architecture in [9] represents a special case, involving only one single content prototype with a standard font. In most cases, the styles are pre-selected and fixed prior to training or utilization.

normalization techniques, such as batch normalization, instance normalization, layer normalization, and adaptive instance normalization (AdaIN), are employed to improve style transfer performance. Based on the *W-Net* framework [9], the **Generalized W-Net** introduces multiple content prototypes with different styles, enabling the generation of characters that combine these prototypes non-linearly to achieve desired styles. Additionally, AdaIN is utilized in different parts of the framework to further enhance style transfer. Extensive experiments demonstrate the effectiveness and capability of the **Generalized W-Net** both visually and statistically.

2   Preliminaries

Notations and preliminaries necessary to demonstrate the proposed **Generalized W-Net** architecture are specified in this section. Firstly, X is denoted as a character dataset, consisting of J different characters with in total I different fonts.

Definition 2.1. *Let x_j^i be a specific sample in X, regarded as the real target. Following [7], the superscript $i \in [1, 2, ..., I]$ represents i-th character style, while the subscript $j \in [1, 2, ..., J]$ denotes the j-th character content.*

Definition 2.2. *Denote $x_j^{c_m}, c_m \in [1, 2, ..., I], m = 1, 2, ..., M$ be the set of content prototypes. It describes a content of the j-th character, the same as the content of x_j^i defined in Definition 2.1.*

Particularly, M different styles are to be pre-selected before the model is optimized. Commonly, they are out of the fonts for the real target ($[1, 2, ..., I]$).

Definition 2.3. *Denote $x_{s_n}^i, s_n \in [1, 2, ..., J], n = 1, 2, ..., N$ be a set of style references with the i-th style, identical to the style of x_j^i defined in Definition 2.1.*

Note that i and c_m are generally different, while j and s_n also differ. The number of *style references* N should be determined before training, but can vary during testing[2]. In our model, each x_j^i combines the j-th content information from $x^{c_m}j$ prototypes and the i-th writing style learned from $x^i s_n$ references. Here, $m \in [1, 2, ..., M]$ and $n \in [1, 2, ..., N]$ follow the definitions in Definition 2.2 and Definition 2.3, respectively.

Definition 2.4. *The proposed **Generalized W-Net** model will synthesize the generated character by taking content prototypes ($x_j^{c_1}, x_j^{c_2}, ..., x_j^{c_M}$, as defined in Definition 2.2) and style references ($x_{s_1}^i, x_{s_2}^i, ..., x_{s_N}^i$, as defined in Definition 2.3) simultaneously. The corresponding generated character is denoted as $G(x_j^{c_1}, x_j^{c_2}, ..., x_j^{c_M}, x_{s_1}^i, x_{s_2}^i, ..., x_{s_N}^i)$[3].*

---

[2] In the proposed **Generalized W-Net** and *W-Net* architecture [9], N can be changed during testing.

[3] The *generated character* will be noted as $G(x_j^{c_m}, x_{s_n}^i)$ for simplicity.

The objective of training is to make the generated character $G(x^{c_m}j, x^i s_n)$ resemble the real target x^i_j in content and style. In the few-shot setting, a set of content prototypes $(x^{c_m}q, m = 1, 2, ..., M)$ is combined with the style reference encoder's outputs $(Enc_r(x^h p_1, x^h_{p_2}, ..., x^h_{p_L}))$ to produce characters $(G(x^{c_m}q, x^h p_l))$ with the desired style. Each value of q generates a different character that imitates the given styles $(x^h_{p_l}, l = 1, 2, ..., L)$. The values of p_l and q can vary within specific ranges, and h can be outside certain ranges to produce the desired character variations given a few styles[4].

3   The Generalized W-Net Architecture

Figure 3 illustrates the structure of the proposed **Generalized W-Net**. It includes the content prototype encoder (Enc_p), the style reference encoder (Enc_r), the feature mixer, and the decoder (Dec).

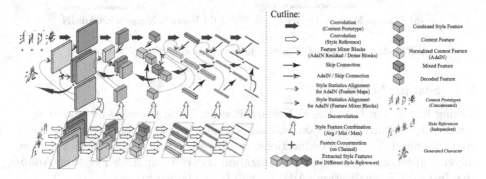

Fig. 3. Model Architecture: **Generalized W-Net**

3.1   Encoders

The Enc_p and Enc_r are formed using convolutional blocks, each comprising convolutional filters (5 × 5 kernel size with stride 2), normalization[5], and ReLU activation. With this configuration, M 64 × 64 prototypes $x^{c_m}_j$ ($c_m \in [1, 2, ..., I], m = 1, 2, ..., M$) and N references $x^i_{s_n}$ ($s_n \in [1, 2, ..., J], n = 1, 2, ..., N$) are both mapped to 1 × 512 feature vectors: $Enc_p(x^{c_1}_j, x^{c_2}_j, ..., x^{c_M}_j)$ and $Enc_r(x^i_{s_1}, x^i_{s_2}, ..., x^i_{s_N})$ respectively.[6]

The single-channel *content prototypes* are concatenated channel-wise, resulting in an M-channel input for Enc_p. The hyper-parameter M remains fixed during training and real-world application. On the other hand, N input *style*

---

[4] The output of the style reference encoder $Enc_r(x^h_{p_1}, x^h_{p_2}, ..., x^h_{p_L})$ is connected to the *Dec* using shortcut or residual/dense block connections (see Sect. 3.2).

[5] The specific normalization method may vary across implementations (see Sect. 3.3).

[6] The encoded outputs will be referred to as $Enc_p(x^{c_m}_j)$ and $Enc_r(x^i_{s_n})$.

references generate N features using shared weights from Enc_r. The combined style features are obtained by averaging, maximizing, and minimizing these N style features. They are then concatenated channel-wise to produce $Enc_r(x^i_{s_n})$. Therefore, the choice of N (or L during testing) can vary due to weight sharing.

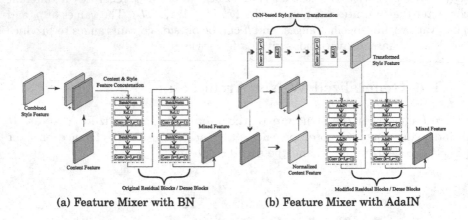

(a) Feature Mixer with BN   (b) Feature Mixer with AdaIN

Fig. 4. Examples of Chinese characters and calligraphic work.

3.2   Feature Mixer

Taking inspiration from the U-Net architecture [8,9], the encoder and character decoder feature mutual connections between equivalent layers for combining content and style. These connections, discussed in Sect. 3.4, involve a feature mixer that incorporates batch normalization (BN). An upgraded version, Adaptive Instance Normalization (AdaIN), will be explained in Sect. 3.3.

Feature maps of deeper layers' from both encoders are concatenated channel-wise, while only the content feature is used in shallower levels, resulting in combined features. In shallower layers, these combined features are processed through multiple dense blocks [4] or residual blocks [3] for further enhancement. Each block consists of batch normalization, ReLU activation, and convolutional filters, as shown in Fig. 4(a). The concatenated feature in deeper layers serves as the mixed feature, simplifying to the original shortcut connection used in the U-Net architecture [8].

3.3   AdaIN in the Feature Mixer

The AdaIN [5] has been proven effective for enhancing style transfer performance in various relevant research studies, such as those mentioned in [6,16,17]. It is defined by Eq. (1), where the statistics $\sigma(\beta)$ and $\mu(\beta)$ of the style features are computed across spatial locations to normalize the content feature α. The calculations for $\sigma(\beta)$ and $\mu(\beta)$ are given by Eqs. (2) and (3), respectively.

$$\mathbf{AdaIN}(\alpha, \beta) = \sigma(\beta) \cdot \left(\frac{\alpha - \mu(\alpha)}{\sigma(\alpha)} \right) + \mu(\beta) \tag{1}$$

$$\sigma_{nc}(\beta) = \sqrt{\frac{1}{HW} \sum_{h,w}^{H,W} (\beta_{nchw} - \mu_{nc}(x))^2 + \epsilon} \qquad (2)$$

$$\mu_{nc}(\beta) = \frac{1}{HW} \sum_{h,w}^{H,W} \beta_{nchw} \qquad (3)$$

N, C, H, and W represent the batch size, number of channels, spatial height, and spatial width of the style feature map β. It signifies the statistical alignment normalization between a content feature map α and the style feature map β. In the proposed feature mixer of the **Generalized W-Net**, AdaIN can be positioned in two locations, as shown in Fig. 4(b).

AdaIN on the Extracted Content Feature. The extracted content features $(f_p^\gamma(x_j^{c_m}))$ in the γ-th layer of the content encoder are normalized by the statistics of the combined style feature $(f_r^\gamma(x_{s_n}^i))$ on the equivalent layer of the style encoder. It is performed by Eq. (1), Eq. (2), and Eq. (3), where $\alpha = f_p^\gamma(x_j^{c_m})$, $\beta = f_r^\gamma(x_{s_n}^i)$, $H = H_\gamma$, $W = W_\gamma$ (H_γ and W_γ are the corresponding feature sizes on the γ-th layer).

AdaIN in the Dense/Residual Blocks. BNs in the residual/dense blocks are replaced with AdaINs. Inspired by [6], multiple Convolutional Neural Network (CNN)-based $Conv(3 \times 3) - ReLU$ transformations process the combined style feature, resulting in the transformed style feature $g(f_r^\gamma(x_{s_n}^i))$ at the γ-th layer. Within these blocks (h), AdaIN replaces BN, as specified by Eq. (1), Eq. (2), and Eq. (3), using the statistics of the transformed style feature $g(f_r^\gamma(x_{s_n}^i))$ instead of the style feature $f_r^\gamma(x_{s_n}^i)$.

Other Corporations. When AdaIN is not used, both the encoders and decoder utilize Batch Normalization (BN). However, as proposed in [6], when AdaIN is employed, BN in the style encoder is omitted. Additionally, in the content encoder, Layer Normalization (LN) is used, and in the decoder, Instance Normalization (IN) is used. The summary of these settings can be found in Table 1.

Table 1. Normalization utilized in the **Generalized W-Net** Architecture

		BN-based	AdaIN-based
Content Encoder		BN	LN
Style Encoder		BN	N/A
Feature Mixer	Content Feature Alignment	N/A	AdaIN
	Residual/Dense Blocks	BN	AdaIN
Character Decoder		BN	IN

3.4 Character Decoder

The *Dec* in the **Generalized W-Net** is designed to be similar to the decoder in the *W-Net* [9]. It consists of deconvolutional blocks connected to the mixed feature. Each block in the *Dec* includes deconvolutional filters (kernel size: 5×5, stride: 2), normalization, and a non-linear activation function. The Leaky ReLU is used for all layers except the last one [12]. The final output, referred to as the *generated character* [1], is obtained by applying the *tanh* nonlinearity.

3.5 Training Strategy and Optimization Losses

The proposed **Generalized W-Net** [9] is trained using the Wasserstein Generative Adversarial Network with Gradient Penalty (W-GAN-GP) framework [2]. It serves as the generative model, denoted as G, taking content prototypes and style references as inputs to generate characters. The generation process is represented by $G(x^{cm}j, x^i s_n) = Dec(Enc_p(x^{cm}j), Enc_r(x^i s_n))$, where *Dec* and *Enc* refer to the decoder and encoder, respectively. The objective is to optimize this formulation to make it similar to x^i_j using reconstruction, perceptual, and adversarial losses, determined by the discriminative model D within the WGAN-GP framework.

Training Strategy. In each alternative training iteration, G and D are optimized in an alternative manner. G optimizes $\mathbb{L}_G = -\alpha \mathbb{L}_{adv-G} + \beta \mathbb{L}_{ac} + \lambda_{pixel}\mathbb{L}_{pixel} + \mathbb{L}_{\phi total} + \psi_p \mathbb{L}_{Const_p} + \psi_r \mathbb{L}_{Const_r}$. Simultaneously, D minimizes $\mathbb{L}_D = \alpha \mathbb{L}_{adv-D} + \alpha_{GP} \mathbb{L}_{adv-GP} + \beta \mathbb{L}_{ac}$.

Adversarial Loss. The adversarial losses are given by $\mathbb{L}_{adv-G} = D(x^{c_{m'}}_j, G(x^{cm}_j, x^i_{s_n}), x^i_{s_{n'}})$ and $\mathbb{L}_{adv-D} = D(x^{c_{m'}}_j, x^i_j, x^i_{s_{n'}}) - D(x^{c_{m'}}_j, G(x^{cm}_j, x^i_{s_n}), x^i_{s_{n'}})$ for G and D respectively. m' and n' are randomly sampled from $[1, 2, ..., M]$ and $[1, 2, ..., N]$ respectively for each training example x^i_j. The gradient penalty $\mathbb{L}_{adv-GP} = ||\nabla_{\widehat{x}} D(x^{c_{m'}}_j, \widehat{x}, x^i_{s_{n'}}) - 1||_2$ [2] is an essential part to make D satisfy the Lipschitz continuity condition required by the Wasserstein-based adversarial training. $\widehat{x}$ is uniformly interpolated along the line between x^i_j and $G(x^{cm}_j, x^i_{s_n})$.

Categorical Loss of the Auxiliary Classifier. As notified in [11], the auxiliary classifier on the discriminator is optimized by $\mathbb{L}_{ac} = \left[\log C_{ac}(i|x^{c_{m'}}_j, x^i_j, x^i_{s_{n'}})\right] + \left[\log C_{ac}(i|x^{c_{m'}}_j, G(x^{cm}_j, x^i_{s_n}), x^i_{s_{n'}})\right]$.

Constant Losses of the Encoders. The constant losses [14] are employed to better optimize both encoders. They are given by $\mathbb{L}_{Const_p} = ||Enc_p(x^{cm}_j) - Enc_p(G(x^{cm}_j, x^i_{s_n}))||^2$ and $\mathbb{L}_{Const_r} = ||Enc_r(x^i_{s_n}) - Enc_r(G(x^{cm}_j, x^i_{s_n}))||^2$ respectively for Enc_p and Enc_r.

Pixel Reconstruction Losses. It represents the discrepancy of pixels from two comparing images, namely, the *generated character* and the corresponding ground truth It is specified by the L1 difference $\mathbb{L}_{pixel} = ||(x_j^i - G(x_j^{c_m}, x_{s_n}^i))||_1$.

Deep Perceptual Losses. The deep perceptual loss aims to minimize the variation between the generated character and the corresponding input images by calculating differences in high-level features. The total perceptual loss, denoted as $\mathbb{L}\phi total$, is composed of three components: $\mathbb{L}\phi real$, $\mathbb{L}\phi content$, and $\mathbb{L}\phi style$, each incorporating mean square error (MSE) discrepancy and von-Neumann divergence [15].

The loss function $\mathbb{L}\phi real$ compares the generated character $G(x^{c_m}j, x^i s_n)$ with the corresponding input x_j^i. It utilizes a pre-trained VGG-16 network, denoted as ϕ_{real}, which is used for character content and writing style classification. Different convolutional feature variations, ϕ_{1-2}, ϕ_{2-2}, ϕ_{3-3}, ϕ_{4-3}, and ϕ_{5-3}, are considered, representing features from specific layers and blocks in the VGG-16 Network.

For $\mathbb{L}\phi content$, the goal is to differentiate the generated character $G(x^{c_m}j, x^i s_n)$ in style while preserving the same character content $x^{c_{m'}}j$. It relies on the $\phi content$ network, trained exclusively for character content classification. To minimize differences in high-level abstract features, only the ϕ_{4-3} and ϕ_{5-3} features are considered.

Similarly, $\mathbb{L}\phi style$ aims to differentiate the generated character $G(x^{c_m}j, x^i s_n)$ in character content while preserving the same writing style $x_{s_{n'}}^i$. The ϕ_{style} network is trained by minimizing cross-entropy for accurate style classification. Again, only the ϕ_{4-3} and ϕ_{5-3} features are utilized to ensure similarity in high-level patterns.

4 Experiments

We only report a few visual examples that are able to demonstrate the effectiveness of the proposed **Generalized W-Net** in this section. Detailed objective results will be in the future work.

Fig. 5. The input few brush-written examples of actual Chinese characters.(These brush-written simplified Chinese character are written by Dr. Fei CHENG from School of Advanced Technology, Xi'an Jiaotong-Liverpool University.)

Most of the characters in the eastern Asian languages including Chinese (traditional or simplified), Korean, and Japanese are constructed by rectangular shapes (known as the *block characters* [13]). In this sense, it is an interesting evaluation to make the well-optimized W-Net model to generate characters of these three kinds of languages. The training of the proposed *Generalized W-Net* follows the description in Sect. 3.5 where only the simplified Chinese characters are available. For each of the generation processes of paragraphs in these kinds of languages, one randomly selected brush-written character listed in Fig. 6 will be specified as the one-shot style reference. Figure 6 illustrates the generated result of the corresponding a traditional Chinese poetry, a Korean lyric, and a Japanese speech. It can be obviously found that the style tendency given in Fig. 5 is well kept in all the three generated outputs.

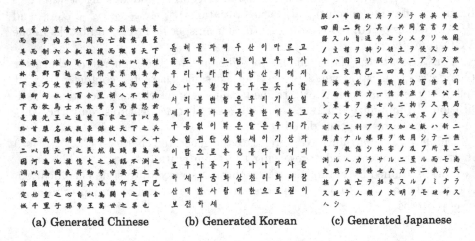

(a) Generated Chinese   (b) Generated Korean   (c) Generated Japanese

Fig. 6. The W-Net (trained with only simplified Chinese characters) generated essay of traditional Chinese, Korean, and Japanese characters from the one selected style reference shown in Fig. 5

A more interesting finding is that the proposed **Generalized W-Net** model trained with simplified Chinese characters is always capable to synthesize the *circular radicals* that are commonly seen in Korean but rarely found in Chinese. As seen in Fig. 7, the *circular radicals* are mostly preserved and recovered in the generated characters (Fig. 7(b)) when compared with the corresponding content prototypes shown in Fig. 7(a). It further demonstrates that the proposed W-Net is capable of being generalized to the useful knowledge in strokes and radicals that are absent from the training data.

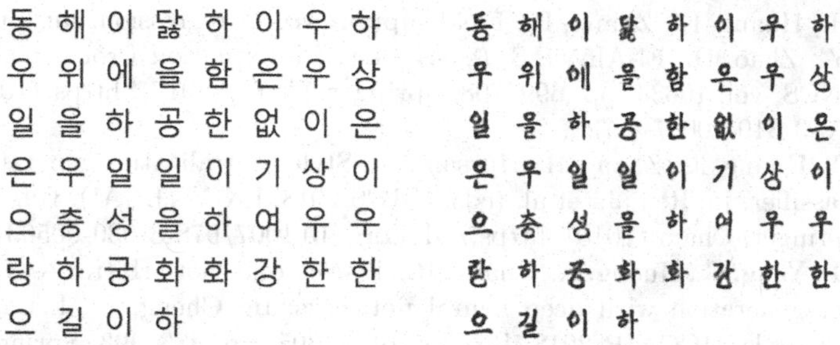

동 해 이 닳 하 이 우 하   동 해 이 닳 하 이 우 하
우 위 에 을 함 은 우 상   우 위 에 을 함 은 우 상
일 을 하 공 한 없 이 은   일 을 하 곰 한 없 이 은
은 우 일 일 이 기 상 이   은 우 일 일 이 기 상 이
으 충 성 을 하 여 우 우   으 충 성 을 하 여 우 우
랑 하 궁 화 화 강 한 한   랑 하 곰 화 화 감 한 한
으 길 이 하   으 길 이 하

(a) Content Prototypes   (b) Generated Characters

Fig. 7. Some Korean characters with *circular radicals*. The characters with circular radicals are selected from the ones given in Fig. 6(b)

5   Conclusion

A novel generalized framework **Generalized W-Net** is introduced in this paperto achieve Few-shot Multi-content Arbitrary-style Chinese Character Generation (FMACCG) task. Specifically, the proposed model, composing of two encoders, one decoder, and a feature mixer with several layer-wised connections, is trained adversarially based on the Wasserstein GAN scheme with the gradient penalty. It enables synthesizing any arbitrary stylistic character by transferring the learned style information from one single style reference to the input single content prototype. Extensive experiments have demonstrated the reasonableness and effectiveness of the proposed **Generalized W-Net** model in the few-shot setting.

References

1. Goodfellow, I., et al.: Generative adversarial nets. In: Advances in Neural Information Processing Systems, pp. 2672–2680 (2014)
2. Gulrajani, I., Ahmed, F., Arjovsky, M., Dumoulin, V., Courville, A.C.: Improved training of wasserstein gans. In: Advances in Neural Information Processing Systems, pp. 5767–5777 (2017)
3. He, K., Zhang, X., Ren, S., Sun, J.: Identity mappings in deep residual networks. In: Leibe, B., Matas, J., Sebe, N., Welling, M. (eds.) ECCV 2016. LNCS, vol. 9908, pp. 630–645. Springer, Cham (2016). https://doi.org/10.1007/978-3-319-46493-0_38
4. Huang, G., Liu, Z., Van Der Maaten, L., Weinberger, K.Q.: Densely connected convolutional networks. In: Proceedings of the IEEE Conference on Computer Vision and Pattern Recognition, pp. 4700–4708 (2017)
5. Huang, X., Belongie, S.: Arbitrary style transfer in real-time with adaptive instance normalization. In: Proceedings of the IEEE International Conference on Computer Vision, pp. 1501–1510 (2017)
6. Huang, X., Liu, M.Y., Belongie, S., Kautz, J.: Multimodal unsupervised image-to-image translation. In: Proceedings of the European Conference on Computer Vision (ECCV), pp. 172–189 (2018)

7. Jiang, H., Huang, K., Zhang, R.: Field support vector regression. In: Liu, D., Xie, S., Li, Y., Zhao, D., El-Alfy, E.S. (eds.) Neural Information Processing. ICONIP 2017. LNCS, vol. 10634, pp. 699–708. Springer, Cham (2017). https://doi.org/10.1007/978-3-319-70087-8_72

8. Jiang, H., Huang, K., Zhang, R., Hussain, A.: Style neutralization generative adversarial classifier. In: Ren, J., et al. (eds.) BICS 2018. LNCS (LNAI), vol. 10989, pp. 3–13. Springer, Cham (2018). https://doi.org/10.1007/978-3-030-00563-4_1

9. Jiang, H., Yang, G., Huang, K., Zhang, R.: *W-Net*: one-shot arbitrary-style Chinese character generation with deep neural networks. In: Cheng, L., Leung, A.C.S., Ozawa, S. (eds.) ICONIP 2018. LNCS, vol. 11305, pp. 483–493. Springer, Cham (2018). https://doi.org/10.1007/978-3-030-04221-9_43

10. Liu, C.L., Yin, F., Wang, D.H., Wang, Q.F.: CASIA online and offline Chinese handwriting databases. In: International Conference on Document Analysis and Recognition (ICDAR), pp. 37–41, September 2011. https://doi.org/10.1109/ICDAR.2011.17

11. Odena, A., Olah, C., Shlens, J.: Conditional image synthesis with auxiliary classifier GANs. arXiv preprint arXiv:1610.09585 (2016)

12. Radford, A., Metz, L., Chintala, S.: Unsupervised representation learning with deep convolutional generative adversarial networks. arXiv preprint arXiv:1511.06434 (2015)

13. Shieh, J.C.: The unified phonetic transcription for teaching and learning Chinese languages. Turk. Online J. Educ. Technol.-TOJET **10**(4), 355–369 (2011)

14. Taigman, Y., Polyak, A., Wolf, L.: Unsupervised cross-domain image generation. arXiv preprint arXiv:1611.02200 (2016)

15. Yang, X., Huang, K., Zhang, R., Hussain, A.: Learning latent features with infinite nonnegative binary matrix trifactorization. IEEE Trans. Emerg. Top. Comput. Intell. **99**, 1–14 (2018)

16. Zhang, Y., Zhang, Y., Cai, W.: Separating style and content for generalized style transfer. In: Proceedings of the IEEE Conference on Computer Vision and Pattern Recognition, pp. 8447–8455 (2018)

17. Zhang, Y., Zhang, Y., Cai, W.: A unified framework for generalizable style transfer: style and content separation. arXiv preprint arXiv:1806.05173 (2018)

Blind Deblurring of QR Codes with Local Extremum Intensity Prior

Wenguang Wang[1], Rongjun Chen[1,2]($\boxtimes$), and Yongqi Ren[2]

[1] School of Electronics and Information Technology, Sun Yat-Sen University,
Guangzhou 510006, China
`wangwg6@mail2.sysu.edu.cn`, `crj321@163.com`
[2] School of Computer Science, Guangdong Polytechnic Normal University,
Guangzhou 510665, China

Abstract. In recent years, QR code has been widely applied in various industries and has provided great convenience for our daily life. However, due to some inevitable factors such as camera shake and defocus, the images captured by the camera sensor might become blurry, which makes the QR code difficult to be recognized and be decoded. To deal with this problem, we propose a blind image deblurring method based on a novel binary prior named Local Extremum Intensity Prior (LEI) for QR code images. The proposed prior is based on the observation that the LEI of clear QR code images is sparser than that of blurred ones. By introducing a simple yet effective threshold technique to compute the LEI-involved regularization term, together with an effective optimization scheme, our method can recover clean and readable results from the blurred QR code images. Extensive experimental results show that our proposed method has advantages in restoration quality compared with the state-of-the-art image deblurring methods.

Keywords: QR codes · Blind deblurring · Local extremum intensity

1 Introduction

With the development of IoT (Internet of Things), QR (Quick Response) code, one of the most common matrix 2D barcodes, is widely used in many fields. However, the QR code images might become blurry, causing difficulty in recognizing and decoding the QR code. To tackle this problem, lots of research has focused on the deblurring of the blurred QR codes, for example, [2,3,12,15], but it still remains a great challenge.

With the assumption that the blur is uniform and spatially invariant, the blur process can be modeled as,

$$B = L \otimes k + N, \tag{1}$$

where B, L, and N represent blurry image, latent image and additive noise, respectively. $\otimes$ stands for discrete image convolution operations. k is a linear

J. Ren et al. (Eds.): BICS 2023, LNAI 14374, pp. 201–210, 2024.
https://doi.org/10.1007/978-981-97-1417-9_19

shift-invariant point spread function, also known as the blur kernel. Image blind deblurring is to obtain the clear image L and the kernel k while the only known item is the blurred image B.

Many methods make use of the prior information of the latent image. For example, hyper-laplacian priors [13] and extreme channels prior [18] for natural images, the L0-regularized intensity and gradient prior [10] for text images, binary prior [7] for QR codes image. Recently, Wen et al. [16] propose a patch-wise minimal prior for natural images. This algorithm imposes sparsity inducing on the local minimal pixels using a simple threshold method, promoting higher reconstruction quality and computation efficiency.

In this paper, we propose a blind image deblurring algorithm for QR codes based on a local binary prior named Local Extremum Intensity (LEI) Prior. Inspired by the work in [16], we observe that the value of local extremum pixels(pixels with the maximum value and minimum value within a local patch) in the QR code images tends to be averaged after the blurring process. We design a prior based on this property and combine it with the global sparse intensity and gradient prior to constraint the solution of the latent image. We keep the sparsity on the LEI of the clear image by exploiting an L0-regularization term of LEI. And we use a simple threshold technique to optimize the challenging L0-regularized term. With a traditional alternating optimization scheme, our algorithm performs favorably against the state-of-the-art image deblurring methods.

The contribution of our work can be summarized as follows: (1) We propose a novel local binary prior named Local Extremum Intensity (LEI) Prior for QR code images; (2) We adopt L_0 norm on the LEI involved term and provide an effective threshold method to optimize this regularization term; (3) We incorporate the optimization of the regularization term of LEI into a conventional optimization scheme for non-convex L0-minimization and our algorithm performs well.

2   Related Work

Recently, many methods have been proposed to solve the image deblurring problem for QR code images. Liu et al. [8] propose an incremental constrained least squares filter method to estimate the Gaussian kernel, incorporating the bi-level constraint of QR code images. Gennip et al. [15] design a hybrid method based on an anisotropic TV regularization, incorporating the pattern information of the QR code's L-shaped finder. Rioux et al. [12] propose a method based on Kullback-Leibler divergence which takes full advantage of QR code symbology. QR code image deblurring often requires extracting features from QR codes. Many feature extraction methods use a customized model to extract image information, such as [2,5,6,9,14].

3   Local Extremum Intensity Prior

3.1   LEI

In this section, we introduce a novel statistical prior, i.e., LEI prior. This prior is based on the observation that in a binary image like QR code image, the pixels with the maximum value and minimum value within a local patch have a sparse distribution. To better describe this observation, we first define the local minimum intensity and local maximum intensity.

Let a binary image $L \in R^{m \times n}$ be divided into q non-overlapped patches with a patch size of r x r, where q $= \lceil \frac{m}{r} \rceil \cdot \lceil \frac{n}{r} \rceil$, $\lceil \cdot \rceil$ denotes ceiling. The local minimum intensity is defined as

$$F(L)(i) = \min_{(x,y) \in \Omega_i} L(x,y), \tag{2}$$

and the local maximum intensity is defined as

$$G(L)(i) = \max_{(x,y) \in \Omega_i} L(x,y), \tag{3}$$

for i $= 1, 2, \ldots$ q, where x and y denote pixel locations in the image, Ω_i denotes the i-th image patch.

Generally, the LEI of clear QR code images is sparser than those of blurred QR code images. As shown in Fig. 1, most values of local minimal intensity and local maximum intensity in the clear QR code images are centered around 0 and 1, while the corresponding values in the blurred version are distributed ranging from 0 to 1. It shows that the LEI of clear QR code images is sparser than the blurred ones because the blur will increase the values of local minimum intensity and decrease the values of local maximum intensity. With this sparse property, clear and blurred images can be differentiated well.

Furthermore, we introduce a new regularization term to enforce the sparsity of LEI in the latent image. Based on above analysis, we choose the L_0 norm as the constraint to maintain its sparseness during the restoration process. According to the regularized term proposed by Yan et al. [15], the sparse constraint on the local extremum intensity is deployed on the minimum intensity and the local maximum intensity, respectively. The LEI-based regularization term is written as follows,

$$\|E(L)\|_0 = \|F(L)\|_0 + \|1 - G(L)\|_0, \tag{4}$$

where the L_0 norm $\| \cdot \|_0$ counts the number of the nonzero-elements in the matrix. Minimizing the term $\|E(L)\|_0$ will obtain a solution that favors clear images.

4   Proposed Algorithm

In this section, we present a blind image deblurring model and develop an efficient optimization algorithm for kernel estimation. We formulate the deblurring problem within the maximum a posteriori (MAP) framework as,

$$\{\hat{L}, \hat{k}\} = \underset{(L,k)}{\arg\min} \ \ell(L \otimes k, B) + \gamma P(k) + \lambda P(L), \tag{5}$$

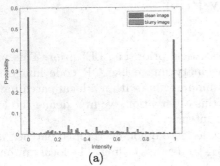 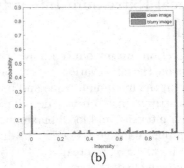

(a)   (b)

Fig. 1. Statistics of the LEI in QR code images. (a) Intensity histograms for local minimum intensity of clear and blurry QR code images. (b) Intensity histograms for local maximum intensity of clear and blurry QR code images.

where, the first item is the data fidelity, which is used to restrict the convolution of recovered latent image L to be consistent with the observed blurred image B, P(k) and P(L) are the priors of the blur kernel and the latent image, respectively.

4.1 Proposed LEI Prior

According to the previous observation and analysis, we propose a L0-regularization of LEI prior to measure the sparsity of QR code images. The regularization term is written as follows:

$$P(L) = \|E(L)\|_0, \tag{6}$$

where $\| \cdot \|_0$ counts the number of nonzero values of $E(L)$.

4.2 Adopted L0-Regularized Intensity and Gradient Prior

As a binary image, the pixel in a clear QR code image can only be black or white. As shown in Fig. 2, the distribution of intensity and gradient of clear QR code images are spares, we adopt the L0 norm proposed by Pan et al. [11] as the regularized constraint. The regularization term is written as follows:

$$P(L) = \sigma\|L\|_0 + \|\nabla L\|_0, \tag{7}$$

where $\| \cdot \|_0$ counts the number of nonzero values of L and ∇L, σ is a weight.

4.3 Objective Function

Our proposed objective function for image deblurring becomes

$$\{\hat{L}, \hat{k}\} = \underset{(L,k)}{\arg\min} \ \|L \otimes k - B\|_2^2 + \gamma\|k\|_2 + \lambda\left(\sigma\|L\|_0 + \|\nabla L\|_0\right) + \eta\|E(L)\|_0, \tag{8}$$

where γ, σ, λ, η are the corresponding weight parameters.

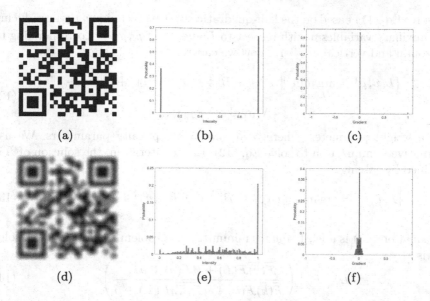

Fig. 2. Statistics of QR code images. (a) clean QR code image. (b) histogram of pixel intensities from (a). (c) histogram of horizontal gradients from (a). (d) blurred QR code image. (e) histogram of pixel intensities from (d). (f) histogram of horizontal gradients from (d)

4.4   Optimization

Since it is difficult to obtain the solution of Eq. (8) directly, we use an alternating minimization algorithm to solve for the latent image L and the kernel k. Thus, the problem is split into two subproblems as follows,

$$\hat{L} = \arg\min_{L} \; \|L \otimes k - B\|_2^2 + \lambda \left(\sigma\|L\|_0 + \|\nabla L\|_0\right) + \eta\|E(L)\|_0, \qquad (9)$$

and

$$\hat{k} = \arg\min_{k} \; \|L \otimes k - B\|_2^2 + \gamma\|k\|_2^2. \qquad (10)$$

Estimating Latent Image. Considering that the L_0 regularization term in Eq. (9) is computationally intractable, we propose an efficient algorithm to tackle it based on the half-quadratic splitting technique and a simple threshold method.

First, we present a simple threshold method to solve for the $\|E(L)\|_0$. We adopt the method in [16], approximately achieving the same sparse constraint as L_0 norm via iteratively thresholding the LEI of the latent image L in the subproblem. Given an image L, let $\lambda_1 > 0$ and $\lambda_2 > 0$ be the threshold parameters, first, in the local minimum intensity of L, we set the pixel value to 0 if it is less than λ_1, and then in the local maximum intensity, we set the pixel value to 1 if it is greater than λ_2. After thresholding, we can rewrite Eq. (9) as

$$\widehat{L} = \arg\min_{L} \; \|L \otimes k - B\|_2^2 + \lambda \left(\sigma\|L\|_0 + \|\nabla L\|_0\right). \qquad (11)$$

We solve Eq. (11) based on the half-quadratic splitting technique. By introducing new auxiliary variables u with respect to L and $\mathbf{g} = (g_h, g_W)^T$ corresponding to horizontal and vertical gradient, now we can rewrite Eq. (11) as

$$\{\hat{L}, \hat{u}, \hat{g}\} = \underset{L,u,g}{\arg\min} \ \|L \otimes k - B\|_2^2 + \beta\|L - u\|_2^2 + \mu\|\nabla L - g\|_2^2$$
$$+ \lambda\left(\sigma\|u\|_0 + \|g\|_0\right), \tag{12}$$

σ is a weight parameter, where β, μ, and λ are penalty parameters. We use alternative minimization to solve Eq. (12). In each iteration, the solution of L is obtained by solving

$$\{\hat{L}, \hat{u}, \hat{g}\} = \underset{L,u,g}{\arg\min} \ \|L \otimes k - B\|_2^2 + \beta\|L - u\|_2^2 + \mu\|\nabla L - g\|_2^2. \tag{13}$$

Equation (13) is a least squares minimization problem, the closed-form solution is

$$\hat{L} = \mathcal{F}^{-1}\left(\frac{\overline{\mathcal{F}(k)}\mathcal{F}(L) + \beta\mathcal{F}(u) + \mu F_G}{\overline{\mathcal{F}(k)}\mathcal{F}(k) + \mu\overline{\mathcal{F}(\nabla)}F(\nabla) + \beta}\right), \tag{14}$$

where $\mathcal{F}(\cdot)$ and $\mathcal{F}^{-1}(\cdot)$ denote the Fast Fourier Transform(FFT) and inverse FFT, respectively; the $\overline{\mathcal{F}^{-1}(\cdot)}$ is the complex conjugate operator, and $F_G = \overline{\mathcal{F}(\nabla_h)}\mathcal{F}(g_h) + \overline{\mathcal{F}(\nabla_v)}\mathcal{F}(g_v)$ where ∇_h and ∇_v denote the horizontal and vertical differential operators, respectively. Given L, the solutions of u and g are obtained based on [17],

$$\hat{u} = \begin{cases} 0, & |L|^2 < \frac{\lambda\sigma}{\beta}, \\ L, & \text{otherwise}, \end{cases} \tag{15}$$

and

$$\hat{g} = \begin{cases} 0, & |L|^2 < \frac{\lambda}{\mu}, \\ \nabla L, & \text{otherwise}. \end{cases} \tag{16}$$

Estimating Blur Kernel. With the given L, Eq. (10) is a least squares minimization problem where a closed-form solution can be computed by FFT. To get a more precise solution, we use the gradient space in [10] to estimate the blur kernel k,

$$\hat{k} = \underset{k}{\arg\min} \ \|\nabla L \otimes k - \nabla B\|_2^2 + \gamma\|k\|_2^2, \tag{17}$$

and the solution can be calculated by FFT:

$$\hat{k} = \mathcal{F}^{-1}\left(\frac{\overline{\mathcal{F}(\nabla L)}\mathcal{F}(\nabla B)}{\overline{\mathcal{F}(\nabla L)}\mathcal{F}(\nabla L) + \gamma}\right). \tag{18}$$

Similar to the state-of-the-art methods, we use the coarse-to-fine framework to execute the kernel estimation with an image pyramid.

5   Experiments Results

In this section, we compare our proposed algorithm with the state-of-the-art deblurring methods on synthetic images. The experiments are carried out on a computer with an Intel i5 processor and 8GB RAM. Matlab-R2022a is used to implement algorithms. The decoders we choose are ZXing (3.3.2), ZBar (0.1.9) and WeChat (4.5.2). The QR codes represent a string composed of ten random characters. In all experiments, we set the parameters as follows: $\lambda = 4e^{-3}, \sigma = 1, \gamma = 2, \beta_{max} = 8$, and $\mu_{max} = 1e^5$, respectively. The size of input images is normalized as 300×300 and the patch size for computing LEI value is fixed as 35×35. The threshold parameters for LEP are set to $\lambda_1 = 0.1$ and $\lambda_2 = 0.9$, respectively. To better evaluate the proposed algorithm, we take the restoration quality metrics and recognition rate as the performance evaluation standards.

(a)   (b)   (c)   (d)   (e)   (f)   (g)   (h)

Fig. 3. Visual Comparisons of state-of-the-art methods on four different blurred QR code images. (a) Input (b) Ground truth (c) Pan et al. [10] (d) Pan et al. [11] (e) Yan et al. [18] (f) Wen et al. [16] (g) Chen et al. [1] (h) ours

5.1   Synthetic Dataset

Reconstruction Quality Metrics. To verify the effectiveness of our method, we randomly select four QR code images with different blurring degrees for quantitative evaluation and compare our method to the state-of-art method. We first give a visual comparison in Fig. 3 and put the responding restoration quality metrics in terms of PSNR and SSIM in Fig. 4. Overall, our method performs well against the state-of-the-art methods. All the methods can recover clear results when the blur is mild. But when it comes to the images with higher blur degrees, our method can generate clear results with higher PSNR and SSIM values than the other four methods. Since the proposed LEI regularization term in our method considers more characteristics of the QR code images.

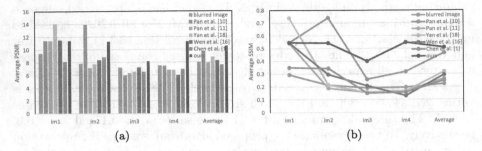

Fig. 4. Quantitative evaluations of different restoration algorithms. (a) Comparisons in terms of PSNR. (b) Comparisons in terms of SSIM

Recognition Rate. In addition, we build up a blurred QR codes image dataset which contains 300 samples of blurred images. From mild to severe, there are totally six different scales of blur. All the blurred images are constructed from the clear ones using the blur function in MATLAB. We further compare the performances of the proposed algorithm against the state-of-art methods [1, 10, 11, 16, 18] by recognition rate. The recognition rate refers to the ratio of the images that can be successfully recognized and decoded by the decoder in the dataset after processing.

Table 1. Comparison of recognition rate.

Algorithms	Recognition Rate		
	ZXing	ZBar	WeChat
None	35.00%	33.00%	43.67%
Pan et al. [10]	92.67%	95.00%	97.33%
Pan et al. [11]	82.67%	84.00%	85.00%
Yan et al. [18]	78.00%	78.67%	78.67%
Wen et al. [16]	82.67%	84.67%	89.67%
Chen et al. [1]	77.67%	78.00%	78.33%
Ours	**97.00%**	**99.67%**	**99.67%**

The responding results are shown in Table 1. The results show that Chen et al. [1] have the least recognition rate at each decoder. Pan et al. [10] perform well at WeChat, with a recognition rate of 97.33%. Our method performs best over all the algorithms at each decoder. The recognition rate of our method can reach up to 99.67% at Zbar and WeChat.

6   Conclusion

In this paper, we propose a blind image deblurring method for QR code images based on the Local Extremum Intensity prior. This prior is motivated by the observation that the distribution of local extremum intensity including local maximum intensity and local minimum intensity will become denser after blurring. And we introduce a simple yet effective threshold method with an efficient optimization scheme to impose the LEI sparsity, which promotes computational efficiency and image restoration quality. Experimental results demonstrate that our method performs favorably against state-of-the-art algorithms. Furthermore, we believe our proposed prior can be extended to other image types such as text images.

Acknowledgments. The work was supported in part by the Special Projects in Key Fields of Ordinary Universities of Guang-dong Province under Grant 2021ZDZX1087, in part by Guangzhou Science and Technology Plan Project under Grant 2023B03J1327 and 202102020857, in part by the Research Projects of Guangdong Polytechnic Normal University under Grant 22GPNUZDJS17 and in part by the Special Project Enterprise Sci-tech Commissioner of Guangdong Province under Grant GDKTP2021033100.

Disclosure of Interests. The authors have no competing interests to declare that are relevant to the content of this article.

References

1. Chen, L., Fang, F., Wang, T., Zhang, G.: Blind image deblurring with local maximum gradient prior. In: Proceedings of the IEEE/CVF Conference on Computer Vision and Pattern Recognition, pp. 1742–1750 (2019)
2. Chen, R., et al.: Rapid detection of multi-qr codes based on multistage stepwise discrimination and a compressed mobilenet. IEEE Internet Things J. (2023)
3. Chen, R., Zheng, Z., Pan, J., Yu, Y., Zhao, H., Ren, J.: Fast blind deblurring of qr code images based on adaptive scale control. Mob. Netw. Appl. **26**(6), 2472–2487 (2021)
4. Chen, R., Zheng, Z., Yu, Y., Zhao, H., Ren, J., Tan, H.Z.: Fast restoration for out-of-focus blurred images of qr code with edge prior information via image sensing. IEEE Sens. J. **21**(16), 18222–18236 (2021)
5. Fu, H., Sun, G., Zhang, A., Shao, B., Ren, J., Jia, X.: Unsupervised 3d tensor subspace decomposition network for hyperspectral and multispectral image spatial-temporal-spectral fusion. IEEE Trans. Geosci. Remote Sens. (2023)
6. Li, Y., et al.: Cbanet: an end-to-end cross band 2-d attention network for hyperspectral change detection in remote sensing. IEEE Trans. Geosci. Remote Sens. (2023)
7. Liu, N., Du, Y., Xu, Y.: QR codes blind deconvolution algorithm based on binary characteristic and l0 norm minimization. Pattern Recognit. Lett. **111**, 117–123 (2018)
8. Liu, N., Zheng, X., Sun, H., Tan, X.: Two-dimensional bar code out-of-focus deblurring via the increment constrained least squares filter. Pattern Recognit. Lett. **34**(2), 124–130 (2013)

9. Ma, P., Ren, J., Sun, G., Zhao, H., Jia, X., Yan, Y., Zabalza, J.: Multiscale superpixelwise prophet model for noise-robust feature extraction in hyperspectral images. IEEE Trans. Geosci. Remote Sens. **61**, 1–12 (2023)
10. Pan, J., Hu, Z., Su, Z., Yang, M.H.: l_0-regularized intensity and gradient prior for deblurring text images and beyond. IEEE Trans. Pattern Anal. Mach. Intell. **39**(2), 342–355 (2016)
11. Pan, J., Sun, D., Pfister, H., Yang, M.H.: Blind image deblurring using dark channel prior. In: Proceedings of the IEEE Conference on Computer Vision and Pattern Recognition, pp. 1628–1636 (2016)
12. Rioux, G., Scarvelis, C., Choksi, R., Hoheisel, T., Marechal, P.: Blind deblurring of barcodes via kullback-leibler divergence. IEEE Trans. Pattern Anal. Mach. Intell. **43**(1), 77–88 (2019)
13. Shan, Q., Jia, J., Agarwala, A.: High-quality motion deblurring from a single image. ACM Trans. Graph. (tog) **27**(3), 1–10 (2008)
14. Sun, G., et al.: Spassa: superpixelwise adaptive ssa for unsupervised spatial-spectral feature extraction in hyperspectral image. IEEE Trans. Cybern. **52**(7), 6158–6169 (2021)
15. Van Gennip, Y., Athavale, P., Gilles, J., Choksi, R.: A regularization approach to blind deblurring and denoising of qr barcodes. IEEE Trans. Image Process. **24**(9), 2864–2873 (2015)
16. Wen, F., Ying, R., Liu, Y., Liu, P., Truong, T.K.: A simple local minimal intensity prior and an improved algorithm for blind image deblurring. IEEE Trans. Circuits Syst. Video Technol. **31**(8), 2923–2937 (2021)
17. Xu, L., Lu, C., Xu, Y., Jia, J.: Image smoothing via l 0 gradient minimization. In: Proceedings of the 2011 SIGGRAPH Asia Conference, pp. 1–12 (2011)
18. Yan, Y., Ren, W., Guo, Y., Wang, R., Cao, X.: Image deblurring via extreme channels prior. In: Proceedings of the IEEE Conference on Computer Vision and Pattern Recognition, pp. 4003–4011 (2017)

Image Enhancement for UAV Visual SLAM Applications: Analysis and Evaluation

Yikun Tian[1], Hong Yue[1(✉)], and Jinchang Ren[2]

[1] Department of Electronic and Electrical Engineering, University of Strathclyde,
Glasgow G1 1XW, UK
{yikun.tian,hong.yue}@strath.ac.uk
[2] National Subsea Centre, Robert Gordon University, Aberdeen AB21 0BH, UK
j.ren@rgu.ac.uk

Abstract. Although simultaneous localisation and mapping (SLAM) has been widely applied in a wide range of robotics and navigation applications, its applicability is severely affected by the quality of the acquired images, especially for those in unmanned aerial vehicles (UAV). In this paper, comprehensive analysis and evaluation of the methods for enhancement of the UAV images are focused, especially the models for denoising of the UAV images using spatial-domain analysis, transform domain analysis and deep learning. Experiments on publicly available datasets are conducted for performance evaluation, along with both qualitative and quantitative results. Surprisingly, deep learning-based approaches did not perform particularly well as these did in other computer vision tasks such as object detection and recognition. Useful discussions are suggested how to further explore this interesting topic.

Keywords: Unmanned Aerial Vehicle (UAV) · visual SLAM · image enhancement · denoising · dehazing

1 Introduction

1.1 A Subsection Sample

With the rapid development of the unmanned aerial vehicle (UAV) techniques, there is a growing trend to apply it to a wide range of applications, including but not limited to survey, surveillance and inspection, supporting various industrial, civil, agriculture, energy, transportation and military tasks. Within these tasks, autonomous visual navigation is a fundamental requirement to enable the automated pilot and survey, for which the simultaneous location and mapping (SLAM) has been widely applied for decades.

There are two major tasks in UAV based visual SLAM, which are constructing a map of the surrounding environment and accurately estimating the motion trajectory of the UAV itself. However, when the UAV system works outdoors, affected by factors such as weather and poor/inconsistent light, the quality of the aerial images on which its visual SLAM environment map construction relies will be seriously degraded. Actually, the quality of the images acquired from UAV platforms may severely affect the accuracy and efficacy of SLAM based navigation tasks. As a result, it will not be possible to accurately estimate the UAV's motion attitude trajectory through aerial images, and it will not be possible to complete the construction of the visual SLAM environment map.

For accurately constructing the surrounding environment map, denoising and enhancement of the image collected by the UAV system becomes essential, before constructing the environment map of visual SLAM. To date, many different models and approaches have been proposed for denoising of aerial images. It is our aim to provide a useful survey and comprehensive evaluation of these models, which will provide a strong base for researchers working in the area to choose the best models accordingly.

2 Image Denoising Models

For denoising of aerial images, numerous researchers have proposed a number of image denoising methods, primarily categorized into traditional image denoising and deep learning-based image denoising methods. Traditional denoising methods can be further subdivided into spatial domain denoising methods and transform domain denoising methods, as detailed below.

2.1 Spatial-Domain Denoising Models

Spatial domain methods primarily utilize filters for denoising. They process the neighbourhood of each pixel in the image using a filter, iterating through the entire image. Spatial domain denoising methods can be classified based on the linearity of the filters into linear filtering methods and non-linear filtering methods [1].

In linear filtering methods, the most common one is the mean filter. For a pixel contaminated by noise, the mean filter calculates the average value of all the pixels in its neighbourhood and assigns this average value to the contaminated pixel. Non-linear filtering methods typically include the median filter and bilateral filter. The median filter initially sorts the pixels around a particular pixel, resulting in an ordered data sequence. Then, it assigns the median value from this sequence to the pixel, effectively removing low and high-frequency components in noisy images. Thus, it is commonly used for eliminating salt-and-pepper noise. However, it has the drawback of potentially causing image discontinuities.

The bilateral filter [2] considers both the grayscale similarity and spatial position relationships between pixels. It assigns higher weight values to pixels that are both close to the center pixel and have similar grayscale values, while giving lower weight values to pixels that are farther away or have dissimilar grayscale values. The advantage of the bilateral filter lies in its ability to preserve more edge information, but it requires further improvement in protecting image texture and detail information.

Local filters can effectively remove noise when the noise level is relatively low but are less effective at higher noise levels. To address this issue, the Non-Local Means (NLM) [3] denoising algorithm leverages the self-similarity and redundancy in the image's structure for denoising. Danbov et al. [4] introduced the Block Matching 3D (BM3D) denoising algorithm, which involves finding a series of similar image blocks and grouping them to obtain multiple three-dimensional blocks. Filtering is then performed in three-dimensional space, followed by using a three-dimensional inverse transform to produce the denoised result. Compared to NLM, this algorithm achieves a higher peak signal-to-noise ratio (PSNR) but comes with higher complexity.

2.2 Transform-Domain Denoising Models

Transform domain methods exploit the distinctive characteristics of images and noise in the transform domain to perform denoising. In the early stages of development for transform domain denoising methods, Fourier transformation was used to remove noise from images. Fourier transformation converts data from the time domain to the frequency domain, where noise in the frequency domain often appears in high-frequency regions. Noise removal can be achieved through low pass filtering in the frequency domain. However, this process also eliminates the texture and detail information in the image.

In addition to Fourier transformation, wavelet transformation has also been employed for image denoising. Denoising methods utilizing wavelet transformation process noise removal based on the differences between image features and noise after undergoing wavelet transformation. The advantage of wavelet-based denoising is that it can simultaneously preserve both frequency and spatial information in the image. However, its drawback lies in its weaker directionality, as it can only extract limited directional information.

2.3 Deep Learning Based Denoising Models

In recent years, deep learning has gained the favor of many researchers due to its powerful feature capturing capabilities and flexible network architectures. Burger et al. employed a Multi-Layer Perceptron (MLP) [5] to learn the mapping from noisy images to clean images, achieving performance comparable to BM3D. Chen et al. [6] designed a trainable Nonlinear Reaction Diffusion (TNRD) denoising model. However, MLP and TNRD can only handle images with fixed noise levels and may not yield ideal results when applied to datasets with varying noise levels.

To enhance the model's ability to handle varying levels of noise, Zhang et al. [7] introduced the Denoising Convolutional Neural Network (DnCNN) model. This model not only addresses images with different noise levels but also utilizes residual learning and batch normalization techniques to expedite model training. Subsequently, Zhang et al. proposed the Fast and Flexible Denoising Network (FFDNet) model [8], which builds upon DnCNN by including noise levels as an additional input to the model. FFDNet is capable of handling spatially correlated noise and plays a crucial role in balancing noise reduction and image detail preservation.

However, these aforementioned models do not yield satisfactory results for real image denoising. To tackle this issue, Guo et al. [9] introduced the Convolutional Blind Denoising Network (CBDNet) and constructed a new noise model to simulate real noise. CBDNet consists of a denoising sub-network and a noise level estimation sub-network, enhancing the network's performance and generalization capacity by introducing an asymmetric loss function. Nevertheless, CBDNet's network structure is complex and comes with a significant computational cost, making it less suitable for practical applications. Therefore, Anwar and Barnes [10] proposed the Real Image Denoising Network (RIDNet) to address real-world denoising scenarios. RIDNet adopts a modular structure for the denoising network and introduces a channel attention mechanism for adaptive channel weight adjustment. The introduction of CBDNet and RIDNet has driven the development of image denoising research in real-world settings.

3 Datasets and Evaluation Criteria

3.1 Datasets Description

For image denoising, the CBSD68 [11] and the SIDD [12] datasets were used. The CBSD68 dataset consists of 68 colour images of varying sizes. The SIDD dataset includes approximately 30,000 noisy images captured under different lighting conditions using five representative smartphones, along with corresponding "noise-free" ground truth images. In addition, an own dataset including 2035 virtual simulation scene images was also used for testing the effect of denoising.

For image dehazing assessment, the SOTS-outdoor [13] public dataset is used. The SOTS dataset is a synthetic dataset consisting of 1000 test images, divided into indoor and outdoor categories, each containing 500 images.

3.2 Evaluation Metrics

3.2.1 Peak Signal-to-Noise Ratio (PSNR)

Given a hazy image I and a haze-free image K both of size M*N, the Mean Squared Error (MSE) is defined as:

$$MSE = \frac{1}{MN} \sum_{j=0}^{M-1} \sum_{j=0}^{N-1} \left[I(i,j) - K(i,j) \right]^2 \tag{1}$$

where $I(i,j)$ and $K(i,j)$ represent the grayscale values of pixels at location (i,j) in the hazy and haze-free images, respectively. The Peak Signal-to-Noise Ratio (PSNR) is then defined as

$$PSNR = 10 log_{10} \left(\frac{(maxI)^2}{MSE} \right) \tag{2}$$

where MAX represents the maximum possible pixel value in the image. This formula is commonly used for grayscale images. For colour images with three channels (RGB), the MSE is calculated separately for each channel, and the resulting MSE values are used to compute the PSNR for each channel. The final PSNR for the color image is obtained by taking the average of the PSNR values across all channels.

3.2.2 Structural Similarity (SSIM)

SSIM is used to measure the structural similarity between two images, which compares the structure, luminance, and contrast of two images. For the two given images x and y, the structural similarity between the two images can be computed as follows [14].

$$SSIM(x, y) = \frac{(2\mu_x\mu_y + c_1)(2\sigma_{xy} + c_2)}{(\mu_x^2 + \mu_y^2 + c_1)(\sigma_x^2 + \sigma_y^2 + c_2)} \tag{3}$$

where μ_x represents the average value (mean) of x, μ_y is the average value (mean) of y, σ_x^2 is the variance of x, σ_y^2 is the variance of y. The constants $c_1 = (k_1L)^2$ and $c_2 = (k_2L)^2$ are used to maintain stability and are typically set to small values like 0.01 and 0.03, respectively. The range of the Structural Similarity (SSIM) metric is from -1 to 1, with larger values indicating less distortion. When two images are identical, the SSIM value is 1.

4 Results and Analysis

4.1 Compared Methods

Based on the recommended good performance, the following models are selected for evaluation in our experiments.

- **BM3D** [4], as a novel image denoising method, BM3D is based on an enhanced sparse representation in the transform domain. The enhanced sparsity is achieved by grouping similar 2-D fragments into 3-D data arrays namely "groups", followed by collaborative filtering being applied to these 3-D groups. The collaborative filtering has helped to reveal the finest details shared by grouped fragments whilst preserving the essential unique features of each individual fragment.
- **Weighted Nuclear Norm Minimization** (WNNM) [15]: The image is modeled as Y = X - N, where Y is also composed of samples with noise, forming a sample matrix. X and N are the corresponding noise-free sample matrices and noise, respectively. The given constraint is that X is a low-rank matrix. Because the matrix composed of similar samples exhibits low-rank characteristics, while the noise does not have low-rank characteristics, image denoising can be achieved through low-rank clustering.
- **Variational Denoising Network** (VDN) [16]: This model is capable of simultaneous image denoising and noise estimation. In typical work, Gaussian white noise is assumed to be present in the image, but this model is not limited to that. The proposed generative model exhibits strong generalization capabilities and performs well even for noise not encountered in the test set. The model provides an explanation for the overfitting phenomenon often observed in deep learning methods trained using MSE loss. This issue is attributed to overfitting the prior of the underlying clean image while neglecting variations in noise. This model explicitly models the generation of noise, thereby avoiding this drawback of deep learning methods.

- **FFDNet** [8]: The adjustable noise level mapping, denoted as M, is used as input to provide flexibility to the denoising model regarding noise levels. An invertible downsampling operator is introduced to reshape the input image of size $W \times H \times C$ into four subsampled sub-images of size $4W/2 \times H/2 \times 4C$, where C represents the number of channels. To ensure that noise level mapping robustly controls the trade-off between denoising and detail preservation without introducing visual artifacts, an orthogonal initialization method is applied to the convolution filters.
- **NAFNet** [17]: Taking inspiration from the Transformer architecture, the use of Layer Normalization (LN) is incorporated to facilitate smoother training. NAFNet also introduces LN operations, leading to significant performance gains on image denoising and deblurring datasets. In the Baseline approach, ReLU is jointly replaced with GELU and CA. GELU helps maintain denoising performance while significantly enhancing deblurring performance. Two new attention module compositions are proposed, namely CA (Channel Attention) and SCA (Spatial-Channel Attention).
- **CycleISP** [18]: The images captured by the camera initially exist as RAW-RGB images, where each pixel contains only one of the three-color channels: R, G, or B. These RAW images are then processed through the camera's ISP (Image Signal Processing), which includes operations like noise reduction, white balance adjustment, gamma transformation, tone mapping, and more, resulting in sRGB images (standard-RGB with three channels). Two neural networks have been employed to simulate this process in both forward and reverse directions. In other words, these networks can transform sRGB images into RAW images and vice versa.

4.2 Results and Analysis

Table 1 below presents a comparison of image denoising experiment results using different algorithms on publicly available datasets. The BM3D [4], WNNM [15], VDN [16], FFDNet [8], and NAFNet [17] were tested on the CBSD68 dataset, while the CycleISP [18] method was applied to the SIDD [12] dataset. The PSNR and SSIM values in the table above represent the peak performance achieved by the respective algorithms on this dataset. From the results, it can be observed that the BM3D and NAFNet algorithms perform well in denoising based on publicly available datasets.

As seen, BM3D apparently outperforms WNNM, thanks for considering the local self-similarity, which has greatly improved the performance of denoising, including those using deep learning models. The relatively poor performance from deep learning can be due to two reasons, i.e. insufficient training and inconsistency between the training and testing samples. The latter may be caused by the random characteristics of the noise within the image, which has potentially affected the learning-based approaches. Nevertheless, in the best deep learning model, NAFNet, the transformer architecture has somehow mitigated such limitations, which can be further explored.

Table 1. Comparison of Image Denoising Methods Using Public Datasets.

Methods	Results	Original Image	Resulted Image
BM3D	PSNR: 41.63dB SSIM: 0.9936		
WNNM	PSNR: 39.38dB SSIM: 0.9750		
VDN	PSNR: 30.83dB SSIM: 0.8533		
FFDNet	PSNR: 28.98dB SSIM: 0.7969		
NAFNet	PSNR: 40.30dB SSIM: 0.9621		
CycleISP	PSNR: 34.70dB SSIM: 0.9822		

5 Conclusions

In this paper, a survey of the denoising models for UAV images in SLAM implementation is focused, followed by a comprehensive evaluation. Six models are selected for both qualitative and quantitative assessment, including conventional approaches in the spatial domain and transform domain as well as deep learning models. By benchmarking on the publicly available datasets, it is found that the BM3D model outperforms all others, even the deep learning approaches, owing mainly to the local-similarity being used in modelling.

This one hand shows the great potential of conventional vision - based perception models in image denoising. On the other hand, it indicates the potential limitations of the deep learning models in this context, due mainly to the ill-posed problem in training the models. Furthermore, the great potential of the NAFNet has suggested that the transformer architecture can help to mitigate the limitations here and improve the modeling thus is worth further investigation.

As the degradation process of UAV images can be much more complicated [19], this paper only covers a small portion, where many other useful topics have not been covered, such as image dehazing, deblurring and normalisation of the lighting effects et al. These will be our future work, along with the integration of other challenging models.

References

1. Cai, R.: Research progress in image denoising algorithms based on deep learning. J. Phys. **1345**, 042055 (2019)
2. Banterle, F., Corsini, M., Cignoni, P., Scopigno, R.: A low-memory, straightforward and fast bilateral filter through subsampling in spatial domain. Comput. Graph. Forum **31**(1), 19–32 (2011)
3. Buades, A.: A non-local algorithm for image denoising. In: Proceedings of the IEEE Conference on Computer Vision and Pattern Recognition (CVPR'05), pp. 60–65 (2005)
4. Dabov, K., Foi, A., Katkovnik, V., Egiazarian, K.: Image denoising by sparse 3-D transform-domain collaborative filtering. IEEE Trans. Image Process. **16**(8), 2080–2095 (2007)
5. Burger, H.C., Schuler, C.J., Harmeling, S.: Image denoising with multi-layer perceptrons, part 1: comparison with existing algorithms and with bounds. Comput. Sci. 8–30 (2012)
6. Chen, W., Huang, Z., Tsai, C., et al.: Learning multiple adverse weather removal via two-stage knowledge learning and multi-contrastive regularization: toward a unified model. In: Proceedings of the IEEE/CVF Conference on Computer Vision and Pattern Recognition, pp. 17653–17662 (2022)
7. Zhang, K., Zuo, W., Chen, Y., et al.: Beyond a Gaussian denoiser: residual learning of deep CNN for image denoising. IEEE Trans. Image Process. **26**(7), 3142–3155 (2017)
8. Zhang, K., Zuo, W., Zhang, L.: FFDNet: toward a fast and flexible solution for CNN-based image denoising. IEEE Trans. Image Process. **27**(9), 4608–4622 (2018)
9. Guo, S., Yan, Z., et al.: Toward convolutional blind denoising of real photographs. In: Proceedings of the CVPR, pp. 1712–1722 (2019)
10. Anwar, S., Barnes, N.: Real image denoising with feature attention. In: Proceedings of the ICCV, IEEE, pp. 3155–3164 (2019)
11. Martin, D., Fowlkes, C., et al.: A database of human segmented natural images and its application to evaluating segmentation algorithms and measuring ecological statistics. In: Proceedings of the 8th IEEE International Conference on Computer Vision, pp. 416–423 (2001)
12. Abdelhamed, A., Lin, S., Brown, M.S.: A high-quality denoising dataset for smartphone cameras. In: Proceedings of the IEEE Conference on Computer Vision and Pattern Recognition, pp. 1692–1700 (2018)

13. Hu, X., Fu, C., Zhu, L., et al.: Direction-aware spatial context features for shadow detection and removal. IEEE Trans. Pattern Anal. Mach. Intell. **42**(11), 2795–2808 (2019)
14. Wang, Z., Bovik, A.C., Sheikh, H.R., Simoncelli, E.P.: Image quality assessment: from error visibility to structural similarity. IEEE Trans. Image Process. **13**(4), 600–612 (2004)
15. Gu, S., Zhang, L., et al.: Weighted nuclear norm minimization with application to image denoising. In: Proceedings of the IEEE Conf. on Computer Vision and Pattern Recognition, pp. 2862–2869 (2014)
16. Yue, Z., Yong, J., Zhao, Q., Meng, D., Zhang, L.: Variational denoising network: toward blind noise modeling and removal. Adv. Neural Inf. Process. Syst. **32** (2019)
17. Chen, L., Chu, X., Zhang, X., Sun, J.: Simple baselines for image restoration. In: Avidan, S., Brostow, G., Cissé, M., Farinella, G.M., Hassner, T. (eds.) Computer Vision – ECCV 2022. ECCV 2022. LNCS, vol. 13667, pp. 17–33. Springer, Cham (2022). https://doi.org/10.1007/978-3-031-20071-7_2
18. Zamir, S.W., et al. :CycleISP: real image restoration via improved data synthesis. In: IEEE/CVF Conference on Computer Vision and Pattern Recognition, pp. 2693–2702 (2020)
19. Fu, H., et al.: Three-dimensional singular spectrum analysis for precise land cover classification from UAV-borne hyperspectral benchmark datasets. ISPRS J. Photogram. Remote Sens. **203**, 115–134 (2023)

Segmentation Framework for Heat Loss Identification in Thermal Images: Empowering Scottish Retrofitting and Thermographic Survey Companies

Md Junayed Hasan[1] , Eyad Elyan[2] , Yijun Yan[1] , Jinchang Ren[1(✉)] , and Md Mostafa Kamal Sarker[3]

[1] National Subsea Centre, Robert Gordon University, Aberdeen AB21 0BH, UK
j.ren@rgu.ac.uk
[2] School of Computing, Robert Gordon University, Aberdeen AB10 7AQ, UK
[3] Institute of Biomedical Engineering, University of Oxford, Oxford OX3 7DQ, UK

Abstract. Retrofitting and thermographic survey (TS) companies in Scotland collaborate with social housing providers to tackle fuel poverty. They employ ground-level infrared (IR) camera-based-TSs (GIRTSs) for collecting thermal images to identify the heat loss sources resulting from poor insulation. However, this identification process is labor-intensive and time-consuming, necessitating extensive data processing. To automate this, an AI-driven approach is necessary. Therefore, this study proposes a deep learning (DL)-based segmentation framework using the Mask Region Proposal Convolutional Neural Network (Mask RCNN) to validate its applicability to these thermal images. The objective of the framework is to automatically identify, and crop heat loss sources caused by weak insulation, while also eliminating obstructive objects present in those images. By doing so, it minimizes labor-intensive tasks and provides an automated, consistent, and reliable solution. To validate the proposed framework, approximately 2500 thermal images were collected in collaboration with industrial TS partner. Then, 1800 representative images were carefully selected with the assistance of experts and annotated to highlight the target objects (TO) to form the final dataset. Subsequently, a transfer learning strategy was employed to train the dataset, progressively augmenting the training data volume and fine-tuning the pre-trained baseline Mask RCNN. As a result, the final fine-tuned model achieved a mean average precision (mAP) score of 77.2% for segmenting the TO, demonstrating the significant potential of proposed framework in accurately quantifying energy loss in Scottish homes.

Keywords: Infrared thermographic testing · instance segmentation · Mask RCNN · thermal images · transfer learning

1 Introduction

Fuel poverty is predicted to affect about 39% of Scottish households after a significant increase in energy prices was announced in April 2023. Therefore, the Scottish government has made it a top priority to reduce carbon emissions from homes and businesses to

J. Ren et al. (Eds.): BICS 2023, LNAI 14374, pp. 220–228, 2024.
https://doi.org/10.1007/978-981-97-1417-9_21

mitigate the fuel poverty. The Scottish Fuel Poverty Act (SFPA) aims to increase energy efficiency and reduce carbon footprint in all infrastructures by 2040 [1]. Additionally, it is imperative to prioritize the energy efficiency of buildings, as it aids in securing funding for emerging technologies like heat pumps, insulation, and retrofitting. Specialized companies that focus on retrofitting and conducting thermographic surveys (TSs) collaborate with social housing providers to ensure compliance with the Energy Efficiency Standard for Social Housing (EESSH) policy [2]. They detect buildings that require retrofitting by using thermal images gathered by ground-level infrared (IR) camera-based-TSs (GIRTSs) [3]. However, the current challenge is the need to manually analyze those images after collection to identify sources of heat loss and eliminate obstructive objects. This process is labor-intensive, time-consuming, and relies heavily on domain experts. As a result, this hinders scalability and limits the utilization of cloud-based statistical thermal profile analyzer toolkits. Therefore, this research aims to develop a deep learning (DL) based-automated solutions to identify the target objects (TO) related to the actual heat loss in GIRTS-thermal images, streamlining data analysis and reducing the mentioned labor-intensive tasks.

Researchers have addressed similar problem types by utilizing vehicle and drone-mounted IR cameras to capture thermal images of building facades and applying various machine learning (ML) algorithms to detect thermal bridges. Macher et al. [4] used a vehicle-mounted camera to create a thermographic 3D point cloud and successfully detect thermal bridges under balconies and between levels. However, this method has limitations such as difficulty in extracting ground-level and basement windows and the inability to detect windows hidden by foliage or other objects. Using drones for thermographic assessments has become more popular as they allow for the entire outside of a building to be captured and there is less interference caused by obstructions. Rakha et al. [5] proposed a thermal drone-based system to locate thermal anomalies in building envelopes and claimed 75% precision, while Mirzabeigi et al. [6] developed a computer vision algorithm and drone flight path to detect thermal anomalies but lack quantitative data on the effectiveness. Kim et al. [7]used neural networks to identify thermal bridges in terrestrial thermographic images with an average precision of 89% and recall of 87%. This study examines the difficulties encountered when using thresholding and histogram methods (as discussed above) to identify TO in non-stationary thermography research in panorama settings. However, these studies face two main challenges. Firstly, the use of aerial thermographic survey (ATS) and vehicle-based moving thermographic survey (VMTS) makes it difficult to capture suitable vantage points for identifying thermal bridges and sources of heat loss. These approaches cover multiple infrastructures and encounter varying conditions, such as weather, lighting, and pose fluctuations, making it challenging to accurately identify the sources of heat loss amid obstructive objects. Secondly, to address the complexity resulting from the first challenge, existing research has employed numerous manual thresholding techniques to remove obstructive objects from ATS/VMTS-thermal images and identify sources of heat loss. However, due to the varying conditions present in these thermal images, the accuracy of thresholding is not optimal. Consequently, ML solutions developed to identify the sources of heat loss struggle to perform effectively and robustly.

To tackle the challenges associated with vantage points in ATS/VMTS-based-thermal image analysis, we conducted the GIRTS in partnership with an industrial TS partner. As a result, we obtained a dataset of 2500 thermal images, from which we carefully selected 1800 images with the help of domain experts and annotated the target objects (TO) to create a customized dataset called GIRTSD. Then, to automate the identification, detection, and removal of potential sources of heat loss as well as obstructive objects, we employed the Mask Region Proposal Convolutional Neural Network (Mask RCNN [8]), a deep learning (DL)-based segmentation algorithm, eliminating the need for manual thresholding and enhancing accuracy. However, since the GIRTSD might not be sufficient for training due to variations in object shapes, forms, and weather conditions, we employed transfer learning strategies. Specifically, we fine-tuned a pre-trained Mask RCNN model from the Microsoft Common Object in Context (MSCOCO) dataset [9], which includes 80 object categories. After preparing the fine-tuned model using a limited set of samples from our GIRTSD, we progressively expanded the training data, integrated diverse image augmentation techniques, and employed transfer learning strategies to improve the model's performance [10, 11]. Thus, after conducting extensive ablation studies, we selected the optimal fine-tuned model, which achieved an impressive mean average precision (mAP) of 77.2% for segmenting the TO. The main contributions of this study are highlighted below.

1. We introduced the ground-level infrared camera-based thermographic survey (GIRTS) for collecting thermal images, resulting in a custom dataset named GIRTSD. This dataset comprises 1800 annotated thermal images representing 7 distinct target objects (TO), including possible sources of heat loss and obstructive objects. Unlike aerial/vehicle-based moving thermographic survey, GIRTS accurately captures various TO regardless of object shapes, forms, and weather conditions.
2. To overcome the limitations of existing research in thermal image-based heat loss source detection, specifically manual thresholding, we propose a Mask RCNN-based framework. This framework detects and crops potential heat loss sources while eliminating obstructive objects, resulting in reduced manual labor and improved efficiency in identifying energy losses in buildings.

The rest of the paper is organized as follows: Sect. 2 refers to the experimental set up and methodology, Sect. 3 discusses the experimental finding, and finally, Sect. 4 concludes the paper.

2   Experimental Setup and Framework Strategy

Figure 1 depicts the complete framework for the validation of our case study. The process begins with data collection by GIRTSs, followed by dataset creation with the help of our industrial TS partner. The GIRTSD creation is the initial step, followed by offline training-testing to refine the proposed Mask RCNN based segmentation framework. Once satisfactory performance is achieved, the optimal model is selected for online evaluation.

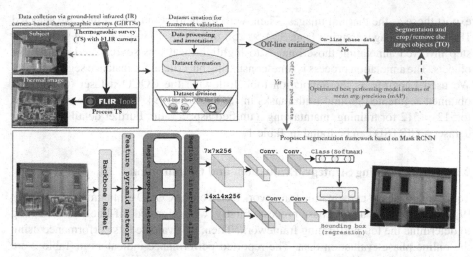

Fig. 1. Proposed segmentation framework for heat loss identification in thermal images.

Table 1. Details of our custom thermal image dataset, GIRTSD

Total 1800 thermal images, with 7 target objects (TO)				
		Off-line phase		On-line phase
		Train	Test	Evaluation set
	TO	Number of instances/TO		
Heat loss sources	Window	807	89	99
	Door	278	31	34
Obstructive objects	Fence	61	7	8
	Tree	85	9	10
	Bin	116	13	14
	Road	111	12	13
	Other	851	94	105

2.1 Data Collection and Dataset Creation

TSs are non-invasive methods used to identify insulation issues and heat loss in buildings [12]. These surveys utilize thermal images to detect air leakage, moisture infiltration, and structural defects, improving energy efficiency and reducing environmental impact. Trained professionals with expertise in both technology and building science are required for accurate surveys. In our study, we collaborated with an industrial TS partner in Scotland and collected thermal images using GIRTSs and FLIR E60bx cameras [13]. The high-resolution infrared detectors and multi-spectral dynamic imaging feature of these cameras provide detailed thermal images with a temperature range of $-20\,°C$ to $+350\,°C$. FLIR Tools software is used to process, and analyse these surveys, and export those as the thermal images. Then, with the help of our TS partner, we selected 1800 diverse images ranging from 320 × 256 to 1920 × 1536 pixels. The subsequent step involved annotating those images to highlight heat loss sources and obstructive objects. The annotation process is time-consuming but crucial for instance segmentation. We used LabelMe as the annotation tool, following the COCO dataset format, and obtained segmented ground truth masks in JSON format [9]. Each image was resized to 512 × 512 for training, maintaining a unified aspect ratio. Further details about our dataset – GIRTSD can be found in Table 1.

2.2 Model Training Strategy, Evaluation, and Optimization

To identify the best segmentation framework based on the COCO evaluation metric (mAP [9]), we conducted a two-phase process. Firstly, we performed off-line phase training to determine the top-performing framework. Then, we evaluated its performance using the online phase-evaluation data. The reported performance indicators in Table 2 are calculated during the off-line phase training and test data.

Given the limited number of samples, constructing a generalized DL model posed challenges. As a solution, we adopted the Mask RCNN model, denoted as Mb, which had been pretrained on the 80-class MS-COCO dataset, as our framework's baseline. While preserving the initial layers that capture low-level features, we retrained the remaining components responsible for high-level features, effectively producing a modified version of Mb. To customize Mb, we leveraged TensorFlow, Keras, OpenCV, and Python 3, utilizing the open-source MMDetection toolbox [14].

Table 2. Ablation study during off-line phase.

Models	Backbone - base model	Data Aug	Training data vol	mAP$^{50-95}$ Train	mAP$^{50-95}$ Test
M1/M2	R50-Mb	No/Yes	20%	89.2/91.1	59.2/60.1
M3/M4	R101-Mb	No/Yes	20%	92.2/94.1	60.9/61.4
M5/M6	R101-M4	No/Yes	40%	94.5/95.1	64.3/65.9
M7/M8	R101-M4	No/Yes	60%	95.7/96.3	67.9/69.2
M9/M10	R101-M4	No/Yes	80%	96.8/97.1	70.7/72.9
M11/**M12**	R101-M4	No/**Yes**	**100%**	97.5/**98.2**	75.8/**78.7**

For the classification tasks within the segmentation pipeline, we employed ResNet-50 (R50) and ResNet-101(R101) as backbone models, using stochastic gradient descent (SGD) [15] and Adam optimizers [16]. This resulted in the generation of four models (M1 - M4) to explore the retraining of Mb on our custom dataset while considering the inclusion or exclusion of image augmentation techniques. The objective was to determine the optimized framework [17]. During this phase, we only utilized 20% of the training data from the offline phase. Once the best-performing model was identified, it was subjected to additional ablation studies and fine-tuning.

In our designed ablation experiments, we aimed to demonstrate the impact of increasing training data on the performance of our model [18]. We initially used only 20% of the data from Table 1 for training, gradually increasing the training data volume by 20% in subsequent experiments. By observing the performance of our model on the off-line phase training and test data, we assessed whether increased data volume resulted in improved model performance.

We evaluated different models (M5- M12) by considering both with and without image augmentation techniques, using *mAP*. The mask head training region of interest (RoI) was set to 200, while the maximum number of ground truth instances per image was raised from 100 to 512. Furthermore, the incremental training RoI parameter was set to 512. Detailed information regarding our various ablation studies can be found in Table 2.

3   Result Analysis and Discussion

Table 2 presents the mAP^{50-95} results for both off-line phase training and testing. The mAP^{50-95} is calculated by averaging accuracy over the intersection over union (IoU) range from 0.5 to 0.95, with a step size of 0.05. Among the models M1-M4, it was observed that M4, with the R101-Mb combination as the backbone-base model along with data augmentation, performed the best. Subsequently, four pairs of models were gradually built from M4, increasing the training data volume to observe performance improvements during the offline training phase. In all experiments, data augmentation consistently led to better performance. Additionally, as the training data volume was incrementally increased by 20%, a gradual improvement in performance was observed. These findings clearly highlight the significance of data diversity and the quantity of training samples in building a generalized and robust model. It is worth noting that as the trained model improved, the performance on the offline test data also improved.

In Table 1, it can be observed that the number of instances/TO varies significantly, which impacts the overall model performance. To validate this observation, a closer examination was conducted on the best performing model from Table 2 (M12), specifically analyzing its off-line phase test mAP^{50-95}/TO. Detailed analysis can be found in Table 3, and Fig. 2.

Based on Table 3, it is evident that windows and doors achieved the best results in terms of bounding box detection (BBOX) and segmentation (SEG). These two object categories have a relatively higher number of training instances. On the other hand, the category labeled as 'other', which includes various miscellaneous objects like benches, chairs, flower gardens, and staircases, had the highest number of instances but did not perform well. This can be attributed to the significant variation in pose and shape among these objects, making it challenging for the model to achieve consistent performance. Therefore, along with a larger number of instances, maintaining a unified object shape and category is crucial. For instance, the 'bin' category, despite having only 116 instances, performed impressively due to the relatively consistent shape of the objects within that category.

Having reached this stage, we have chosen the best-performing model, M12, from our offline phase. To assess the overall performance of our framework, we utilized

Table 3. Off-line phase test performance of M12/TO.

mAP$^{50-95}$	BBOX	SEGM
Window	78.9	80.3
Door	74.8	77.5
Fence	59.1	62.3
Tree	57.2	47.3
Bin	71.2	72.4
Road	59.0	60.8
Other	66.7	65.3

Fig. 2. Snapshot of segmentation results.

M12 for TO classification and segmentation first. Then, the unwanted objects will be removed based on the segmentation results. During this online phase with the test data, M12 achieved a *mAP$^{50-95}$* of 77.2%, which is remarkably close to our performance in the offline phase. Alongside the development of our prototype, we have designed a user-friendly web interface for end-users to evaluate the functionality. This interface allows users to upload GIRTS-thermal images and visually inspect the detected/segmented regions for possible heat loss sources or obstructive objects. It also displays the cropped and cleaned versions of the images, enabling users to save the thermal information from the processed images. Figure 3 illustrates the activities and workflow of our proposed solution within the web user interface.

Fig. 3. Snapshot of our developed web-interface.

4 Conclusions

In this research, we propose an AI-driven solution for ground level infrared camera-based thermographic surveys using thermal images. We implement a Mask RCNN-based deep learning framework to identify, segment, and remove heat sources and unwanted objects. As a result, we created a new thermographic dataset. To overcome the issue of limited thermal images for training, we employed transfer learning strategies and various image augmentation techniques. Through fine-tuning experiments and parameter adjustments, our final model achieved a mAP score of 77.2%. In future research, we aim to improve the training dataset by utilizing generative deep architectures [19]. We also plan to address the challenge of dense clusters of distracting objects that can potentially lead to the model overlooking certain segments [20]. Furthermore, we aim to demonstrate the practical implementation of this solution by developing an open-source tool specifically designed for industries involved in thermographic surveys. Additionally, we have plans to publicly release the GIRTS dataset, enabling further exploration of the potential outcomes achieved through this dataset.

Acknowledgement. We are thankful to IRT surveys for the thermograph survey support and Data Lab for the funding of this project.

References

1. Scotland Government: Home energy and fuel poverty policy of Scotland Government. https://www.gov.scot/policies/home-energy-and-fuel-poverty/fuel-poverty/. Accessed 16 June 2023
2. The Scottish Government: The Energy Efficiency Standard for Social Housing post 2020 (EESSH2). https://www.gov.scot/policies/home-energy-and-fuel-poverty/. Accessed 22 Jan 2023
3. Yan, Y., et al.: Cognitive fusion of thermal and visible imagery for effective detection and tracking of pedestrians in videos. Cogn. Comput. **10**, 94–104 (2018)
4. Macher, H., Landes, T., Grussenmeyer, P.: Automation of thermal point clouds analysis for the extraction of windows and thermal bridges of building facades. Int. Arch. Photogramm. Remote. Sens. Spat. Inf. Sci. **43**, 287–292 (2020)
5. Rakha, T., Liberty, A., Gorodetsky, A., Kakillioglu, B., Velipasalar, S.: Heat mapping drones: an autonomous computer-vision-based procedure for building envelope inspection using unmanned aerial systems (UAS). Technol. Archit. Des. **2**, 30–44 (2018)
6. Mirzabeigi, S., Razkenari, M.: Automated vision-based building inspection using drone thermography. In: Construction Research Congress 2022, pp. 737–746 (2022)
7. Kim, C., Choi, J.-S., Jang, H., Kim, E.-J.: Automatic detection of linear thermal bridges from infrared thermal images using neural network. Appl. Sci. **11**, 931 (2021)
8. He, K., Gkioxari, G., Dollár, P., Girshick, R.: Mask r-cnn. In: Proceedings of the IEEE International Conference on Computer Vision, pp. 2961–2969 (2017)
9. Lin, T.-Y., et al.: Microsoft coco: common objects in context. In: Fleet, D., Pajdla, T., Schiele, B., Tuytelaars, T. (eds.) Computer Vision – ECCV 2014. ECCV 2014. LNCS, vol. 8693, pp. 740–755. Springer, Cham (2014). https://doi.org/10.1007/978-3-319-10602-1_48
10. Shorten, C., Khoshgoftaar, T.M.: A survey on image data augmentation for deep learning. J. Big Data **6**, 1–48 (2019)

11. Zhao, Z.-Q., Zheng, P., Xu, S., Wu, X.: Object detection with deep learning: a review. IEEE Trans. Neural Netw. Learn. Syst. **30**, 3212–3232 (2019)
12. U.S Department of Energy: Thermographic Inspections
13. TELEDYNE FLIR: TELEDYNE FLIR E60bx. https://www.flir.co.uk/support/products/e60bx. Accessed 22 Jan 2023
14. OpenMMLab: Openmm Detection Mask-R-CNN. https://openmmlab.com/. Accessed 22 Jan 2023
15. Bottou, L.: Stochastic gradient descent tricks. In: Montavon, G., Orr, G.B., Müller, K.R. (eds.) Neural Networks: Tricks of the Trade. LNCS, vol. 7700, pp. 421–436. Springer, Berlin, Heidelberg (2012). https://doi.org/10.1007/978-3-642-35289-8_25
16. Kingma, D.P., Ba, J.: Adam: a method for stochastic optimization (2014). arXiv preprint arXiv:1412.6980
17. Huang, J., et al.: Speed/accuracy trade-offs for modern convolutional object detectors. In: Proceedings of the IEEE Conference on Computer Vision and Pattern Recognition, pp. 7310–7311 (2017)
18. Ma, P., et al.: Multiscale superpixelwise prophet model for noise-robust feature extraction in hyperspectral images. IEEE Trans. Geosci. Remote Sens. **61**, 1–12 (2023)
19. Li, Y., et al.: Cbanet: an end-to-end cross band 2-d attention network for hyperspectral change detection in remote sensing. IEEE Trans. Geosci. Remote Sens. **61** (2023)
20. Sun, M., Li, P., Ren, J., Wang, Z.: Attention mechanism enhanced multi-layer edge perception network for deep semantic medical segmentation. Cogn. Comput.Comput. **15**(1), 348–358 (2023)

Mixed-Precision Collaborative Quantization for Fast Object Tracking

Yefan Xie[1,2], Yanwei Guo[1,2], Xuan Hou[1,2], and Jiangbin Zheng[1,2(✉)]

[1] National Engineering Laboratory for Integrated Aero-Space-Ground-Ocean Big Data Application, School of Computer Science, Northwestern Polytechnical University, Xi'an 710129, Shaanxi, China
zhengjb@nwpu.edu.cn
[2] Shaanxi Provincial Key Laboratory of Speech Image Information Processing, Xian, China

Abstract. To address the non-differentiability of quantizers and inaccurate gradient propagation in training low-bit quantized tracking models, we propose a mixed-precision collaborative quantization method for fast object tracking that combines a full-precision auxiliary module and low-bit quantization blocks through parameter sharing. Specifically, our approach constructs a partial full-precision auxiliary module that receives multiple intermediate outputs from the low-bit module, allowing the parameters of the low-bit model to combine gradient information from itself and the auxiliary module via gradient averaging. Additionally, the multi-branch feature enhancement block is utilized to extract different features from different branches, enabling diverse feature representations in the low-bit quantization tracking network. Extensive experiments are conducted to validate the effectiveness of the proposed mixed-precision collaborative quantization approach compared to existing quantization methods, demonstrating the superior performance of our quantization framework on object tracking networks.

Keywords: Visual Object Tracking · Model Quantization · Mixed-Precision Quantization

1 Introduction

Object tracking is a fundamental task in the field of computer vision, widely applied in various domains including autonomous driving and surveillance. Its primary objective is to accurately locate and track objects of interest within video frames. However, when performing object tracking in dynamic background conditions, a significant challenge arises: the computational speed is directly influenced by the hardware performance of the deployed models. It is observed that tracking models with higher hardware capabilities exhibit faster inference speeds, while their deployment feasibility on resource-constrained platforms becomes weaker.

© The Author(s), under exclusive license to Springer Nature Singapore Pte Ltd. 2024
J. Ren et al. (Eds.): BICS 2023, LNAI 14374, pp. 229–238, 2024.
https://doi.org/10.1007/978-981-97-1417-9_22

Therefore, there is a pressing need to research a low-bit quantized tracking network within the context of object tracking. Such a network not only enables the deployment of sophisticated networks on low-computing-power devices but also reduces network computation, leading to lower energy consumption. This research endeavor holds significant industrial demand and research value. By addressing these challenges, advancements in low-bit quantized tracking networks can revolutionize object tracking applications and contribute to the development of more efficient and accessible computer vision systems.

Aiming to address the non-differentiability of quantizers and inaccurate gradient propagation in training low-bit quantized tracking models, we propose a mixed-precision collaborative quantization for fast object tracking that combines a full-precision auxiliary module and low-bit quantization blocks through parameter sharing. Specifically, our approach constructs a partial full-precision auxiliary module that receives multiple intermediate outputs from the low-bit module. During the training of the tracking model, a joint optimization of the low-bit network and the mixed-precision model is performed, allowing the parameters of the low-bit model to incorporate gradient information from both itself and the hybrid model through gradient averaging, thereby addressing the challenge of gradient propagation in low-bit quantization. By combining it with the full-precision auxiliary module, a novel learning strategy for training low-bit object tracking networks is established. Furthermore, a feature enhancement module utilizing asymmetric convolutions is introduced to extract diverse features and enhance the expressive power of the network.

The main contributions of our work are summarized as follows:

1) A collaborative training method involving full-precision modules is proposed to address the non-differentiability of quantized networks. It achieves smaller auxiliary space utilization during the training phase without requiring additional hyper-parameters.
2) A multi-branch feature enhancement block is designed specifically for object tracking networks, which can narrow the quantization gap of quantized networks when combined with the collaborative training method.
3) Extensive experiments are conducted to validate the effectiveness of the proposed mixed-precision collaborative quantization approach compared to existing quantization methods, demonstrating superior performance on object tracking networks.

2   Related Work

2.1   Model Quantization

The quantization scheme can be classified into two categories: post-training quantization (PTQ) and quantization-aware training (QAT). Post-training quantization (PTQ) methods are mainly used in scenarios where there is no training data but only a model, and their work focuses on the optimization of the truncation value [1,12]. Quantization-aware training (QAT) methods have been proposed to address the limitations of post-training quantization, improving the performance of quantization networks. DoReFa [17] applies the tanh function to restrict parameters within the range of [0, 1]. PACT [4] introduces parameterized truncation values and initializes them based on the real-valued distribution during network training. DSQ [7] employs multi-segment tanh functions with regularization terms to fit the quantization values and progressively approximate the quantization function. LSQ+ [3] utilizes iterative parameter optimization to determine the optimal discrete value step. It is important to note that the studies mentioned above have focused on fixed quantization bit widths.

2.2 Object Tracking

The existing research on object tracking algorithms could be categorized into four types: generative tracking algorithms [5], discriminative tracking algorithms [14], collaborative tracking algorithms [15], and deep learning-based tracking algorithms. The deep learning-based tracking algorithms have demonstrated superior performance compared to traditional tracking algorithms. MDNet [13] combines multi-domain learning and convolutional neural networks to learn a shared representation model for object tracking. SiamFC [2] deviates from the traditional online learning approach of tracking models. It achieves higher operational speed by employing twin networks for more efficient spatial search and making effective use of available data. SiamPRN++ [9] enhances model accuracy by aggregating multi-layer features and replacing the upper channel cross-correlation layer in SiamPRN [10] with a depthwise cross-correlation layer.

3 Methodology

3.1 Overall Architecture

The overall training framework of our Mixed-Precision Collaborative Quantization is outlined in Fig. 1. In this framework, the lower part of the network is denoted as Q, representing the low-bit network to be learned in this chapter. The upper part is the partially full-precision auxiliary module designed in this chapter, denoted as F.

During the training phase, the video frame input is fed into Q and generates two outputs in the intermediate E-Block: one is the normal output of the low-bit network E-Block, which is connected with the residual from Q and input to the next E-Block; the other output is input into the auxiliary module F for training. Q and F together form a mixed-precision network Q∘F, where the auxiliary module F receives the intermediate outputs from the low-bit module Q. This allows the gradient propagation in the mixed-precision network Q∘F to have a gradient component that influences the low-bit network Q. Consequently, the gradient of the output loss from the full-precision module is backpropagated to the low-bit model, and averaging the gradients between the two modules ensures more accurate parameter updates for each E-Block. This effectively mitigates the difficulty of gradient propagation in quantized models.

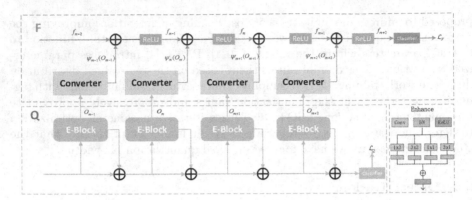

Fig. 1. Mixed-Precision Collaborative Quantization Training Framework Based on Full-Precision Auxiliary Modules.

3.2   Full-Precision Auxiliary Module

The full-precision auxiliary module achieves parameter sharing by receiving intermediate outputs from the low-bit tracking network, thereby averaging the gradients of the mixed-precision network and the original low-bit tracking network during back-propagation, guiding more accurate updates of the parameters in the low-bit tracking network.

The sub-modules of the mixed-precision network Q∘F consist of a group of Converters and a group of Connectors. The partial-precision auxiliary module F receives a collection of feature output branches $O_m{}_{m=1}^{M}$ from the E-Block, which is transformed by the feature enhancement module, in the low-bit object tracking module. For the m-th sub-module, the Converter takes the feature output branch O_m from the low-bit tracking module and produces a matching feature $\Psi_m(O_m)$ that can align with the aggregated output features of the previous group of sub-modules in the full-precision auxiliary module. By transforming and fusing the intermediate features from the low-bit tracking network, a mixed-precision network with rich features is formed. The formulaic expression for the m-th sub-module of the mixed-precision network Q∘F is given by Eq. 1.

$$f_m = ReLU(\Psi_m(O_m) + f_{m-1}) \tag{1}$$

where the Converter component receives the feature output branch O_m from the low-bit tracking network, which is then fused with the aggregated output feature f_{m-1} from the previous group of sub-modules. This fusion is followed by a non-linear activation function $ReLU(\cdot)$, resulting in the current module's fused feature f_m.

All the sub-modules, as defined by Eq. 1, eventually undergo classification prediction and mixed-precision loss calculation in the final layer of the partial full-precision auxiliary module, which is expressed by Eq. 2.

$$\min_{\{\theta^Q, \theta^F\}} \sum_{N}^{i=1} L_Q(F(x_i; \theta^Q), y_i) + L_F((Q \circ F)(x_i; \theta^F, \theta^Q), y_i) \qquad (2)$$

where the training samples $\{x_i, y_i\}_{i=1}^{N}$ are inputted into both the low-bit tracking network Q and the mixed-precision network Q∘F. During the forward propagation, the tracking loss L_Q of the low-bit network and the task loss L_F of the partial full-precision network are computed. During the back-propagation, the parameters of both the low-bit tracking network θ^Q and the full-precision auxiliary module θ^F are updated.

By introducing partially full-precision auxiliary modules, the gradients of the mixed-precision network and the original low-bit tracking network are averaged, ensuring more accurate updates for the gradient branches of the low-bit tracking network. This strategy provides richer and more accurate gradient information and updates directions for the low-bit tracking network, thereby compensating for the information loss issue that may arise from using the Straight-Through Estimator (STE) and Learned Step Size Quantization (LSQ) in the tracking network.

3.3 Multi-branch Feature Enhancement Block

The multi-branch structure is utilized to extract different features from different branches, enabling diverse feature representations in the tracking network. To increase the network's receptive field for different features, a non-symmetric convolution approach is employed. This approach decomposes a standard $d \times d$ convolution layer into two sequential convolution layers with sizes of $d \times 1$ and $1 \times d$, respectively.

To ensure the fusion of convolution branches in the feature module, it is necessary to design multiple compatible 2D convolutions on the same input. All branches in the feature enhancement module adopt the same stride, allowing them to produce output feature maps of the same size. The result of adding these feature maps is equivalent to the point-wise summation of the convolution kernels in each branch, as shown in Fig. 2.

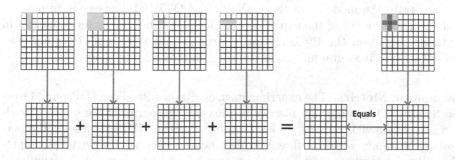

Fig. 2. The illustration of multi-branch 2D convolution equivalent transformation

Given four convolution branches, namely 1×1, 1×3, 3×1, and 3×3, on the same input, the sum of the features from these branches is equivalent to convolving conv1, conv2, conv3, and conv4 at corresponding positions and obtaining the same feature. This mathematical expression is represented by Eq. 3.

$$I \otimes K^{(1)} + I \otimes K^{(2)} + I \otimes K^{(3)} + I \otimes K^{(4)} = I \otimes (K^{(1)} \oplus K^{(2)} \oplus K^{(3)} \oplus K^{(4)}) \quad (3)$$

where the combination of the results of convolving the four branches, denoted as $K^{(1)}$, $K^{(2)}$, $K^{(3)}$, and $K^{(4)}$, on the processed matrix I, is equal to the point-wise summation of the convolution kernels $K^{(1)} \oplus K^{(2)} \oplus K^{(3)} \oplus K^{(4)}$ followed by convolution to obtain the output.

To maintain the original network structure during inference, it is necessary to merge the outputs of the four branches into a single output corresponding to the original convolution kernel. According to the homogeneity property of convolution, the result of branch fusion is expressed as Eq. 4.

$$O^{(i)} = I \otimes \left(\frac{\gamma_i}{\sigma_i} F^{(i)} \oplus \frac{\bar{\gamma}_i}{\bar{\sigma}_i} \bar{F}^{(i)} \oplus \frac{\hat{\gamma}_i}{\hat{\sigma}_i} \hat{F}^{(i)} \oplus \frac{\tilde{\gamma}_i}{\tilde{\sigma}_i} \tilde{F}^{(i)} \right)$$
$$- \frac{\mu_i \gamma_i}{\sigma_i} - \frac{\bar{\mu}_i \bar{\gamma}_i}{\bar{\sigma}_i} - \frac{\hat{\mu}_i \hat{\gamma}_i}{\hat{\sigma}_i} - \frac{\tilde{\mu}_i \tilde{\gamma}_i}{\tilde{\sigma}_i} + \beta_i + \bar{\beta}_i + \hat{\beta}_i + \tilde{\beta}_i \quad (4)$$

where the convolutional kernels represented by the branches of 1×1, 1×3, 3×1, and 3×3 are denoted as $F^{(i)}$, $\bar{F}^{(i)}$, $\hat{F}^{(i)}$, and $\tilde{F}^{(i)}$ respectively. $F^{(i)} \in \mathbb{R}^{H \times W \times C}$, and the input data is denoted as I. The convolution operation $\otimes$ yields the result $O^{(i)}$ for the i-th convolution. μ_i and σ_i donate the mean and standard deviation of the current channel, respectively. γ_i and β_i donate scale factors and bias terms learned by linear combination, respectively.

4   Experiments

4.1   Implementation Details

Datasets and Implementation Settings. The datasets used in this experiments are the GOT-10k [8], UAV123 [11], and OTB100 [16] datasets, respectively. The OTB100 dataset has different tracking scenarios but unifies the location and range of common ground truth objects. The GOT-10k training set was used to train the model, and the UAV123 and OTB100 datasets were used to test the performance of the experimental methods. The code is implemented in Python3, based on the PyTorch deep learning framework, cuda10.0, and two NVIDIA GTX 1080 graphics cards are used in the training phase.

Evaluation Metrics. The experiments use Distance Precision (DP), and Overlap Success (OS) evaluation metrics to measure the performance of the mixed-precision quantized tracking model. Distance Precision (DP) represents the percentage of frames within a threshold distance between the target location predicted by the tracking model and the target labeled frame, which is formulated as Eq. 5.

$$DP = \frac{1}{K} \sum_{j=1:k} \frac{\sum_{i=1:N_j} l(|p_{j,i} - g_{j,i}|_2 < 20)}{N_j} \tag{5}$$

where $p_{j,i}$, $g_{j,i}$ denote the number of frames that have been tracked in the current tracking sequence and the data label box respectively, $l(\cdot)$ denotes the number of frames that have been successfully tracked within a threshold of 20 pixels for the center position error, N_j denotes the number of frames that have been tracked in the current tracking sequence, and K denotes the total number of video tasks in the tracking model in the dataset.

Overlap Success (OS) indicates the percentage of frames where the overlap between the target position outer frame and the target labeled frame predicted by the tracking model is greater than a threshold value, which is formulated as Eq. 6.

$$OS = \frac{1}{K} \sum_{j=1:k} \frac{\sum_{i=1:N_j} s(IoU(p_{j,i} - g_{j,i}) > 0.5)}{N_j} \tag{6}$$

where $p_{j,i}$, $g_{j,i}$ denote the tracking result of the completed tracking video frame and the external frame of the data label respectively in the tracking sequence, $IoU(\cdot)$ calculates the intersection ratio of the frame object tracking result and the external frame of the data label. The function $s(\cdot)$ indicates whether the intersection ratio result is greater than the threshold value. It returns 1 if the condition is satisfied, and 0 otherwise.

4.2   Experimental Results

The full-precision DiMP18 tracking model based on ResNet18 was trained on the GOT-10k dataset and compared the performance of our method with existing quantization methods on the UAV123 and OTB100 datasets. Specifically, the following quantization methods were compared in the experiment: 1) DoReFa [17]; 2) PACT [4]; 3) LSQ [6] 4) LSQ+ [3]. The DiMP18 object tracking network was reconstructed using the LSQ [6] as a 4-bit baseline.

On the UAV123 dataset, the performance of LSQ+, DoReFa, PACT, and LSQ (as a low-bit benchmark) was evaluated in terms of two performance metrics, DP and OS, in the DiMP18 tracking network. These performance comparisons are illustrated in Fig. 3. It is observed that the performance comparison of different quantization methods is shown in Fig. 3(a). Compared to the full-precision tracking network, LSQ+, DoReFa, and PACT experience performance degradation of 8.2%, 6.7%, and 6.3%, respectively. Even the LSQ method, serving as a low-bit benchmark, which uses trainable quantization scales to adjust the quantization errors in weights and activations, cannot avoid the errors introduced. This is reflected in a 5.1% performance decrease compared to the full-precision network. Our proposed method, compared to the LSQ 4-bit baseline, achieves a 3.0% performance improvement and reduces the performance gap with the full-precision tracking network to only 2.1%, which is significantly smaller

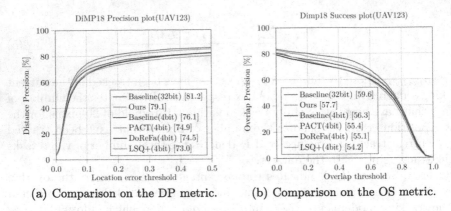

(a) Comparison on the DP metric.   (b) Comparison on the OS metric.

Fig. 3. The performance comparison of different methods in the UAV123 dataset.

than the fixed-bit quantization methods. Similar performance differences are shown in Fig. 3(b) for the OS metric. Our method achieves a 1.4% performance improvement compared to the 4-bit baseline, with a reduced performance gap of 1.9% with the full-precision network.

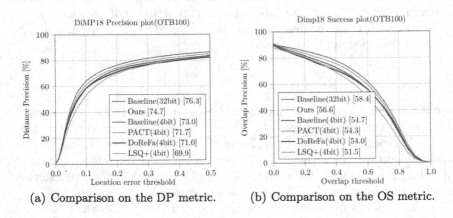

(a) Comparison on the DP metric.   (b) Comparison on the OS metric.

Fig. 4. The performance comparison of different methods in the OTB100 dataset.

The performance comparison of the full-precision network, LSQ+, DoReFa, PACT, LSQ (as 4-bit baseline quantization method), and Ours in the DiMP18 tracking network on the OTB100 dataset is shown in Fig. 4. From the results in Fig. 4(a), it can be observed that LSQ+, DoReFa, and PACT experience performance degradation of 6.4%, 5.3%, and 4.6%, respectively, compared to the full-precision tracking network. Among the quantization methods, the LSQ 4-bit baseline network also exhibits a performance decrease of 3.3% compared to the full-precision network. Our method, after incorporating the feature enhancement module and the full-precision auxiliary module, achieves a 1.7% performance improvement compared to the low-bit baseline on the OTB100 dataset. The performance gap with the full-precision tracking network is reduced to 1.6%, which is smaller than all the quantization methods. Figure 4(b) shows that LSQ+, DoReFa, and PACT methods exhibit DP and OS metrics of 51.5%, 54.0%, and 54.3%, respectively, which corresponds to performance decreases of 6.9%, 4.4%, and 4.1%, respectively, compared to the full-precision tracking network. Our method still demonstrates significant improvements on the OTB100 dataset, with a performance increase from the low-bit baseline of 54.7% to 56.6%, reducing the performance gap with the full-precision network to 1.8%. The experimental results verify the effectiveness of the full-precision auxiliary module and multi-branch feature enhancement block we introduced in assisting gradient update and improving the feature expression of the quantization model.

Moreover, we validate the experimental results of tracking average latency on two sub-sequences, "bird1" from the UAV123 dataset and "Dog" from the OTB100 dataset. Compared with the full-precision network, the single-frame processing time of our quantization method is reduced from 35.56 ms to 31.24 ms, achieving a speedup of about 14%. Comparing our method with a 4-bit quantization baseline, our method achieves better performance while barely compromising tracking speed, increasing single-frame processing time by only 0.31 ms. It offers a compelling solution for fast object tracking, making it particularly advantageous in resource-constrained scenarios.

5   Conclusion

To overcome the problems of non-differentiability and inaccurate gradient propagation in training low-bit quantized tracking models, we propose a mixed-precision collaborative quantization for fast object tracking networks. Based on the full-precision auxiliary module, our quantization algorithm constructs an improved mixed-precision model by combining partially full-precision auxiliary modules with low-bit modules. In addition, we introduce a multi-branch feature enhancement block specifically designed for object-tracking networks. This scheme uses asymmetric convolutions to extract different features, thereby improving both the feature representation capability and the feature extraction performance of the quantized network. We also compare and analyse the performance differences of fixed-point quantization methods (DoReFa, PACT, LSQ, and LSQ+) in the DiMP18 tracking networks on the UAV123 and OTB100 datasets. Through experimental results and comparisons, we demonstrate the effectiveness and advantages of our method in the object tracking network.

References

1. Banner, R., Nahshan, Y., Soudry, D.: Post training 4-bit quantization of convolutional networks for rapid-deployment. In: Advances in Neural Information Processing Systems 32 (2019)

2. Bertinetto, L., Valmadre, J., Henriques, J.F., Vedaldi, A., Torr, P.H.S.: Fully-convolutional Siamese networks for object tracking. In: Hua, G., Jégou, H. (eds.) ECCV 2016. LNCS, vol. 9914, pp. 850–865. Springer, Cham (2016). https://doi.org/10.1007/978-3-319-48881-3_56

3. Bhalgat, Y., Lee, J., Nagel, M., Blankevoort, T., Kwak, N.: LSQ+: improving low-bit quantization through learnable offsets and better initialization. In: Proceedings of the IEEE/CVF Conference on Computer Vision and Pattern Recognition Workshops, pp. 696–697 (2020)

4. Choi, J., Wang, Z., Venkataramani, S., Chuang, P.I.J., Srinivasan, V., Gopalakrishnan, K.: PACT: parameterized clipping activation for quantized neural networks. arXiv preprint arXiv:1805.06085 (2018)

5. Dekel, T., Oron, S., Rubinstein, M., Avidan, S., Freeman, W.T.: Best-buddies similarity for robust template matching. In: Proceedings of the IEEE Conference on Computer Vision and Pattern Recognition, pp. 2021–2029 (2015)

6. Esser, S.K., McKinstry, J.L., Bablani, D., Appuswamy, R., Modha, D.S.: Learned step size quantization. arXiv preprint arXiv:1902.08153 (2019)

7. Gong, R., et al.: Differentiable soft quantization: bridging full-precision and low-bit neural networks. In: Proceedings of the IEEE/CVF International Conference on Computer Vision, pp. 4852–4861 (2019)

8. Huang, L., Zhao, X., Huang, K.: GOT-10k: a large high-diversity benchmark for generic object tracking in the wild. IEEE Trans. Pattern Anal. Mach. Intell. **43**(5), 1562–1577 (2019)

9. Li, B., Wu, W., Wang, Q., Zhang, F., Xing, J., Yan, J.: SiamRPN++: evolution of Siamese visual tracking with very deep networks. In: Proceedings of the IEEE/CVF Conference on Computer Vision and Pattern Recognition, pp. 4282–4291 (2019)

10. Li, B., Yan, J., Wu, W., Zhu, Z., Hu, X.: High performance visual tracking with Siamese region proposal network. In: Proceedings of the IEEE Conference on Computer Vision and Pattern Recognition, pp. 8971–8980 (2018)

11. Mueller, M., Smith, N., Ghanem, B.: A benchmark and simulator for UAV tracking. In: Leibe, B., Matas, J., Sebe, N., Welling, M. (eds.) ECCV 2016. LNCS, vol. 9905, pp. 445–461. Springer, Cham (2016). https://doi.org/10.1007/978-3-319-46448-0_27

12. Nagel, M., van Baalen, M., Blankevoort, T., Welling, M.: Data-free quantization through weight equalization and bias correction. In: Proceedings of the IEEE/CVF International Conference on Computer Vision, pp. 1325–1334 (2019)

13. Nam, H., Han, B.: Learning multi-domain convolutional neural networks for visual tracking. In: Proceedings of the IEEE Conference on Computer Vision and Pattern Recognition, pp. 4293–4302 (2016)

14. Ning, J., Yang, J., Jiang, S., Zhang, L., Yang, M.H.: Object tracking via dual linear structured SVM and explicit feature map. In: Proceedings of the IEEE Conference on Computer Vision and Pattern Recognition, pp. 4266–4274 (2016)

15. Sui, Y., Tang, Y., Zhang, L., Wang, G.: Visual tracking via subspace learning: a discriminative approach. Int. J. Comput. Vision **126**, 515–536 (2018)

16. Wu, Y., Lim, J., Yang, M.H.: Object tracking benchmark. IEEE Trans. Pattern Anal. Mach. Intell. **37**(9), 1834–1848 (2015)

17. Zhou, S., Wu, Y., Ni, Z., Zhou, X., Wen, H., Zou, Y.: DoReFa-Net: training low bitwidth convolutional neural networks with low bitwidth gradients. arXiv preprint arXiv:1606.06160 (2016)

Pre-diagnosis for Autism Spectrum Disorder Using Eye-Tracking and Machine Learning Techniques

Mustafa Mehmood[1], Hafeez Ullah Amin[2]([✉]), and Po Ling Chen[1]

[1] University of Nottingham Malaysia, 43500 Semenyih, Selangor, Malaysia
[2] Edge Hill University, Ormskirk L39 4QP, UK
aminh@edgehill.ac.uk

Abstract. This study explores the potential of utilizing machine learning in conjunction with gaze-tracking data to facilitate early or pre-diagnosis of ASD which can be cost-effective and beneficial to people with limited access to healthcare resources. A dataset comprising gaze-tracking information mapped onto images to differentiate between control subjects and autistic individuals is utilized and treated as an image classification problem. Two machine learning frameworks were employed for model training and testing: (1) a fast approach using principal component analysis (PCA) on the images followed by conventional machine learning algorithms such as ANN, Decision Tree, and support vector machines (SVM), which yielded an accuracy of 78% and an AUC of 0.82; and (2) a deep learning approach that involved a custom convolutional neural network (CNN) model, achieving an accuracy of 92% and an AUC of 0.96. Several transfer learning models were also evaluated, with the ResNet50 model providing the best results (accuracy: 0.86, AUC: 0.94). These findings demonstrate the viability of these methods for the pre-diagnosis of autism.

Keywords: Autism Spectrum Disorder · Mental Disease Diagnosis · Deep Learning · Transfer Learning · Principal Component Analysis · Feature Extraction

1 Introduction

Autism Spectrum Disorder, a complex neurodevelopmental condition causing difficulties with social interaction, communication, and repetitive behaviors, demands early diagnosis and intervention for better outcomes, including specialized treatment and support that can enhance the quality of life. The availability of skilled professionals frequently limits today's diagnostics, which often rely on expert assessments that can be time-consuming, subject to availability, and random. Thus, quantitative, objective, and powerful analytic approaches are in high demand. The goal is to discover autism early, in order to transform the lives of those affected and their families.

Eye tracking technology, which captures and analyzes gaze patterns, has shown potential in differentiating individuals with autism from typically developing individuals. Research suggests that the gaze patterns of individuals with ASD exhibit distinct

J. Ren et al. (Eds.): BICS 2023, LNAI 14374, pp. 239–250, 2024.
https://doi.org/10.1007/978-981-97-1417-9_23

characteristics, reflecting the unique social and cognitive processing of this population. Using eye-tracking data, the aim is to create new diagnostic methods to differentiate between autistic and control subjects through the application of machine learning models. By overcoming the shortcomings of current methods, it should have the potential to advance in the field and explore innovative techniques.

The importance of this project lies in its potential to aid in the early identification of ASD, which will subsequently improve access to early intervention and support. The research can also assist in reducing the subjectivity associated with current diagnostic procedures, leading to more accurate and reliable diagnoses, by developing a data-driven and objective diagnostic tool.

A dataset of gaze-tracking information is mapped onto images to differentiate between control subjects and autistic individuals. Two machine learning frameworks are trained and compared - a fast approach using principal component analysis (PCA) and support vector machines (SVM), and a deep learning approach employing a custom convolutional neural network (CNN) model and transfer learning models. The results highlight the viability of these methods for pre-diagnosing autism. By examining the potential of gaze tracking data in the early diagnosis of autism and providing a platform for ongoing research, this study contributes to the development of innovative, objective, and efficient diagnostic tools that can improve the lives of individuals with ASD and their families.

2   Literature Review

Various studies have been conducted to explore the potential of eye-tracking technology in diagnosing autism. (Klin et al. 2002) investigated the visual fixation patterns of individuals with autism while they viewed naturalistic social situations. The researchers found that the gaze patterns of autistic individuals differed significantly from those of typically developing individuals, suggesting that eye-tracking technology could be used to identify autism.

(Chita-Tegmark 2016) conducted a review and meta-analysis of eye-tracking studies focused on attention allocation in individuals with an autism spectrum disorder. The review highlights the differences in gaze patterns between individuals with ASD and typically developing individuals, emphasizing the potential of eye-tracking technology for an autism diagnosis. (Carette et al. 2018) employed an LSTM-RNN-based ML classifier on eye-tracking data from children aged 8 to 10 and achieved an average accuracy of over 90%. However, they acknowledged the possibility of model overfitting due to the small dataset. In contrast, (Tao and Shyu 2019) proposed a framework called SP-ASDNet for autism screening based on gaze scan paths. They used a CNN-LSTM architecture to extract features from the saliency map and achieved an accuracy of 74.22% on the validation dataset.

More recently, (Oliveira et al. 2021) collected eye-tracking data from 106 subjects using Tobii Pro TX300 device. They used a genetic algorithm to identify ASD-relevant features and tested two classifiers, SVM and ANN. The SVM model scored an AUC of 77.5%, while the ANN model achieved an AUC of 82.2%. Despite the promising results, the authors encountered challenges in obtaining annotated datasets from experts.

(Nayar et al. 2022) utilized eye-tracking technology to study the gaze patterns of toddlers with autism spectrum disorder during a dynamic social interaction. Their findings indicated that autistic toddlers demonstrated reduced monitoring of social situations, further emphasizing the potential for eye tracking in autism diagnosis.

This highlights the potential of eye-tracking technology in ASD diagnosis, with various machine-learning models demonstrating promising results. However, a common limitation among these studies is the lack of a comprehensive eye-tracking dataset well suited for machine learning-based research work. Additionally, many authors reported the need for expensive equipment in their research, indicating a potential barrier to the widespread application of these methods. However, more research is needed to fully explore the potential of eye-tracking and machine learning in the diagnosis of mental conditions and Neurodevelopmental disorders.

3 Dataset

A public dataset from research by (Carette et al. 2019) was used that aimed to see if eye-tracking technology could help with the early screening of Autism Spectrum Disorder (ASD). The dataset is made for experimenting with eye-tracking in ASD research. The interesting thing about Carette et al.'s approach is that they turned numerical eye-tracking data into images, which means diagnosing ASD can be treated like an image classification task. This creative method allows for using different machine-learning models for image classification in ASD diagnosis.

There were 59 children in the study, and they focused on including young participants to help with the early detection and diagnosis of autism. The subjects were shown a series of autism-related visual stimuli while their eye movements were recorded using an SMI RED eye tracker. To make the images, the researchers took three dynamic components from the eye-tracking data: velocity, acceleration, and jerk. They then put these onto three equal-sized layers that matched the red, green, and blue (RGB) colors of an image. The values of these three components changed the colors in each layer, creating a unique picture of each child's eye-tracking path.

By using this dataset, the study wants to build on the existing research and further explore how eye-tracking technology and machine learning models can help with early diagnosis of autism spectrum disorder. The summarized important info about the dataset is in Table 1, and you can see examples of images from both groups (TD and ASDs) in Fig. 1.

3.1 Image Augmentation

A significant challenge faced during this study was the limited amount of data available for training the model. To tackle this issue and lower the risk of overfitting, image data augmentation was used, which is a common method to increase dataset size and enhance model generalization.

In this study, data augmentation was applied to create variations of the eye-tracking scan path visualizations. Starting with an initial dataset of 547 images, an additional 2,735 samples were generated, making five synthetic images for each original visualization.

This process effectively increased the dataset size by five times, providing a stronger foundation for training the machine learning models.

A straightforward Python script was created to carry out the data augmentation, performing various image transformations like zooming, shearing, and rotation. By creating these variations, the goal was to make a more diverse dataset that would help the models better understand the patterns in the eye-tracking scan paths. Ultimately, this should lead to improved performance in early diagnosis of autism spectrum disorder.

Table 1. Summary of Eye-tracking Scan-Path Images Dataset

Number of participants	59
Gender Distribution M: F	39: 21
No. of Control subjects	30
No. of ASD subjects	29
Age (Mean/median) years	7.8/8.1
CARS Score (Mean)	32.97
Original Image Resolution	640x480x3
Total Images	547
Images from Ctrl subjects	328
Images from ASD subjects	219

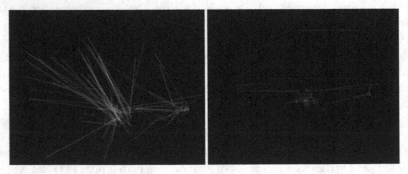

Fig. 1. Visual representation of eye-tracking scan paths. The left-sided image is ASD-diagnosed participants, while the right-sided one is for non-ASD when given the same Visual Stimuli

4 Framework 1

4.1 Methodology

The first framework shown in Fig. 2 used for model development began with the processing of the augmented image dataset. Initially, the images were resized to a uniform size of 224 × 224 pixels to expedite processing times. The images were then flattened to create a one-dimensional representation of the pixel values. The next step in the framework was to extract features from the flattened images using Principal Component Analysis (PCA).

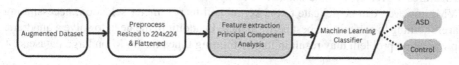

Fig. 2. Principal Component Analysis Framework

Principal Component Analysis (PCA)

Principal component analysis (PCA) is a statistical technique widely employed for feature extraction and dimensionality reduction in various fields, including image processing, bioinformatics, and finance. Transforming the high-dimensional image data into a lower-dimensional space, PCA helped mitigate the challenges posed by the curse of dimensionality, such as overfitting and lengthy computation times. It allows us to retain the most significant features of the eye-tracking scan path visualizations while reducing the complexity of the dataset, thereby enhancing the performance and interpretability of the machine-learning models.

The original set of variables can be transformed into a fresh set of independent variables, known as principal components. These chief elements are linear combinations of the initial variables and are arranged in order of the variance they account for. By electing a selection of the principal components that cover the majority of the variance in the information, we can effectively truncate the dimensionality of the dataset while maintaining maximum information. The optimal number of components was determined by examining the explained variance. This allowed us to strike a balance between reducing the dimensionality of the dataset and preserving the most significant features for the machine-learning models. Once the features were, various machine learning classifiers were trained and tested, including Random Forest, Logistic Regression, ANNs, Decision Jungle, and Support Vector Machines (SVM).

To evaluate and compare the performance of each classifier, several evaluation metrics were employed, such as the Receiver Operating Characteristic (ROC) curve, confusion matrix, accuracy, precision, recall, and F1 score. The SVM classifier and Decision Jungle were chosen based on their performance.

Support Vector Machines

SVM (Chita-Tegmark 2016) aims to find the optimal hyperplane that best separates the data points belonging to different classes. The algorithm focuses on maximizing the margin between the nearest data points, called support vectors, and the decision boundary. This approach enhances the model's generalization ability and minimizes the risk of overfitting. In cases where the data is not linearly separable, it employs kernel functions, such as polynomial or radial basis function (RBF) kernels, to project the data into a higher-dimensional space so a linear decision boundary can be found.

SVM has several key advantages, including its ability to handle high-dimensional data, its robustness against overfitting, and its capacity to identify complex decision boundaries. Additionally, SVM is particularly effective in scenarios where the number of features is greater than the number of samples, as the model is less likely to overfit. SVM, however, can be delicate to the kernel and its parameter selection, necessitating careful tuning to achieve the best performance. In addition, the need to solve a quadratic optimization problem may result in longer training times for SVM on large datasets.

Decision Jungle

The decision jungle (Criminisi and Shotton 2013) approach uses a number of decision trees and mixes their predictions to get better accuracy and stability. It works by making lots of decision trees, each one trained on a random bit of the input data, with some over-laps. The trees in a Decision Jungle are not very deep, which makes the model work faster and keeps it from overfitting. When predicting outcomes, the algorithm puts together the outputs of all the trees in the group, often using majority voting or weighted averaging, to get the final answer.

It can deal with messy or unbalanced data, not over-fitting and able to handle complex decision boundaries by putting together a few simple trees. Moreover, the algorithm can be split up to work faster on big datasets, and it is easy to understand, which is helpful for different uses. Decision Jungle might not do so well if the trees are too shallow or if there is not enough variety in the group. Also, it can be a bit picky in choosing the hyper-parameters, like the number of trees, tree depth, and how the data is sampled. Therefore, extra precautions are needed to get the best results.

Artificial Neural Networks

Artificial Neural Networks (ANNs) imitate how the brain processes information. Imagine a vast network of neurons that are connected to one another and send electrical signals to one another. In a similar way, ANNs are made up of layers of synthetic neurons, or nodes, that communicate with one another. These nodes are arranged into three layers: input, hidden, and output, with each layer carrying out particular functions. The nodes transform the input as data moves through the network by applying mathematical operations, which ultimately produces the desired output.

Training an ANN involves adjusting the weights and biases of the connections between nodes to minimize the error in the network's predictions. This is achieved through a process called back-propagation. During training, the network compares its predictions to the actual target values and calculates the error. Then, in order to reduce the error, it moves backward from the input layer to the output layer while adjusting the weights and biases. Over time, the ANN becomes more adept at making accurate predictions, much like a student learning from their mistakes and improving their understanding of a subject.

4.2 Experimental Results

The top performers, ANN, SVM, and Decision Jungle, were the subjects of further experimentation to determine how well they would perform. To demonstrate how they performed, we used various metrics, tables, and graphs, including ROC curves, AUC, confusion matrices, and other metrics. The confusion matrices for original dataset are shown in Fig. 4 along with a comparison of the ROC curves in Fig. 3.

For framework 1, the SVM produced some fairly good outcomes, with an accuracy of 77%, precision of 75%, recall of 0.77, F1 score of 0.76, and AUC of 0.82. Decision Jungle produced an accuracy of 73%, a precision of 71%, a recall of 0.72, an F1 score of 0.73, and an AUC of 0.78. Finally, the best-performing version of ANN using PCA and trained for 30 epochs achieved an accuracy of 71%, precision of 67%, recall of 0.54, F1 score of 0.67, and AUC of 0.81.

In comparison, SVM came out on top among these methods in Framework 1. Although, Framework 2 performed better in terms of classification accuracy, but Framework 1 is efficient and required less computational resources. There is a trade-off between performance and speed when choosing between the two frameworks.

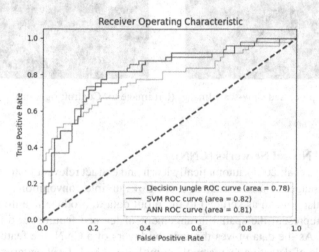

Fig. 3. ROV Curves (1) Blue – SVM, (2) Yellow – Decision Jungle, (3) Green -ANN (Color figure online)

5 Framework 2

5.1 Methodology

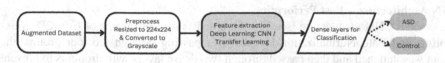

Fig. 4. Methodology for Framework 2

Framework 2 is slower than Framework 1, but it produces more consistent results. The process for implementing Framework 2 can be seen in Fig. 4. First, the augmented dataset was resized to 224 × 224 and converted to grayscale (see Fig. 5) as follows.

$$\text{New grayscale image} = (0.3 * R) + (0.59 * G) + (0.11 * B). \qquad (1)$$

This step was performed to speed up the model training time, as working with grayscale images requires fewer computational resources compared to colored images. After preprocessing, the images were fed into deep-learning models for feature extraction. The first model tested was a custom CNN architecture.

Fig. 5. Pre-processed Gray-scale Images (left image for Control; right image for ASD)

Convolutional Neural Networks (CNN)

CNNs, (Oliveira et al. 2021) automatically learn and extract relevant features from input data, such as images. These extractors consist of multiple convolutional layers, which apply filters (also known as kernels) to the input data to produce feature maps. These feature maps represent the spatial arrangement of learned features at different levels of abstraction. As the data moves through the layers of a CNN, the feature extractors become capable of detecting increasingly complex and high-level patterns.

In the initial layers of a CNN, the feature extractors usually identify simple patterns, such as edges and textures. As the data moves deeper into the network, the feature extractors learn to recognize more complex structures, like object parts or even entire objects. The learned features are then combined and used by subsequent layers in the network, such as fully connected layers, to make predictions or perform other tasks e.g. object detection and segmentation. The ability of CNN feature extractors to learn hierarchical representations of input data ensures their efficacy for various computer vision tasks, such as image classification, object detection, and semantic segmentation.

CNN Architecture – Best Performing

The custom CNN architecture (see Fig. 6), consists of four groups of Conv2D, Max-Pooling, and BatchNormalization layers. After passing through the four groups, the output was flattened and then passed through two dense layers with dropouts to prevent overfitting. Finally, a sigmoid function was used to obtain probabilities.

Transfer Learning

In addition to the custom CNN, the preprocessed dataset was tested on transfer learning

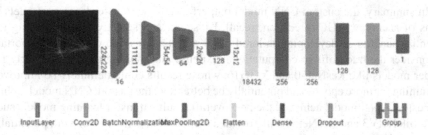

Fig. 6. Architecture of CNN

models. Transfer learning (Arai 2021) is a technique where a pre-trained model, usually on a large-scale dataset, is fine-tuned to perform a new task. It leverages the knowledge acquired by the model during its initial training and can lead to faster convergence and improved performance on the new task. Three popular transfer learning models were tested: VGG16, ResNet50 (Su 2022), and MobileNetV3 (Zhang et al. 2023). VGG16 is a deep CNN architecture known for its simplicity and strong performance on image classification tasks. It consists of 16 layers, including 13 convolutional layers and three fully connected layers. ResNet50 is another deep CNN architecture that introduced the concept of residual learning, which allows for training very deep networks without suffering from vanishing or exploding gradients. MobileNetV3 is a more lightweight CNN architecture that focuses on efficiency, making it suitable for mobile and embedded applications. It employs depthwise separable convolutions and other optimizations to reduce computational complexity while maintaining good performance.

The top layers were excluded, and the pre-trained layers were frozen during model training followed by a 1024 nodes dense layer and finally sigmoid to get predictions.

5.2 Experimentation Results

The results for Framework 2, which focused on using CNN and transfer learning models, showed a higher performance compared to Framework 1. First, the custom CNN model was trained for 15 epochs using the Adam optimizer and binary cross-entropy as the loss function. This model achieved an impressive accuracy of 92% and an AUC of 96. Additionally, it had a precision of 0.91, a recall of 0.90, an F1 score of 0.89, and a sensitivity of 0.81. The ROC curve for this model can be viewed in Fig. 7. Next, the MobileNetV3 model, which is known for its lightweight architecture and efficiency, was trained for 20 epochs. It obtained an accuracy of 86% and a ROC of 94. Furthermore, it had a precision of 0.86, recall of 0.81, F1 score of 0.77, and sensitivity of 0.68. The ResNet50 model, which utilizes residual connections to address the vanishing gradient problem, was trained for 10 epochs due to its computational demands and our limited resources. It achieved an accuracy of 87% and a ROC of 93. Moreover, it demonstrated a precision of 0.86, a recall of 0.85, an F1 score of 0.82, and a sensitivity of 0.79. Finally, the VGG16 model, known for its deeper architecture, was trained for only 5 epochs because of its high computational requirements. It managed to achieve an accuracy of 81% and a ROC of 0.88. The model also had a precision of 0.91, recall of 0.79, F1 score of 0.74, and sensitivity of 0.66.

In summary, the custom CNN model outperformed the transfer learning models in terms of accuracy, AUC, precision, recall, F1 score, and sensitivity, highlighting the potential benefits of developing tailored CNN architectures. However, it is important to consider the trade-offs in computational expense and training time, particularly for deeper models like ResNet50 and VGG16, whose results could definitely be improved by training for more epochs and potentially be better than the custom CNN model. While the custom CNN model achieved the best overall results, transfer learning models such as MobileNetV3 and ResNet50 still provided competitive performance with potentially faster training times and reduced computational resources.

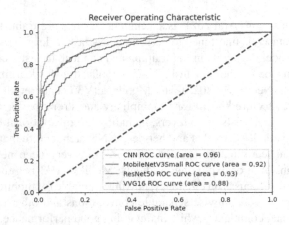

Fig. 7. ROC curve Deep Learning Models

6 Discussion: Analytical Comparison of Two Frameworks

When comparing the performance of Framework 1 and Framework 2 (see Table 2), it is evident that Framework 2 outperformed Framework 1 in terms of the obtained evaluation metrics. Framework 1, which focused on using PCA for feature extraction and traditional machine learning classifiers, achieved an accuracy of up to 77% (using SVM), while Framework 2, which utilized CNN and transfer learning models, reached a maximum accuracy of 92% (with the custom CNN model). The superior performance of Framework 2 can be attributed to the ability of deep learning models to capture more complex and abstract features in the input data, which in turn allows for better discrimination between the classes. In the case of Framework 1, PCA was used to reduce the dimensionality of the data and extract important features. However, this approach might not capture all the relevant information present in the eye-tracking scan path images. On the other hand, Framework 2 leverages the power of deep learning models, which can automatically learn hierarchical feature representations from the data. This allows the models to capture more intricate patterns in the images, leading to improved classification performance.

Table 2. Comparing evaluation metrics for the two Frameworks

Model	Accuracy	AUC	Precision	Recall	F1 Score
SVM (F1)	77%	0.82	0.75	0.77	0.76
Decision Jungle (F1)	73%	0.78	0.71	0.72	0.73
ANN (F1)	71%	0.81	0.67	0.54	0.67
CNN(F2)	92%	0.96	0.91	0.89	0.83
MobileNetv3(F2)	86%	0.92	0.86	0.81	0.77
ResNet(F2)	87%	0.93	0.86	0.85	0.82
VVG16(F2)	81%	0.88	0.91	0.79	0.74

7 Conclusion

The experiments showed that the performance of traditional machine learning methods was inferior to that of custom CNNs, and transfer learning models based on deep learning. However, traditional machine-learning methods are much faster and require far fewer computational resources (Yan et al. 2023), (Ma et al. 2023), (Liu et al. 2021). In order to improve the accuracy of these models for detecting autism in individuals, further investigation is necessary. By identifying individuals with autism early and offering the appropriate support, eye-tracking, and machine learning can assist those individuals and their families. This eye-tracking technique in conjunction with deep learning may be helpful not only for autism but also for other mental illnesses like ADHD (Zhao et al. 2022), (Padfield et al. 2021), (Li et al. 2023). Further studies may reveal how eye- tracking can be used to diagnose various ailments and facilitate patient access to necessary care. Overall, this research demonstrated that eye-tracking may be an excellent method for making an early diagnosis of autism. This approach might become more accurate and effective with additional studies. It should pave the way for more accurate methods of diagnosing Neurodevelopmental disorder conditions.

References

Arai, K.: Intelligent systems and applications: Proceedings of the 2021 intelligent systems conference (IntelliSys), vol. 1. Springer, Heidelberg (2021)

Carette, R., Cilia, F., Dequen, G., Bosche, J., Guerin, J.-L., Vandromme, L.: Automatic autism spectrum disorder detection thanks to eye-tracking and neural network-based approach. In: Ahmed, M.U., Begum, S., Bastel, J.-B. (eds.) HealthyIoT 2017. LNICSSITE, vol. 225, pp. 75–81. Springer, Cham (2018). https://doi.org/10.1007/978-3-319-76213-5_11

Carette, R., Elbattah, M., Cilia, F., Dequen, G., Guérin, J.-L., Bosche, J.: Learning to predict autism spectrum disorder based on the visual patterns of eye-tracking scanpaths. In: Proceedings of the 12th International Joint Conference on Biomedical Engineering Systems and Technologies (2019). https://doi.org/10.5220/0007402601030112

Chita-Tegmark, M.: Attention allocation in ASD: a review and meta-analysis of eye-tracking studies. Rev. J. Aut. Dev. Disord. 3(3), 209–223 (2016). https://doi.org/10.1007/s40489-016-0077-x

Criminisi, A., Shotton, J.: Classification forests. In: Decision Forests for Computer Vision and Medical Image Analysis, pp. 25–45. Springer, London (2013). https://doi.org/10.1007/978-1-4471-4929-3_4

Figure 4: (A) Architecture of the original VGG16, (B) VGG16 architecture with the strategy applied.https://doi.org/10.7717/peerj-cs.451/fig-4

Klin, A., Jones, W., Schultz, R., Volkmar, F., Cohen, D.: Visual fixation patterns during viewing of naturalistic social situations as predictors of social competence in individuals with autism. Arch. Gen. Psychiatry **59**(9), 809 (2002). https://doi.org/10.1001/archpsyc.59.9.809

Nayar, K., Shic, F., Winston, M., Losh, M.: A constellation of eye-tracking measures reveals social attention differences in ASD and the broad autism phenotype. Molec. Aut. **13**(1) (2022). https://doi.org/10.1186/s13229-022-00490-w

Oliveira, J.S., et al.: Computer-aided autism diagnosis based on visual attention models using eye tracking. Sci. Rep. **11**(1) (2021). https://doi.org/10.1038/s41598-021-89023-8

Tao, Y., Shyu, M.-L.: SP-ASDNet: CNN-LSTM based ASD classification model using observer ScanPaths. In: 2019 IEEE International Conference on Multimedia & Expo Workshops (ICMEW) (2019). https://doi.org/10.1109/icmew.2019.00124

Zhang, Y., Xu, C., Du, R., Kong, Q., Li, D., Liu, C.: MSIF-MobileNetV3: an improved MobileNetV3 based on multi-scale information fusion for fish feeding behavior analysis. Aquacult. Eng. **102**, 102338 (2023). https://doi.org/10.1016/j.aquaeng.2023.102338

Su, Q.: Brief analysis of resnet50. Comput. Sci. Appl. **12**(10), 2233–2236 (2022). https://doi.org/10.12677/csa.2022.1210227

Yan, Y., et al.: PCA-domain fused singular spectral analysis for fast and noise-robust spectral-spatial feature mining in hyperspectral classification. IEEE Geosci. Remote Sens. Lett. **20**, 1–5 (2023)

Ma, P., et al.: Multiscale superpixelwise prophet model for noise-robust feature extraction in hyperspectral images. IEEE Trans. Geosci. Remote Sens. **61**, 1–12 (2023)

Liu, Q., et al.: EACOFT: An energy-aware correlation filter for visual tracking. Pattern Recogn. **112**, 107766 (2021)

Zhao, H., et al.: SC2Net: a novel segmentation-based classification network for detection of COVID-19 in chest X-ray images. IEEE J. Biomed. Health Inform. **26**(8), 4032–4043 (2022)

Padfield, N.: Sparse learning of band power features with genetic channel selection for effective classification of EEG signals. Neurocomputing **463**, 566–579 (2021)

Li, Y., et al.: CBANet: an end-to-end cross band 2-d attention network for hyperspectral change detection in remote sensing. IEEE Trans. Geosci. Remote Sens. **61**, 1–11 (2023)

HRMOT: Two-Step Association Based Multi-object Tracking in Satellite Videos Enhanced by High-Resolution Feature Fusion

Yuqi Wu[1,2], Xiaowen Zhang[1,2], Qiaoyuan Liu[1(✉)], Donglin Xue[1(✉)], Haijiang Sun[1], and Jinchang Ren[3]

[1] The Changchun Institute of Optics, Fine Mechanics and Physics, Chinese Academy of Sciences, Changchun 130033, China
{liuqy,xuedl,sunhj}@ciomp.ac.cn
[2] The University of Chinese Academy of Sciences, Beijing 100049, China
{wuyuqi21,zhangxiaowen22}@mails.ucas.ac.cn
[3] The National Subsea Centre, Robert Gordon University, Aberdeen AB21 0BH, U.K.
jinchang.ren@ieee.org

Abstract. Multi-object tracking in satellite videos (SV-MOT) is one of the most challenging tasks in remote sensing, its difficulty mainly comes from the low spatial resolution, small target and extremely complex background. The widely studied multi-object tracking (MOT) approaches for general images can hardly be directly introduced to the remote sensing scenarios. The main reason can be attributed to: 1) the existing MOT approaches would cause a significant missed detection of the small targets in satellite videos; 2) it is difficult for the general MOT approaches to generate complete trajectories in complex satellite scenarios. To address these problems, a novel SV-MOT approach enhanced by high-resolution feature fusion (HRMOT) is proposed. It is comprised of a high-resolution detection network and a two-step based association strategy. In the high-resolution detection network, a high-resolution feature fusion module is designed to assist the detection by maintaining small object features in forward propagation. Based on high-quality detection, the densely-packed weak objects can be effectively tracked by associating almost every detection box instead of only the high score ones. Comprehensive experiment results on the representative satellite video datasets (VISO) demonstrate that the proposed HRMOT can achieve a competitive performance on the tracking accuracy and the frequency of ID conversion with the state-of-the-art (SOTA) methods.

Keywords: Multi-object Tracking · Satellite Video · High-resolution Feature Fusion · Data association

1 Introduction

With the rapid development of remote sensing earth observation technology: the successive launch of the video satellites such as Jilin-1, SkySat. etc. [1, 2], remote sensing video data is becoming increasingly easy to obtain and play an important role in the tasks of military precision strike, civilian traffic flow estimation, fire detection, etc. [3–5]. Video satellites finish dynamic ground observation by gazing at specific areas for a certain period of time, could provide richer temporal information and a larger observation area. Therefore SV-MOT has become one of the hotspots in the satellite video processing and analysis. Different from the general MOT approaches, SV-MOT would have to face more challenges such as low resolution, complex background, especially targets in very small sizes [6–9].

In recent decades, general MOT have made significant breakthroughs, which could be divided into two branches: detection-based tracking (DBT) and joint track and detection (JDT). Specifically, the DBT approaches would first get detection boxes in every frame and then do association over time with the detection and association modules have no affection between each other, whose tracking performance is mainly depended on the quality of detection. However, the independent structure has low tracking efficiency and it could not make the joint optimization together, they can achieve a competitive performance easily but with a heavy training cost; while the JDT approaches would do detection and tracking with one network, so as to estimate the objects and learn re-ID features simultaneously. These models are more efficient and fast because of considering detection and association at the same time, but it is difficult to maintain a balance between detection and association, which may lead to a tracking performance decrease. SORT [10] are the most representative DBT methods, The SORT algorithm is based on a Kalmanfilter to establish an observation model foundation, which can accurately predict the position and motion speed of the target, and achieve tracking of the target. In order to reduce the number of ID conversions, Deepsort [11] introduced ReID features on the basis of SORT, which can better distinguish different feature targets in videos. Bytetrack first associates high score detection boxes and then associate low score detection boxes, finally distinguishes between real targets and false alarms through the similarity of trajectories. In the JDT branches, FairMOT [12] optimizes both the detection model and the ReID model, and achieves a balance between detection and ReID tasks through some strategies. CenterTrack [13] proposes an integrated network of detection and tracking based on key point. In the data association stage, CenterTrack performers data association based on the distance between the predicted offset of the center point and the center point of the tracked target in the previous frame. Since there is a big difference between the general videos and the satellite videos, directly apply any methods discussed above to satellite videos, there will be a significant performance degradation. This is because compared to natural scenes, the proportion of target pixels in satellite videos is low, the object size in satellite videos is about 10–100 pixels, increasing the difficulty of detection and the background of satellite videos is complex such as widespread cloudiness, reflective noise, which may cause more missed detection and false alarm.

To fill the gap between SV-MOT and MOT, a few tracking methods for satellite videos are proposed one after another. SV-MOT methods could also be divided into the branches of DBT and JDT. The DBT method such as CKDNet-SMTNet [14] proposed a spatial motion information-guided network for tracking performance enhancement, extracting spatial information of targets in the same frame and motion information of targets in consecutive frames. However, its double LSTM structure leads the approach unable to track online. Since the JDT methods such as TGraM [15] proposed a graph based spatiotemporal inference module to explore potential high-order correlations between video frames, modeling multi-object tracking as a graph information inference process from the perspective of multi task learning. CFTracker [16] proposed a cross frame feature update module and a method for cross frame training. But their network cannot handle occlusion issues well and training is very time-consuming.

It is worth mentioning that although some progresses have been made by existing SV-MOT methods, The high-frequency tracking miss and ID switches while dealing with dense targets occlusion are still the key issues which have not been well handled. In our opinion the missing tracking problem mainly comes from a poor detection on small objects, and the high frequency of ID switches mainly come from a weak data association. Therefore, this paper focuses on both the detection and association parts, and proposes a two-step association based multi-object tracking approach for satellite video enhanced by high-resolution feature fusion (HRMOT). For a better tracking performance, we follow the branch of DBT, which do detection at first and then association. In detail, a novel detection module with high-resolution feature fusion for small targets is introduced, combined with a two-step association for tracklets prediction. Comprehensive experiment results on VISO datasets [17] demonstrate, the proposed SV-MOT can achieve a comparable performance with the SOTA methods.

2   Method

2.1   High-Resolution Detection for Network

In the satellite videos, the appearance information of object is weak because the object is very small in satellite scenarios and the background is complex, such as cloud interference, which result in a great deal of missed detection. A large-scale of missed detection is one of the main reasons that general MOT cannot be introduced into satellite videos. If there are not enough detection boxes generated at the beginning, any data association have to face a tracking failure. Therefore, in this paper the general detection module is replaced by a small object detection module to adapt the characteristics of satellite video. At present, most small target detectors obtain strong semantic information through downsampling, and then upsampling to recover high-resolution location information. However, this approach can lead to the loss of a large amount of effective information during the continuous upsampling and downsampling process. HRNet achieves the goal of strong semantic information and precise positional information by parallelizing multiple resolution branches and continuously exchanging information between different branches. So the most representative concept of multi-resolution is introduced for an essential improvement in tracking performance. Specifically, the most representative detection framework Yolov5 [18] is set as the baseline, and a high-resolution feature fusion strategy is designed to make full use of features in muti-resolution, whose network structure is shown in Fig. 1.

The idea of HRNet [19] is introduced to Yolov5 for high-resolution feature fusion. However, different from the original HRNet which only uses the highest resolution feature map as the output of the model, HRNetv2 [20] added a concatenate operation for feature maps with different resolutions, and then subsampled the combined feature maps with average pooling to obtain multiple feature maps with different resolutions. Therefore, in this paper an operation similar to HRNetv2 is adopted. In specific, the HRNet retains high-level semantic information by adopting a parallel structure with different resolutions. Within the forward propagation, multiple resolutions interact between different branches, which further reduce the impact caused by the decreasing channel dimensions. The HRNet composes of basic blocks and fuse layers. The input of the basic block is $x_{in} \in \mathbb{R}^{H \times W \times C}$, and the output $x_{out} \in \mathbb{R}^{H \times W \times C}$ of each basic block can be expressed as:

$$x_{out} = x_{in} + f(f(x_{in})), \tag{1}$$

where $f(\cdot)$ denotes the convolutional layer with a Batch Normalization layer (BN) and ReLU activate function layer. As for the fuse layer, the input is combined with multi-resolution features, which are concatenated in channel dimension. Then the concatenated feature goes through an operation $f(\cdot)$ to generate the output, where the output channel varies with the position of the fuse layer.

Besides, to reduce the information loss of feature maps with different resolutions in the sampling process and reduce the calculation amount, the feature maps with the current resolution are also concatenated. Finally, three feature maps with different resolutions and channel numbers are obtained. These feature maps are then fed into the detection head for further prediction.

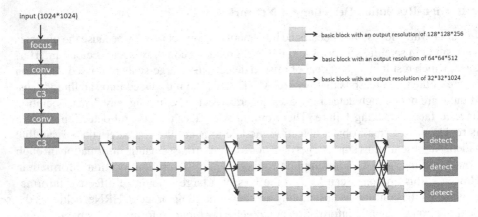

Fig. 1. Structure of the proposed high-resolution detection network.

2.2 Two-Step Association Based SV-MOT

Targets in satellite videos either with small size or affected by complex background would both easily get a low detection confidence. Common MOT associations would always first filter out the detection boxes with low confidence, and then do association. But in satellite scenarios, the real targets with small size cannot always be detected with a high confidence. So, the filter operation based on confidence may cause a lot of small objects lost. But associating all the detection boxes may result in tracking a lot of false alarm. To tackle this issue, the similarity of the trajectories is applied to distinguish the false alarms of correct targets. Specifically, in order to make more objects in satellite video be associated to form a trajectory, a two-step association method is proposed which firstly associates the high-score detection frames then low-score detection frames, finally filters out false alarms through the similarity with trajectory. The flowchart can be seen from Fig. 2.

Below, we will specifically discuss data association methods. (1)According to an adaptive threshold, the detection boxes are divided into high-score detection boxes and low-score detection boxes. (2) the trajectory would be predicted in the new frame. (3) the high-score detection boxes in new frame are associated with the existing trajectory, The similarity between the predicted trajectory and the high-score detection boxes is calculated by IOU. The Hungarian algorithm [22] was used to solve the optimal matching between the track and the detection boxes. (4) the unmatched trajectories are associated with the low-score detection boxes in order to achieve better tracking performance on scale change, occlusion and motion ambiguity, the false alarms are distinguished from the real target by the similarity with trajectories. (5) the high-score detection boxes that have no match is initialized to a new track, and the tracklets which have not been matched over a certain number of frames would be deleted.

Here we will introduce our object motion model, Kalman filter [21], which is an algorithm that utilizes linear system state equations to perform optimal estimation of system state through system input and output observation data. Due to the inclusion of noise and interference in the observed data, the optimal estimation can also be seen as a filtering process. As long as it is a dynamic system with uncertain information, Kalman filter can make informed speculations about what the system will do next. Even with noise information interference, Kalman filter can usually figure out what is happening and find imperceptible correlations between images. Therefore, Kalman filtering is very suitable for constantly changing systems, and its advantages include small memory footprint (only retaining the previous state), fast speed, making it an ideal choice for real-time problems and embedded systems. In the paper, the state vector of each object was chosen to be a eight-tuple:

$$X = [x, y, a, h, \dot{x}, \dot{y}, \dot{a}, \dot{h}] \qquad (2)$$

where x and y represent the horizontal and vertical coordinates of the object, while a and h represent the aspect ratio and the height of object's bounding box respectively. The last four parameters represent their rate of change, respectively.

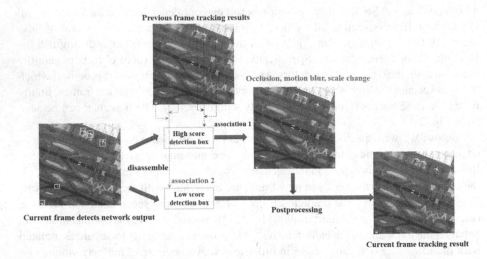

Fig. 2. Flowchart of the two-step association.

3 Experiments

In this section, we firstly introduce the details of implementation, including the dataset, metrics and experimental setup. Then ablation experiments are performed to verify the effectiveness of each module proposed. Finally, performance of HRMOT and start-of-art methods are compared, experimental results demonstrate the superiority of the proposed HRMOT.

3.1 Datasets

We validate our method on VISO dataset [17] which consists of 47 satellite videos captured by the Jilin-1 satellite. Jilin-1 satellite is a video satellite system launched by Chang Guang Satellite Technology Co., Ltd. The satellite video obtained by Jilin-1 satellite consists of a series of true color images. The coverage area of real scenes can reach several square kilometers. The majority of instances in VISO have a size smaller than 50 pixels and the videos cover different types of city-scale elements, such as roads, bridges, lakes, and a variety of moving vehicles. VISO is a diverse and comprehensive dataset which consist of different traffic situations such as dense lanes and traffic jams. Therefore, the satellite videos in dataset cover a wide range of challenges, including complex background, illumination variations, dense targets and so on. Every video is in the size of 1024 × 1024. The frame numbers are different from 324 - 326. We randomly selected 20 videos as training set and 5 unrepeated videos as testing set from the VISO dataset.

3.2 Evaluation Metrics

To evaluate the performance of different methods and the module that we proposed in multi-object tracking in satellite video. Six representative metrics are utilized for evaluation, including false positives (FP), false negatives (FN), ID switches (IDs), ID F1 score (IDF1), multi-object tracking precision (MOTP) and multi-object tracking accuracy (MOTA), the discrimination of the metrics are as follow:

(1) FP: the number of false positives in the whole video.
(2) FN: the number of false negatives in the whole video.
(3) IDs: the total number of ID switches.
(4) IDF1: the ratio of correctly identified tests to the average true and calculated number of tests.
(5) The MOTP is defined as follow, in which $d_i{}^t$ is the euclidean distance between the i-th target and the assumed position in frame t, c_t is the number of matches in frame t:

$$MOTP = \frac{\sum d_t^i}{\sum c_t} \tag{3}$$

(6) The MOTA is defined as follow, in which GT is the ground truth:

$$MOTA = 1 - \frac{FN + FP - IDs}{GT} \tag{4}$$

3.3 Experimental Setup

In our HRMOT, the detection part with high-resolution feature fusion is first trained with the VISO dataset, then followed by the two-step association strategy to get the trajectories of multi-objects in the satellite video. 20 videos datasets in VISO are trained on GPU 3090. During the training process, random gradient descent (SGD) is selected as the optimizer, the initial learning rate is 1e–2, and the weight attenuation parameter is 5e–4. In the tracking process, the maximum number of lost track frames is set to 30 frames, and the matching threshold of tracking is set to 0.8. We compare our method with the state-of-the-art trackers in natural and satellite scenarios, such as SORT, FairMOT, TGraM and Bytetrack. All methods use the same satellite video training and all experiments were conducted on RTX 3090 GPU to ensure a fair comparison.

3.4 Ablation Experiments

To verify the effectiveness of separate modules proposed in HRMOT, 5 representative videos in VISO datasets (as shown in Fig. 3) [17] with challenges like serious occlusion, complex background are selected as testing set. Yolov5 + SORT was selected as our baseline methods. Each module is integrated independently into the baseline for an effective evaluation. We compare the proposed HRNet + baseline method, YOLOX + baseline method, two-step association + baseline method and HRNet + two-step association + baseline method(HRMOT) with the baseline method under the above 5 test videos. The experimental result of adding each module to baseline are shown in Table 1, the tracking performance of the 5 tested videos produced by HRMOT are shown in Table 2.

It can be seen from Table 1 that compared with the baseline, compared with the baseline, Yolov5 + Byte (two-step association strategy added) has achieved a great tracking enhancement by improving MOTA by 36% and IDF1 by 45.3%, at the same time IDs has also been significantly reduced, which verifies the effectiveness of the two-step association. We also try a latest detection model Yolox [23] for evaluation, compare to Yolox + Byte, Yolov5 + Byte achieves better performance by increasing the MOTA by 2.7% and IDF1 by 0.6%, which double confirmed the effective of Yolov5 in multi-object tracking in satellite videos. Therefore, the high-resolution feature fusion module is integrated into the framework of Yolov5, after integrating HRNet into Yolov5, We have achieved improvements in tracking performance by improving MOTA by 4.1% and IDF1 by 1%. It can be seen that the performance of HRMOT (Yolov5 + HRNet + Byte) is far superior to other methods, the MOTA reached 74% and the IDF1 reached 79.6%.

| (a) | (b) | (c) | (d) | (e) |

Fig. 3. Five selected test satellite videos in VISO

Table 1. Ablation study for different detection headers and data association methods.

	IDF1	FP	FN	IDs	MOTA	MOTP
Yolov5 + SORT (baseline1)	33.9%	16011	11248	14990	33.9%	0.38
Yolov5 + Byte	79.2%	1098	16206	199	72.6%	0.296
Yolox + Byte	78.6%	1405	17681	161	69.9%	0.307
HRMOT	79.6%	1020	15395	208	74%	0.294

Table 2. Quantitative results of our method on VISO test set.

	IDF1	FP	FN	IDs	MOTA	MOTP
1	94.0%	165	639	11	92.7%	0.265
2	80.0%	151	5240	42	69.5%	0.329
3	88.6%	146	294	14	89.2%	0.205
4	69.5%	378	6268	84	66.6%	0.305
5	76.3%	180	2954	57	69.6%	0.303
overall	79.6%	1020	15395	208	74.0%	0.294

3.5   Comparison with Other Methods

In this section, SOTA approaches including FairMOT [12], SORT [10], Bytetrack [24] and TGraM [15] are compared with the proposed method on the test of VISO, and the experimental results are shown in Table 3. It can be seen from Table 3 that we have achieved the best results on MOTA, IDF1 and other metrics, the proposed HRMOT achieved 74% of MOTA and 79.6% of IDF1, much better than the previous work. Specifically, for the most representative MOT approach-Bytetrack, we can achieve 4.1% higher on MOTA and 1% higher on IDF1. The superiority of the proposed method not only reflected in accuracy but also in speed. Although our tracking method is based on the paradigm of DBT, our tracking part does not have a deep network model and the detection network is simple, with fast inference speed. Our tracking speed can reach over 100 FPS under satellite video, making it suitable for in orbit tracking.

Somevisualization results of the proposed method were shown in Fig. 4, Fig. 5 and Fig. 6. The tracked trajectories comparison between the proposed method and the SOTA multi-object tracking methods were shown in Fig. 4 and Fig. 5. it can be concluded that the trajectories of TGraM, FairMOT, SORT and other methods tend to be fragmented and confused. But in Fig. 4(d) (e) and Fig. 5(d) (e) with our two-step association strategy could perform a stable tracking and the tracklets are tend to be more completed, which demonstrate the availability of the two-step association. Finally, in Fig. 4(f) and Fig. 5(f) with high-resolution feature fusion added, the number of tracklets have been increased, which intuitively demonstrating the effectiveness of our detection network, Yolov5 + HRNet. Key frames of the representative visual experimental results on the challenges of occlusion and intensive targets are shown in Fig. 6. It can be seen in Fig. 6(a) that when two cars meet and block each other, the IDs of the two cars remain unchanged and there is no missed tracking, when a vehicle crosses the bridge, the ID information of the vehicle remains unchanged before and after crossing the bridge, which indicates that our method can effectively handle the problem of background occlusion, and mutual occlusion of targets in complex satellite scenes. It can be seen in Fig. 6(b) that our method can maintain simultaneous and stable tracking of multiple targets in satellite scenes with dense targets. Overall, Facing rather complex tracking challenges, our HRMOT can achieve more stable tracking, forming more complete trajectories with fewer ID conversions and higher accuracy.

Table 3. Comparison of the state-of-the-art methods on VISO test set.

	IDF1	FP	FN	IDs	MOTA	MOTP
FairMOT	22.3%	4014	51454	3036	8.4%	0.523
Baseline1	33.9%	16011	11248	14990	33.9%	0.38
TGraM	32.6%	3240	45621	2548	13.3%	0.475
Yolox + Byte	78.6%	1405	17681	161	69.9%	0.307
Yolov5 + Byte	79.2%	1098	16206	199	72.6%	0.296
HRMOT	79.6%	1020	15395	208	74%	0.294

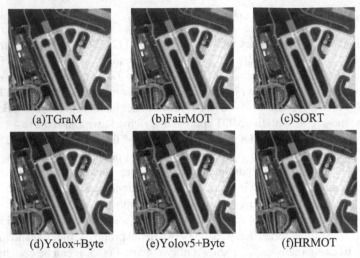

Fig. 4. The tracked trajectories comparison between the proposed method and the SOTA multi-object tracking methods. Different colors represent different trajectories(video1).

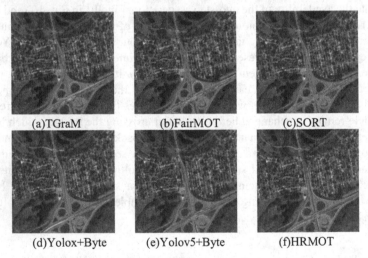

Fig. 5. The tracked trajectories comparison between the proposed method and the SOTA multi-object tracking methods. Different colors represent different trajectories(video2).

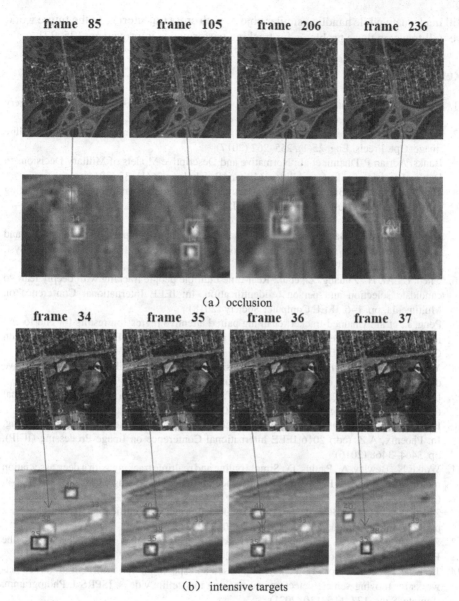

(a) occlusion

(b) intensive targets

Fig. 6. Detail experimental results on the challenges of occlusion and intensive targets.

4 Conclusion

In this paper, a novel multi-object tracking approach for satellite videos called HRMOT is proposed. Which consist of a high-resolution detection network for small objects and a two-step association. Compared with the SOTA approaches, our method achieves higher tracking accuracy and a lower frequency on ID switches. Nevertheless, our method is still inadequate while handling similar objects with cross trajectories. In the future work, we will focus on these problems for tracking performance improvement [25–27].

References

1. Aoran, X., Zhongyuan, W., Lei, W., et al.: Super-resolution for "Jilin-1" satellite video imagery via a convolutional network. Sensors **18**(4), 1194 (2018)
2. Wei, X.U., et al.: Target fast matching recognition of on-board system based on Jilin-1 satellite image. Opt. Precis. Eng. **25**(1), 255–262 (2017)
3. Banks, Adrian P, Dhami, et al.: Normative and Descriptive Models of Military Decisions to Deploy Precision Strike Capabilities. Military Psychology 26(1), 33 (2014)
4. Toth, C.K., Grejner-Brzezinska, D.: Extracting dynamic spatial data from airborne imaging sensors to support traffic flow estimation. ISPRS J. Photogramm. Remote. Sens. **61**(3–4), 137–148 (2016)
5. Dimitropoulos, K., Barmpoutis, P., Grammalidis, N.: Spatio-temporal flame modeling and dynamic texture analysis for automatic video-based fire detection. IEEE Trans. Circuits Syst. Video Technol. **25**, 339–351 (2015)
6. Chen, L., Ai, H., Zhuang, Z., et al.: Real-time multiple people tracking with deeply learned candidate selection and person re-identification. In: IEEE International Conference on Multimedia, pp. 1–6. IEEE Computer Society (2018)
7. Peng, J., et al.: Chained-tracker: chaining paired attentive regression results for end-to-end joint multiple-object detection and tracking. In: Proceedings of the European Conference on Computer Vision (2020)
8. Yan, Y., Ren, J., Zhao, H., et al.: Cognitive fusion of thermal and visible imagery for effective detection and tracking of pedestrians in videos. Cogn. Comput. **10**, 94–104 (2018)
9. Liu, Q., Ren, J., Wang, Y., et al.: EACOFT: an energy-aware correlation filter for visual tracking. Pattern Recogn. **112**, 0031–3203 (2021)
10. Bewley, A., Ge, Z,. Ott, L., Ramos, F., Upcroft, B.: Simple online and realtime tracking. In: Phoenix, A.Z. (ed.) 2016 IEEE International Conference on Image Processing (ICIP), pp. 3464–3468 (2016)
11. Wojke, N., Bewley, A., Paulus, D.: Simple online and realtime tracking with a deep association metric. In: 2017 IEEE International Conference on Image Processing (ICIP), pp. 3645–3649 (2017)
12. Zhang, Y., et al.: FairMOT: on the fairness of detection and re-identification in multiple object tracking. Int. J. Comput. Vision **129**, 3069–3087 (2021)
13. Zhou, X., Koltun, V., Krähenbühl, P.: Tracking objects as points. In: Proceedings of the European Conference on Computer Vision (2021)
14. Jie Feng, A., et al.: Cross-frame keypoint-based and spatial motion information-guided networks for moving vehicle detection and tracking in satellite videos. ISPRS J. Photogramm. Remote. Sens. **177**, 116–130 (2021)
15. He, Q., Sun, X., Yan, Z., Li, B., Fu, K.: Multi-object tracking in satellite videos with graph-based multitask modeling. IEEE Trans. Geosci. Remote Sens. **60**, 1–13 (2022)
16. Kong, L., Yan, Z., Zhang, Y., Diao, W., Zhu, Z., Wang, L.: CFTracker: multi-object tracking with cross-frame connections in satellite videos. IEEE Trans. Geosci. Remote Sens. **61**, 1–14 (2023)
17. Yin, Q., et al.: Detecting and tracking small and dense moving objects in satellite videos: a benchmark. IEEE Trans. Geosci. Remote Sens. **60**, 1–18 (2022)
18. Mekhalfi, M.L., Nicolò, C., Bazi, Y., Rahhal, M.M.A., Alsharif, N.A., Maghayreh, E.A.: Contrasting YOLOv5, transformer, and efficientdet detectors for crop circle detection in desert. IEEE Geosci. Remote Sens. Lett. **19**, 1–5 (2022)

19. Sun K., Xiao, B., Liu D., Wang, J.: Deep high-resolution representation learning for human pose estimation. In: 2019 IEEE/CVF Conference on Computer Vision and Pattern Recognition (CVPR), Long Beach, CA, pp. 5686–5696 (2019)
20. Sun, K., Zhao, Y., Jiang, B., et al.: High-resolution representations for labeling pixels and regions. arXiv (2019)
21. Welch, G., Bishop, G.: An Introduction to the Kalman Filter. University of North Carolina at Chapel Hill (1995)
22. Mills-Tettey, A., Stent, A., Dias, M. B.: The Dynamic Hungarian Algorithm for the Assignment Problem with Changing Costs. Carnegie mellon university (2007)
23. Ge, Z., Liu, S., Wang, F., Li, Z., Sun, J.: YOLOX: exceeding YOLO series in 2021. arXiv (2021)
24. Zhang, Y.: ByteTrack: multi-object tracking by associating every detection box. In: Proceedings of the European Conference on Computer Vision (2021)
25. Li, Y., et al.: Cbanet: an end-to-end cross band 2-d attention network for hyperspectral change detection in remote sensing. IEEE Trans. Geosci. Remote Sens. **61** (2023)
26. Luo, F., Zhou, T., Liu, J., Guo, T., Gong, X., Ren, J.: Multiscale diff-changed feature fusion network for hyperspectral image change detection. IEEE Trans. Geosci. Remote Sens. **61**, 1–13 (2023)
27. Liu, Q., Ren, J., Wang, Y., Wu, Y., Sun, H., Zhao, H.: EACOFT: an energy-aware correlation filter for visual tracking. Pattern Recogn. **112**, 107766 (2021)

Data Analysis and Machine Learning

Application of Manifold Recognition Target Identification Method in Seismic Exploration

Jing Zhao[1](✉), Haojie Lei[1], Yang Li[1], Fuku Zhang[1], Wenhao Zhou[1], Changrao Tian[1], Fuxiao Zhou[1], Jiale Cui[1], and Daxing Wang[2]

[1] School of Earth Science and Engineering, Xi'an Shiyou University, Xi'an 710065, China
zhaojing@xsyu.edu.cn
[2] Research Institute of E & D, Changqing Oil-Field Company of CNPC, Xi'an 710065, China

Abstract. Currently, there are more than 200 types of seismic attributes extracted from seismic data, the high-dimensional features of the data have become increasingly obvious. Although the increasing number of seismic attribute parameters is beneficial for researchers to understand seismic data, the massive amount of data also leads to redundancy and makes it difficult to further explore deeper buried information in seismic attributes, thereby reducing the accuracy of reservoir prediction. Manifold learning projects high-dimensional data into low-dimensional space by maintaining the local structure of the data, and mines and discovers the inherent characteristics and regularities hidden in the data. It is a new field of seismic attribute optimization research.

This study is based on the application of manifold learning algorithms from cognitive science technology, comparing the advantages and disadvantages of the Isometric Mapping (ISOMAP) and Multidimensional Scaling (MDS) for extracting seismic attribute features, reducing the dimensionality of seismic attributes and optimizing the attributes. Both theoretical model analysis and practical application show that manifold learning has better clustering analysis ability and feature extraction performance in dealing with nonlinear problems, Seismic attributes extracted by ISOMAP are more accurate than those extracted by MDS in characterizing the distribution characteristics of favorable reservoirs, which provide powerful tools for subsequent reservoir characterization, sweet spot identification, and seismic interpretation.

Keywords: Manifold learning · ISOMAP · MDS · Seismic attribute reduction

1 Introduction

Due to the decline in oil and gas reserves, the focus of oil and gas exploration has gradually shifted towards regions with deeper and more complex structures. Considering the complexity of the Earth's structure, seismic events reflecting underground structures often have complex horizons and weak energy [1]. Simultaneously, various stages involved in the generation, propagation, and acquisition of seismic signals are interference by different types of noise, including noise from the surrounding environment, multiple reflections, and coherent noise from surface waves. There are noises in

J. Ren et al. (Eds.): BICS 2023, LNAI 14374, pp. 267–281, 2024.
https://doi.org/10.1007/978-981-97-1417-9_25

the original seismic data, which reducing the signal-to-noise ratio and posing significant challenges in identifying seismic reflection horizons and analyzing the subsurface structure [2].

Facing this challenge, we can explore new seismic research methods. Seismic attributes, with their rich information content, can serve as a starting point for oil and gas geophysical exploration. Seismic attributes can be applied to seismic signal analysis, processing, and the prediction of oil and gas reservoirs. The account of seismic attributes extracted from seismic records can be more than 200, which leads to high-dimensional characteristics of the data. While high-dimensional data contains rich and detailed information about the objective reservoir, its processing can be quite challenging. There exists a nonlinear relationship between seismic attributes and geological features, making it challenging for traditional linear-based dimensionality reduction methods to fully capture this relationship. Additionally, the vast amount of data makes it difficult to further explore the deeper information buried within seismic attributes, thus compromising the accuracy of reservoir prediction. This phenomenon, known as the 'curse of dimensionality' poses a challenge in the field of manifold learning. To addressing this challenge, manifold learning has emerged as a promising field in optimizing seismic attributes. Manifold learning techniques aim to project high-dimensional data onto a lower-dimensional space while preserving the local structure of the data, thereby uncovering and discovering the inherent features and patterns hidden within the data [3].

Manifold-based cognitive method is one of the dimensionality reduction. It aims to uncover the nonlinear structural information of high-dimensional data and find compact embedding of the original high-dimensional data in a lower-dimensional space. The essence of manifold learning lies in capturing the intrinsic geometric distribution patterns of a high-dimensional dataset when it resides in a low-dimensional manifold. By reducing the dimensionality of the sampled data, we can uncover the inherent characteristics of the data and utilize them for classification and recognition purposes. Manifold learning can be categorized into linear and nonlinear methods. Linear dimensionality reduction algorithms primarily include Principal Component Analysis (PCA) [6], Independent Component Analysis (ICA) [7], and Multidimensional Scaling (MDS) [8], among others. These methods can yield favorable results when dealing with high-dimensional seismic attribute parameters that exhibit linear or quasi-linear relationships. Nonlinear dimensionality reduction algorithms mainly include Self-Organizing Map (SOM) [9], Generative Topographic Mapping (GTM) [10], Locally Linear Embedding (LLE) [11], and Isometric Mapping (ISOMAP) [12], among others [13].

In the year 2000, Seung H S and colleagues [14] published a research paper on manifold learning in the journal Science. This research highlighted the inspiration drawn from the natural evolution of biological systems, suggesting that the human brain can utilize nonlinear manifolds to perceive the external world. These high-dimensional pieces of information are often hidden in a lower-dimensional space with a nonlinear manifold structure. In the same year, Tenenbaum [15] and Roweis [16] independently proposed two important manifold learning algorithms: Isometric Mapping (ISOMAP) and Locally Linear Embedding (LLE). These algorithms analyze the potential low-dimensional feature variables in high-dimensional nonlinear data based on the manifold learning. The ISOMAP algorithm encounters several challenges, such as sensitivity to noise, inability to estimate the intrinsic dimensionality of samples with high curvature or sparsity, and low computational efficiency. To tackle these issues, extensive research has been conducted. In 2002, Balasu bramanian et al. [17] conducted an analysis of the topological stability in the ISOMAP algorithm. In 2003, deSilva and Tenenbaum [18] introduced the conformal ISOMAP algorithm, which preserves angles, specifically addressing low-dimensional manifolds with significant intrinsic curvature in the sample data. In the same year, they also proposed the Landmark ISOMAP algorithm, which incorporates landmarks to alleviate the computational efficiency issue of ISOMAP. In addition, in 2005, Yang proposed a method to improve the estimated accuracy of geodesic distance matrix by constructing k-connectivity graphs. In 2007, Choi et al. [16] introduced a technique to eliminate critical estimation points from the sample data, enhancing the topological stability of the ISOMAP algorithm and mitigating its sensitivity to data noise [19]. The Locally Linear Embedding (LLE) algorithm, which is also a kind of nonlinear dimensionality reduction methods, has a lower computational complexity and is capable of handling local features. However, the data manifold cannot be closed and the sample points need to be densely distributed. Moreover, there is a potential issue of inaccurate results after the dimensionality reduction process, especially when dealing with noisy data or samples with low feature correlation. To address these issues, researchers have conducted in-depth analyses of the LLE theory and proposed a series of improved algorithms. For example, NPE (Neighborhood Preserving Embedding) introduced by He in 2005 and ONPP (Out-of-Sample Extension by Nyström with Partial Projections) proposed by Kokiopoulou in 2007 address the problem of outliers in the LLE algorithm. The supervised LLE algorithm, proposed by De Ridder et al., enhances the algorithm's classification recognition capability. Additionally, Chang et al. improved the robustness of the LLE algorithm by incorporating robust principal component analysis techniques. Due to these enhancements, Over the past two decades, the LLE algorithm has been successfully applied in various fields, such as face recognition, emotion analysis, and text classification, information retrieval, remote sensing image processing, data visualization, and computer vision [20], showcasing its powerful dimensionality reduction capabilities. In 2022, Tan JP et al. [21] proposed a novel robust low-rank multi-view diversity optimization model with adaptive-weighting based manifold learning. It is an improved non-negative matrix factorization (NMF), which is used to extract the hidden structure information in sparse decomposition and useful diversity discrimination information in error matrix. Luo FL et al. [22] introduce the hypergraph into semi-supervised.

leraning to reveal the complex multi-structures of HSI, and construct a semi-supervised discriminant hypergraph learning (SSDHL) method by designing an intra-class hypergraph and an interclass graph with the labeled samples. The proposed method can simultaneously utilize the labeled and unlabeled samples to represent the homogeneous properties and restrain the heterogeneous characteristics of HSI. In 2023, Yan Y et al. [23] proposed a PCA domain 2DSSA approach for spectral-spatial feature mining in HSI. By applying 2DSSA only on a small number of PCA components, the overall computational complexity has been significantly reduced whilst preserving the discrimination ability of the features.

In order to address the challenge of dimensionality reduction for high-dimensional nonlinear seismic attribute parameters, researchers have introduced the ISOMAP algorithm for seismic attribute parameter data. By comparing the dimensionality reduction results obtained by ISOMAP with those obtained by linear MDS using wavelet neural network testing, the suitability of the ISOMAP algorithm for dimensionality reduction in seismic attribute parameter processing is discussed. The results demonstrate that the ISOMAP algorithm has stronger dimensionality reduction capability and the ability to discover the inherent structure of the data compared to the MDS algorithm. This study will employ manifold cognitive methods to investigate the application of dimensionality reduction in seismic attributes. The aim is to embed seismic attribute parameters into lower-dimensional spaces, with the goal of achieving effective dimensionality reduction. This provides a new approach for addressing geological data processing issues.

2 ISOMAP Algorithm

Isometric Mapping (ISOMAP) is one of the most renowned methods for nonlinear dimensionality reduction. The main objective of this algorithm is to map high-dimensional data points to a lower-dimensional space in order to better understand the relationships and structures among the data. The core idea of this algorithm is to utilize the nearest neighbor graph to reconstruct the manifold structure of the data samples. Then, by preserving the global information of the distances between data points, a dimensionality reduction operation is performed. These distance information can be obtained by calculating the shortest paths from each point to other points. Based on this, techniques such as Multidimensional Scaling (MDS) or other related methods can be applied to map the data points into a lower-dimensional space while preserving the relative distances between samples. It builds upon MDS by introducing the concept of 'geodesic distance', which directly addresses the problem of MDS being unable to handle nonlinear manifolds using Euclidean distances [22].Geodesic distance is the shortest distance between two points in a high-dimensional manifold. In high-dimensional manifolds, the spatial structure is irregular, which means that the shortest distance may not necessarily be a straight-line distance (Euclidean distance).Similar to how an ant cannot directly traverse a cube but instead crawls from one face to another, the ISOMAP algorithm performs dimensionality reduction by considering the geodesic paths between points. By utilizing MDS with a multi-scale analysis approach, ISOMAP determines the optimal geometric structure, accurately identifying the potential parameter space for data flow [23].The optimal objective of ISOMAP is:

$$f = \arg\min_j \sum_{i,j} \left(d_M\left(x_i, x_j\right) - d\left(f\left(x_i\right), f\left(x_j\right)\right) \right)^2 \tag{1}$$

First, construct the nearest neighbor graph G. Calculating the Euclidean distance between each sample point and other sample points. If the Euclidean distance between sample points x_i and x_j is smaller than a threshold value ε, or if point x_i is k-nearest neighbors of point x_j, then these two points are considered adjacent to each other. In other words, there is an edge connecting these two points in the graph G, and the weight

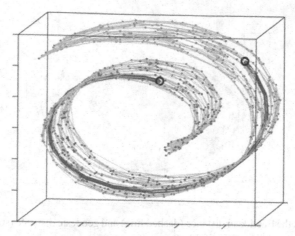

Fig. 1. Selection of neighboring points in the nearest neighbor graph computing the shortest path

of the edge is denoted as $d_X(i, j)$. Figure 1 illustrates the selection of neighboring points in the nearest neighbor graph.

Then, calculate the shortest path. For sample points x_i and x_j, if there is an edge connecting them in graph G, initially set their shortest path as $d_G(i, j) = d_X(i, j)$; otherwise, set it as $d_G(i, j) = \infty$. Then, independently set l as $1, 2, \ldots, n$, where n is the number of sample points, and compute

$$d_G(i, j) = \min\{d_G(i, j), d_G(i, l) + d_G(l, j)\} \tag{2}$$

The shortest path matrix $D_G = \{d_G(i, j)\}$ will contain the shortest path distances between any two points in graph G. The following diagram compares the Euclidean distance and the geodesic distance between any two points in a non-linear structure (Fig. 2).

Finally, computing D-dimensional embedding. We utilize the MDS algorithm for dimensionality reduction and perform the computation of D-dimensional embedding. Let's assume that λ_p is the pth eigenvalue of matrix $\tau(D_G)$ (The eigenvalues are arranged in descending order, $\tau(D_G) = -\frac{1}{2}HSH$, H is a centered matrix, $S_{ij} = d_G^2(i, j)$). If V_p^i is the ith component of the pth eigenvector, then the pth component of the D-dimensional embedding of y_i is $\sqrt{\lambda_p}V_p^i$. The following diagram illustrates the shortest path in the 2D embedding space (Fig. 3).

Figure 4 is the diagram of the method proposed in this paper.

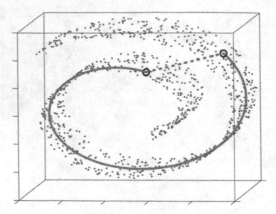

Fig. 2. Compares the Euclidean distance (dashed line) and geodesic distance (solid line) between any two points on a non-linear manifold

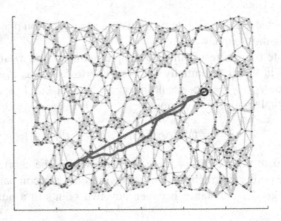

Fig. 3. Illustrates a schematic diagram of the shortest path in the two-dimensional embedding space

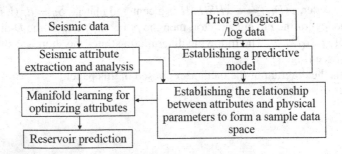

Fig. 4. The diagram of the proposed method

3 Examples

3.1 Synthetic Seismic Model with Five Common Reflection Points

We applied the ISOMAP algorithm and MDS algorithm on the synthetic seismic data. The synthetic seismic data consists of three seismic attributes: instantaneous frequency (IF), instantaneous amplitude (IA), and envelope peak instantaneous frequency (EPIF). These attributes are represented by matrices of size 1600*49. For the purpose of applying the ISOMAP algorithm and MDS algorithm, we extracted the first and the 49th seismic traces from the set of 49 traces and performed dimensionality reduction on them.

In the ISOMAP algorithm used here, the K-Nearest Neighbor (KNN) approach is employed with a default value of 12 for k. This means that each point selects its 12 nearest neighbors as its adjacent points. The geodesic distance between each point is then calculated based on these neighboring points. Choosing a value of k that is too large can lead to inaccurate dimensionality reduction results, losing the nonlinear characteristics of the ISOMAP algorithm. On the other hand, selecting a value of k that is too small introduces larger errors in constructing the distance matrix. In this particular processing, the optimal range for k selection is between 10 and 15. Figure 5 (a) represents the instantaneous frequency profile, Fig. 5 (b) represents the envelope peak instantaneous frequency profile, and Fig. 5 (c) represents the instantaneous amplitude profile.

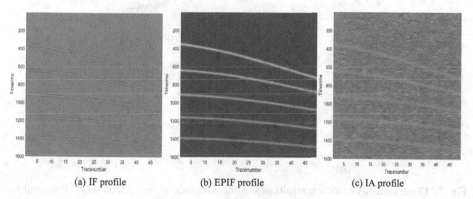

(a) IF profile   (b) EPIF profile   (c) IA profile

Fig. 5. Seismic attribute profiles

3.1.1 First Set of Processing Results

Figure 6 depicts a 3D visualization of seismic attributes for the first trace of seismic section. Since the reduced-dimensional seismic attributes happen to be three in number, they can be visualized in three dimensions. Each axis in the figure represents a distinct seismic attribute. The image also showcases the main view, top view, and left view of the 3D seismic attribute plot. Therefore, each two-dimensional plane in the plot can be considered as a display of the corresponding two seismic attributes. Each coordinate axis represents one of the three seismic attribute parameters of the shot point gather.

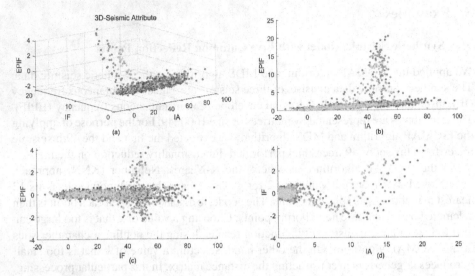

Fig. 6. Synthetic seismic attribute 3D plot

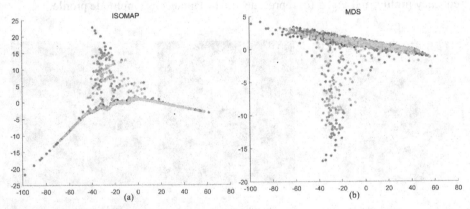

Fig. 7. Dimensionality reduction results of seismic attributes: (a) using the ISOMAP algorithm; (b) using the MDS algorithm

From the first set of data in Fig. 7, it can be observed that both the ISOMAP and MDS algorithms have successfully reduced the dimensionality. When comparing the distribution of data points with the 3D seismic attribute plot, it is evident that the algorithms preserve the structural integrity of the data and effectively showcase local features. Comparing them with the 3D seismic attribute plot, the MDS algorithm largely preserves the distance relationships and relative positions from the original data, consistent with the characteristics of MDS. On the other hand, the ISOMAP algorithm better captures the features of the low-dimensional manifold in the seismic attributes, highlighting the linear characteristics of the data. Based on the data analysis, the reduced-dimensional seismic attributes retain the IA and EPIF profiles.

3.1.2   Second Set of Processing Results

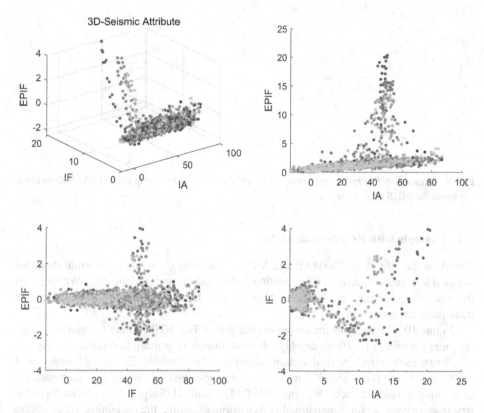

Fig. 8. Synthetic seismic attribute 3D plot

Figure 8 depicts a 3D visualization of seismic attributes for the 49th seismic trace, which shares similarities with the 3D seismic attribute plot for the first trace. This image also showcases the main view, top view, and left view of the 3D seismic attribute plot, making each two-dimensional plane in the plot represent a relative display of two seismic attributes.

From the second set of data in Fig. 9, it is evident that both the ISOMAP and MDS algorithms once again successfully accomplished dimensionality reduction. The reduced-dimensional data points still preserve the distribution and positions of the seismic attributes, without compromising the underlying structure of the data. Local features are effectively showcased in the reduced-dimensional space. Comparing them with the 3D seismic attribute plot, the MDS algorithm largely preserves the distance relationships of the original data, while the ISOMAP algorithm better captures the features of the low-dimensional manifold in the seismic attributes, highlighting the linear structural characteristics of the data. Based on the data analysis, the reduced-dimensional seismic attributes retain the profiles of instantaneous amplitude and envelope peak instantaneous frequency.

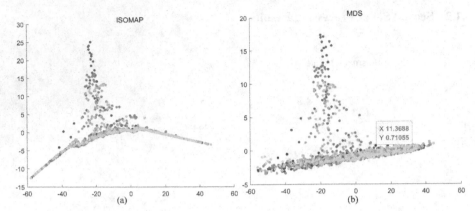

Fig. 9. Dimensionality reduction results of seismic attributes: (a) using the ISOMAP algorithm; (b) using the MDS algorithm

3.2 Example with Real Seismic Data

Based on the effects of ISOMAP and MDS algorithms on synthetic seismic data, we decided to apply dimensionality reduction techniques to real seismic data. We focus on three seismic attributes: instantaneous frequency, instantaneous amplitude, and envelope peak instantaneous frequency.

Figure 10 (a) represents the common shot gather, Fig. 10 (b) shows the instantaneous frequency profile, Fig. 10 (c) displays the instantaneous amplitude profile, and Fig. 10 (d) illustrates the envelope peak instantaneous frequency profile. Each profile consists of a 595 × 1250 matrix. These 595 traces were then grouped into seven sets, with each set containing 85 seismic traces. We applied ISOMAP and MDS algorithms separately to the first trace of each set for dimensionality reduction, assuming the remaining traces within each set to be similar. Finally, the dimensionality reduction and optimal attributes were determined by comparing the results of the seven groups. Since the dimensionality reduction results of the seven groups were quite similar, we will present the dimensionality reduction results of the first two groups for demonstration:

Figure 11 shows the three-dimensional plot of the first trace of the seismic attribute of the first group, along with its main view, top view, and left view. Therefore, each two-dimensional plane can also be viewed as a two-dimensional representation relative to two seismic attributes. Each coordinate axis represents one of the three seismic attributes of the shot point gather.

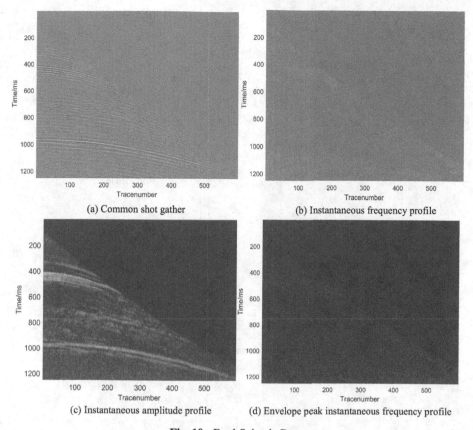

(a) Common shot gather

(b) Instantaneous frequency profile

(c) Instantaneous amplitude profile

(d) Envelope peak instantaneous frequency profile

Fig. 10. Real Seismic Data

By examining the dimensionality-reduced data in Fig. 12 and Fig. 13, we can observe that the distribution pattern of the actual seismic data is less distinct compared to the synthetic seismic data. Both the ISOMAP and MDS algorithms have successfully performed dimensionality reduction on the real seismic data. The two sets of dimensionality-reduced data exhibit well-structured distributions, and when compared to their two-dimensional plots, it is evident that the inherent structure of the data has been preserved, and local features are well represented. The MDS algorithm largely preserves the original data's distance relationships, while the ISOMAP algorithm effectively captures the characteristics of low-dimensional manifolds in the seismic attributes. This can be seen from the linear structural features present in the distribution of the seismic attribute parameters.

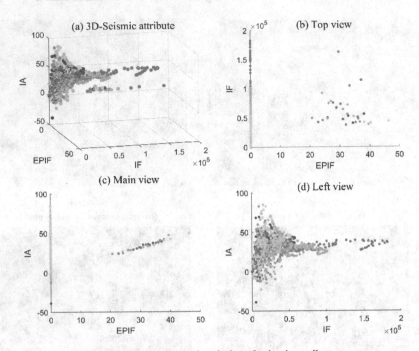

Fig. 11. Three-dimensional plot of seismic attributes

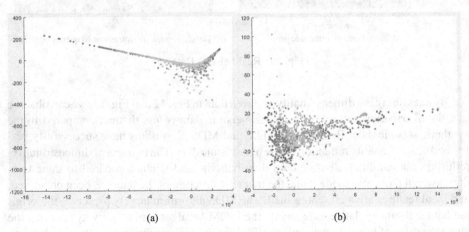

(a)   (b)

Fig. 12. ISOMAP Algorithm and (b) MDS Algorithm Dimensionality Reduction Results of the First Set - Seismic Attributes

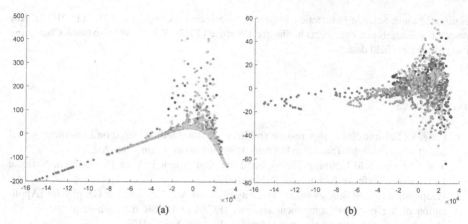

Fig. 13. ISOMAP Algorithm and (b) MDS Algorithm Dimensionality Reduction Results of the second Set - Seismic Attributes

4 Conclusion

Utilizing manifold learning methods for dimensionality reduction of seismic data is a promising research direction. Given the abundance of high-dimensional information in seismic data, it is necessary to perform dimensionality reduction to facilitate subsequent prediction and interpretation using seismic attribute analysis techniques. The ISOMAP algorithm, with its advantage in nonlinear dimensionality reduction, can preserve the topological structure of the data while mapping high-dimensional data into a lower-dimensional space, thus revealing the intrinsic characteristics of the data more effectively. Applying manifold learning methods to dimensionality reduction of seismic data holds the potential to deepen our understanding of the patterns and mechanisms in seismic data, and provide valuable data support for seismic data interpretation.

Both the ISOMAP and MDS algorithms demonstrated their effectiveness in handling real seismic data by successfully accomplishing the task of dimensionality reduction without compromising the overall data structure and local features. They not only reduce the dimensionality of the data but also help reveal the intrinsic dimensionality of the dataset. A fixed dimensionality of 2 was obtained after dimensionality reduction, indicating the inherent structure of the data. The ISOMAP algorithm demonstrates greater robustness compared to the MDS algorithm. It excels at uncovering hidden information in data that contains a certain amount of noise. However, when dealing with data that has minimal noise, the MDS algorithm is more effective in preserving the original structure of the data. Compared to other dimensionality reduction algorithms, the ISOMAP algorithm exhibits superior feature preservation capability and stability. It is effective in handling high-dimensional data and addresses the issue of dangling nodes while preserving the global structure, resulting in more accurate dimensionality reduction results.

Acknowledgment. This work is partially supported by the Natural Science Foundation of Shaanxi Province, China (2021JQ-587), National Natural Science Foundation of China (41604113),

280   J. Zhao et al.

National Nature Science Foundation Project of International Cooperation (41711530128), Key research and development projects in Shaanxi Province (2022GY-148). We also thank Changqing oilfield for their field data.

References

1. Feng, Y.: Seismic Noise Suppression and Interpolation Method Based on Low Dimensional Manifold in Framelet Transform Domain. Jilin University, Changchu (2022)
2. Yin, J.S.: Manifold Learning Theories and the applications in Face Recognition. National University of Defense Technology, Beijing (2007)
3. Geophysics; Recent Studies from University of Siena Add New Data to Geophysics [Application of the principal component analysis (PCA) to HVSR data aimed at the seismic characterization of earthquake prone areas]. J. Technol. Sci. (2017)
4. Zhang, Y.X., Tian, X.M.: Seismic signal denoising method based on improved PSO-ICA. Oil Geophys. Prospect. **47**(01), 56–62+188+194 (2012)
5. Yin, H.Y., Chen, T.J., Song, X.: Methods for predicting the thickness of coal seams based on seismic attribute optimization and machine learning. Coal Geol. Explorat. **51**(05), 164–170 (2023)
6. Liu, Z., Cao, J., Chen, S., Lu, Y., Tan, F.: Visualization analysis of seismic facies based on deep embedded SOM. IEEE Geosci. Remote Sens. Lett. **18**(8), 1491–1495 (2020)
7. Wang, M., Wu, Q.: Research of advanced GTM and its application to gas-oil reservoir identification. Int. J. Pattern Recogn. Artif. Intell. **31**(05), 1750015 (2017)
8. Liu, X.F., Zheng, X.D., Xue, G.C., et al.: Locally linear embedding-based seismic attribute extraction and application. Appl. Geophys. **7**(04), 365–375+400–401 (2010)
9. Xiang, J., Weng, J.G., Zhuang, Y.T., Wu, F.: Ensemble learning HMM for motion recognition and retrieval by Isomap dimension reduction. J. Zhejiang Univ.y-Sci. A **7**(12), 2063–2072 (2006)
10. Liu, X.F., Zheng, X.D., Xu, G.C.: Seismic attribute feature extraction and application based on manifold learning. In: Special issue of the 2010 International Petroleum Geophysical Technology Exchange Conference (2010)
11. Seung, H.S.: The manifold ways of perception. Science **290**, 2268–2269 (2000)
12. Tenenbaum, J.B., de Silva, V., Langford, J.C.: A global geometric framework for nonlinear dimensionality reduction. Science **290**(5500), 2319–2323 (2020)
13. Roweis, S.T., Saul, L.K.: Nonlinear dimensionality reduction by locally linear embedding. Science **290**(5500), 2323–2326 (2000)
14. Balasubramanian, M., Schwartz, E.L., Tenenbaum, J.B., et al.: The isomap algorithmand topological stability. Science **295**, 7 (2002)
15. Vin de Silva, Joshua B. Tenenbaum. Global versus Local Methods in Nonlinear Dimensionality Reduction. Advances in Neural Information Processing Systems.2003(15)
16. He, B.R.: The Advantages and Disadvantages of ISOMAP and LLE in Dimensionality Reduction. Capital University of Economics and Business, Beijing (2016)
17. Zhang, X.P., Nie, R.H.: The application of seismic exploration technology to geological anomalous body interpretation. Chin. J. Eng. Geophys. **10**(04), 465–471 (2013)
18. Ni, Y.: Nonlinear Dimensionality Reduction of Isomap in the Analysis of Seismic Attribute Parameter Data. Chengdu University of Technology, Chengdu (2007)
19. Huang, Y.Y., Chen, T.: Visual analysis of research hotspots in the field of domestic online commentary take 364 Chinese papers in CNKI as an example. **33**(03), 66–70 (2020). https://doi.org/10.14059/j.cnki.cn32-1276n.2020.03.036

20. Yousaf, M.: Research on Nonlinear Isomap Dimension Reduction Method for High-Dimensional Datasets. University of Science and Technology of China, Beijing (2021)

21. Tan, J., Yang, Z., Ren, J., Wang, B., Cheng, Y., Ling, W.K.: A novel robust low-rank multiview diversity optimization model with adaptive-weighting based manifold learning. Pattern Recogn. **122**, 108298 (2022)

22. Luo, F., Guo, T., Lin, Z., Ren, J., Zhou, X.: Semi-supervised hypergraph discriminant learning for dimensionality reduction of hyperspectral image. J. Sel. Topics Appl. Earth Obs. Remote Sens. **13**, 4242–4256 (2020). https://doi.org/10.1109/JSTARS.2020.3011431

23. Yan, Y., Ren, J., Liu, Q., Zhao, H., Sun, H., Zabalza, J.: PCA-domain fused singular spectral analysis for fast and noise-robust spectral–spatial feature mining in hyperspectral classification. IEEE Geosci. Remote Sens. Lett. **20**, 1–5 (2021)

Effects of PCA-Enabled Machine Learning Classification of Stress and Resting State EEGs

Kah Kit Ng and Doreen Ying Ying Sim[✉]

University of Nottingham Malaysia, 43500 Semenyih, Selangor, Malaysia
{hcykn1,Doreen.Sim}@nottingham.edu.my

Abstract. This paper aims to investigate the effects of Principle Component Analysis (PCA) enabled approach for the classification of stress and resting state of Electroenphalogram (EEG) when performing Mental Arithmetic Tasks (MAT) through the usage of signal processing and machine learning classifiers. For this analysis, we investigate our approach on a group of 25 subjects from the SAM-40 dataset and subsequently using the average power computed as a feature for our machine learning classifier. The results have shown the benefits of having PCA-enabled features where an increase of more than 10% is achieved, paving way for the potential future usage as a computer-aided diagnosis tool to assist in the detection of stress.

Keywords: Principal Component Analysis (PCA) · Average Power · Electroencephalogram (EEG)

1 Introduction

Stress research has been ongoing for the longest time as it provides desirable outcomes when endured successfully while detrimental effects when the effects of it becomes uncontrolled. This is shown when eustress, positive stress is known to provide motivation and goals for someone to overcome and allowing them to be more resilient to future challenges of similar nature. However, the problem comes when distress, negative stress becomes prevalent and completely obstructs the desire to improve or overcome the challenges posed, ultimately limiting the potential of what could have been achieved. Further, detrimental effects of stress can lead to numerous cardiovascular [1] and brain related disorders [2]. While several measures such as the Perceived Stress Scale (PSS) is used to measure the stress levels of individuals, it remains to be an unreliable source given the subjective nature of the questions that gives room for bias reporting from both the person administering the test and the one responding to it. Therefore, it is useful to note that a physiological response is used whereby objective measures can be obtained to accurately and reliably determine the presence or absence of negative stress. Consequently, the usage of electroencephalography (EEG) has been employed in several studies for mental stress detection [3], especially with the aid of computer techniques such as machine learning and artificial intelligence.

© The Author(s), under exclusive license to Springer Nature Singapore Pte Ltd. 2024
J. Ren et al. (Eds.): BICS 2023, LNAI 14374, pp. 282–290, 2024.
https://doi.org/10.1007/978-981-97-1417-9_26

Electroencphalogram (EEG) is a scientific technique that measures the action potential generated by the firing of neurons from the chemical reaction in the brain. Such action potentials will produce a potential difference that is then measured using electrodes placed on the scalp of a human. Because of the relatively weak voltage produced, the measuring instrument will amplify the signals received while also measuring its surrounding noisy signals. Therefore, simple pre-processing of the signals is usually done either through eye-inspection of by passing the signals to band filters to remove known noises. Ideally, this will produce an EEG signal corresponding to the brain regions it was measured. It still remains however, a relatively unanswered question as to which part of the physiological response obtained by EEG directly contributes to the detection of stress. We propose a simple machine learning classification for mental stress detection aimed to look at the effect variable between stress and brain physiology provided from the analysis. Further, we have also decided to use Principal Component Analysis (PCA) as some studies have used it to reduce the dimensionality of the features extracted alongside compressing the information for faster processing [4]. By using PCA, we hope to further understand the effects of disregarding specfic and long-held beliefs about certain specific parts of the brain contributing directly to stress by showing the positive improvement when using a PCA-enabled feature for stress detection with the relevant machine learning classifiers.

2 Materials and Methods

2.1 Dataset

The SAM 40 [5] EEG dataset is used for this experiment, available publicly. It is collected using a 32-channel Emotiv Epoc Flex gel kit with a sampling rate of 128Hz where each task lasts for 25 s. The channels considered for recording were Cz, Fz, Fp_1, F_7, F_3, FC_1, C_3, FC_5, FT_9, T_7, CP_5, CP_1, P_3, P_7, PO_9, O_1, Pz, Oz, O_2, PO_{10}, P_8, P_4, CP_2, CP_6, T_8, FT_{10}, FC_6, C_4, FC_2, F_4, F_8, FP_2. This dataset contains 40 participants undergoing the arithmetic task by solving mathematical problems, Stroop colour word task and the identification of symmetric mirror images. Each participant also recorded their resting EEG state in contrast to each of the stress task performed. In each task, the participants conducted three independent trials alongside the resting states. After each of the stress related task, the participants are asked to rate their stress levels on a scale of 1–10. To determine which of the three trials conducted to use for our analysis, we have decided to perform a simple average of the overall stress levels reported to determine the most suitable trial for usage. Given the higher stress levels reported in trial 3 for the arithmetic task, we have decided to only use trial 3 for our analytical experiment. Finally, we separated the participant's stress levels according to the rated stress levels provided by the participants where any values above 5 (6–10) is considered high stress. Thus, the participant's corresponding relaxed state are used as a control measure. Specifically, we used participants 2, 3, 4, 6, 7, 9, 10, 12, 14, 15, 16, 17, 18, 20, 21, 22, 23, 28, 29, 30, 31, 32, 33, 37, 39 that has been labelled as high stress and their relaxed state.

The original dataset employed simple band-pass filtering in the range of 0.5–45 Hz to remove noise from the raw data. After which, Savitzky-Golay filter and wavelet thresholding were used to remove noisy components such as eye moments, eye blinks, muscular activity and etc. Consequently, this experimental set-up uses the pre-processed dataset by the original authors.

2.2 Feature Extraction

The individual bands were first computed using Welsch power spectral density and the window size used is 50% of the sampling rate used to record the EEG dataset [6]. We have decided to use the range of 1 Hz to 4 Hz as the delta band, 4 Hz to 8 Hz as the theta band, 8 Hz to 13Hz as the alpha band, 13 Hz to 30 Hz as the beta band and finally 30 Hz to 45 Hz as the Gamma band. Anything above it is discarded as noise. Following from this. Thus, we can compute the powers of each channel and by applying the mean function provided in MATLAB, the average power is obtained and used as a feature. As a result, we have a total of 20,000 features (i.e. a result of 32 channels $\times$ 25 High Stress $\times$ 25 Resting States).

Next, we applied Principal Component Analysis (PCA) to reduce the number of features by combining the average power features into a new set of features. PCA is a feature reduction and generation technique to reduce large number of features into smaller size by first standardizing all the existing features. Next, the covariance matrix is computed to determine if there is any relationship between features before computing their eigenvectors and eigenvalues to identify the principal components. Principal components are a new set of features that are constructed as linear combinations of the initial set of features such that information is squeezed and compressed into the first few components. We used PCA with 95% variance in the existing feature set for the new features generated. Thus, only one principal component is used out of the 32 feature channels, providing a set of 625 features (1 principal component out of the 32 $\times$ 25 High Stress $\times$ 25 Resting States).

2.3 Machine Learning Classifiers

We used the Classification Learner toolbox provided in MATLAB alongside the features provided. As such, we compared most of the classifiers except for the optimizable classifiers, namely Fine Tree, Medium Tree, Coarse Tree, Linear Discriminant, Quadratic Discriminant, Logistic Regression, Gaussian Naïve Bayes, Kernel Naïve Bayes, Linear Support Vector Machine (SVM), Quadratic SVM, Cubic SVM, Fine Gaussian SVM, Medium Gaussian SVM, Coarse Gaussian SVM, Fine K-Nearest Neighbour (KNN), Medium KNN, Coarse KNN, Cosine KNN, Cubic KNN, Weighted KNN, Boosted Trees, Bagged Trees, Subspace Discriminant, Subspace KNN, Random Undersampling Boosted (RUSBoosted) Trees, Narrow Neural Network, Medium Neural Network, Wide Neural Network, Bilayered Neural Network, Trilayered Neural Network, SVM Kernel and Logistic Regression Kernel.

The basic configurations for each of the classifiers has been used as provided by the toolbox without altering any of its default settings. Each of the classifiers uses the features extracted with and without applying PCA for comparison.

3 Results

MATLAB R2022a was used for this experiment alongside the help of the Classification Learner toolbox from MATLAB. A Windows 10 operating system running with 8 GB RAM and 11th Generation Intel Core @2.40 GHz laptop was used.

To obtain a fair assessment and reliable evaluation, the accuracy obtained was used alongside a 10-fold cross validation to validate the experiment. Formula (1) or Eq. (1) below shows the method to calculate the prediction and/or classification accuracy of the machine learning classification of each classifier.

Classification or Prediction Accuracy. Accuracy is obtained from the sum of true positive and true negative divided with the sum of the true positive, true negative, false positive, and false negative. True positive is when the classifier correctly classifies the data as stress while true negative is when the classifier correctly classifies the data as relaxed. Similarly, false positive is when the classifier falsely classifies the data as stressed when it should be relaxed while false negative is when the classifier incorrectly classifies as relaxed when it should be stressed. This measure allows us to look at the overall correctness of the classifier in determining whether someone is in the stress or relax class. Equation (1) below is the common misclassification equation usually adopted by machine learning researches and this is used by this research to measure the prediction or classification accuracy of each classifier.

$$Accuracy = \frac{True\ Positive + True\ Negative}{True\ Positive + True\ Negative + False\ Positive + False\ Negative} \quad (1)$$

Comparison and Contrast of Each EEG Signaling Band in Tables 1 and 2. Tables 1 and 2 are arranged in the order of the feature band used, followed by an indication of whether PCA was applied and lastly the type of classifiers used. The highest accuracy achieved in each band is bolded alongside their corresponding classifiers. It should be noted that the results for quadratic discriminant without applying PCA is unavailable because of the default settings used by the toolbox where the predictors had zero covariance before applying PCA. However, once PCA is applied, there is a covariance structure that enables the workings of the quadratic discriminant analysis.

Table 1. Machine learning classification of the average prediction or classification accuracy of Alpha and Beta (High Stress versus Relaxed State) EEG signaling bands for each classifier.

Average power **Alpha**		Average power **Beta**		Classifier applied
Without PCA	With PCA	Without PCA	With PCA	Classifier applied for each EEG signaling band
40	50	42	36	Fine Tree
40	50	42	36	Medium Tree
44	50	32	38	Coarse Tree
60	50	44	44	**Linear Discriminant**
–	48	–	36	Quadratic Discriminant
52	52	46	44	**Logistic Regression**
56	54	46	34	Gaussian Naïve Bayes
50	48	52	26	Kernel Naïve Bayes
40	48	30	36	**Linear SVM**
50	48	30	30	Quadratic SVM
46	46	32	38	Cubic SVM
50	46	28	32	Fine Gaussian SVM
40	46	48	30	Medium Gaussian SVM
44	40	38	40	Coarse Gaussian SVM
32	46	28	28	Fine KNN
60	50	46	54	Medium KNN
40	40	40	40	Coarse KNN
50	48	54	52	Cosine KNN
58	48	48	58	**Cubic KNN**
42	54	34	26	Weighted KNN
40	40	40	40	Boosted Trees
40	54	28	24	Bagged Trees
38	52	32	44	Subspace Discriminant
36	48	30	36	Subspace KNN
46	44	40	32	RUSBoosted Trees
36	48	32	34	Narrow Neural network
44	42	36	42	Medium Neural Network
36	42	36	30	Wide Neural Network
40	48	40	26	Bilayered Neural Network
44	42	36	34	Trilayered Neural Network

Table 1. (*continued*)

Average power **Alpha**		Average power **Beta**		Classifier applied
Without PCA	With PCA	Without PCA	With PCA	Classifier applied for each EEG signaling band
44	66	40	44	SVM Kernel
48	58	34	34	Logistic Regression Kernel

Table 2. Machine learning classification of the average prediction or classification accuracy of Delta, Gamma and Theta (High Stress versus Relaxed State) EEG signaling bands for each classifier.

Avg. Power **Delta**		Avg. Power **Gamma**		Avg. Power **Theta**		Classifier applied
Without PCA	With PCA	Without PCA	With PCA	Without PCA	With PCA	Classifier applied for each EEG signaling band
48	30	48	36	42	34	Fine Tree
48	30	48	36	42	34	Medium Tree
48	40	48	48	42	38	Coarse Tree
44	56	**58**	42	40	52	**Linear Discriminant**
–	54	–	42	–	48	Quadratic Discriminant
42	**58**	54	42	44	50	**Logistic Regression**
52	52	38	38	44	46	Gaussian Naïve Bayes
46	34	50	38	44	36	Kernel Naïve Bayes
44	52	30	40	36	**58**	**Linear SVM**
34	34	30	56	34	42	Quadratic SVM
40	40	40	50	40	34	Cubic SVM
46	50	32	42	30	36	Fine Gaussian SVM
44	44	48	38	40	38	Medium Gaussian SVM
54	40	40	40	40	40	Coarse Gaussian SVM

(*continued*)

Table 2. (*continued*)

Avg. Power **Delta**		Avg. Power **Gamma**		Avg. Power **Theta**		Classifier applied
Without PCA	With PCA	Without PCA	With PCA	Without PCA	With PCA	Classifier applied for each EEG signaling band
28	30	24	36	34	38	Fine KNN
52	38	38	48	36	50	Medium KNN
50	40	50	40	50	40	Coarse KNN
46	54	42	44	46	46	Cosine KNN
50	44	48	46	46	46	**Cubic KNN**
40	32	28	34	34	34	Weighted KNN
50	38	50	46	50	40	Boosted Trees
44	32	40	38	38	40	Bagged Trees
38	54	28	42	36	52	Subspace Discriminant
36	38	24	28	46	40	Subspace KNN
50	38	50	38	50	40	RUSBoosted Trees
34	30	30	42	36	34	Narrow Neural network
34	38	34	32	34	38	Medium Neural Network
32	38	32	40	32	32	Wide Neural Network
46	38	32	38	40	42	Bilayered NN
32	36	26	40	32	46	Trilayered NN
52	46	32	40	48	44	**SVM Kernel**
36	32	36	48	38	46	Logistic Regress. Kernel

4 Discussions

From the results, the effects of applying principal component analysis have significantly improved the classification accuracy with a range of 10 to 20 percent increment when looking at the same classifier. The only exception to this is the Gamma band, possibly since it may still contain noise from the range specified when separating the band levels. Whereas alpha, beta and theta band has been associated to mental stress from existing studies [7, 8]. Further evidence suggests that the range specified in our experiment is not useful as there has been reports about the seeming relation between high gamma band and emotions where the range is 50 to 80 Hz [9]. Other notable decrease in performance where PCA is applied is seen in neural networks where conventional networks require a larger set of features than a compressed version. It may instead be a possibility where the PCA applied features could use all the principal components instead of only the ones explained with a 95% variance. Though this may sound contradictory to the purpose of applying PCA in the first place, it may be worth noting that PCA also provides a new set of features that ought to be considered in its entirety before dismissing it entirely. Figure 1 below shows the comparison and contrast between highest prediction or classification accuracies with and without PCA-Enable classifiers.

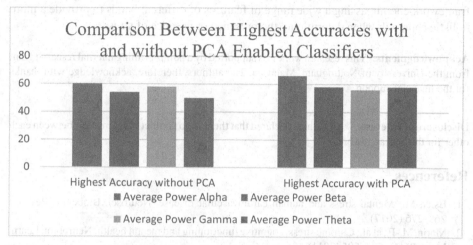

Fig. 1. Comparison and contrast among the highest accuracies obtained by enabling or disabling PCA in the classifiers.

Next, it is surprising to note that despite being unable to determine the exact channels and locations of the associated brain regions relating to stress, the results shown when PCA is applied clearly improves significantly compared to without. This can be shown in Fig. 1 above. This may instead suggest that stress is not limited to any one or few regions in the brain, but a combination of regions combined in a linear manner as indicated in the combination of PCA. Further, the most significant increase of 22% is shown in the average alpha features where PCA is applied in the SVM Kernel classifier. This is because alpha band as a feature in power spectral density has been useful in previous studies for mental stress detection [10].

Best Performance in Alpha, Beta, Delta, Gamma and Theta Band Respectively. Finally, classifiers like linear discriminant, logistic regression, linear SVM, cubic KNN, and SVM Kernel has the best performance in alpha, beta, delta, gamma and theta band respectively. Combined, they have an average accuracy of 59.6% across the bands, suggesting the possible combination of an ensemble network whereby each of the respective classifier is used to predict stress on a particular band before collating the information to further predict the presence or absence of stress. This will pave way for an ensemble of classifiers that is diverse to allow for different aspects to be considered instead of producing similar ones [11].

5 Conclusion

We have successfully demonstrated the effective use of Principal Component Analysis (PCA) when detecting stress from participants undergoing mental arithmetic task with an improvement of 10 to 20 percent depending on the frequency band used during classification. Although the results is still far from clinically approve usage, the application of PCA into the context of machine learning classification for mental stress detection will benefit in computational power where more features are available in future experiments. This also shows the possibility of implementing a type of ensemble machine learning framework encompassing a wide range of features of different bands to provide a more holistic approach when determining the presence or absence of stress.

Acknowledgments. This research work was supported by a pump-priming internal research grant from the University of Nottingham, Malaysia. The authors, therefore, acknowledge with thanks for the financial support.

Disclosure of Interests. All authors declared that there is no conflict of interests in between each other for this research work.

References

1. Esler, M.: Mental stress and human cardiovascular disease. Neurosci. Biobehav. Rev. **74**, 269–276 (2017)
2. Marin, M.-F., et al.: Chronic stress, cognitive functioning and mental health. Neurobiol. Learn. Mem. **96**(4), 583–595 (2011)
3. Katmah, R., Al-Shargie, F., Tariq, U., Babiloni, F., Al-Mughairbi, F., Al-Nashash, H.: A review on mental stress assessment methods using EEG signals. MDPI Sensors **21**, 1–26 (2021)
4. Xia, L., Malik, A.S., Subhani, A.R.: A physiological signal-based method for early mental-stress detection. Biomed. Signal Process. Control **46**, 18–32 (2018)
5. Ghosh, R., et al.: SAM 40: Dataset of 40 subject EEG recordings to monitor the induced-stress while performing Stroop color-word test, arithmetic task, and mirror image recognition task. Data Brief **40**, 107772 (2022)
6. Amin, H.U., Mumtaz, W., Subhani, A.R., Saad, M.N.M., Malik, A.S.: Classification of EEG signals based on pattern recognition approach. Front. Comput. Neurosci. **11**(103), 1–12 (2017)
7. Awang, S.A., Pandiyan, P.M., Yaacob, S., Ali, Y.M., Ramidi, F., Mat, F.: Spectral density analysis: theta wave as mental stress indicator. Signal Process. Image Process. Pattern Recogn. **260**, 103–110 (2011)
8. Hernán, D.M., Cid, F.M., Otárola, J., Roberto, R., Oscar, A., Cañete, L.: EEG Beta band frequency domain evaluation for assessing stress and anxiety in resting, eyes closed, basal conditions. Procedia Comput. Sci. **162**(1), 974–981 (2019)
9. Yang, K., Tong, L., Shu, J., Zhuang, N., Yan, B., Zeng, Y.: High Gamma band EEG closely related to emotion: evidence from functional network, brain imaging and stimulation. Front. Hum. Neurosci. **14**, 1–12 (2020)
10. Al-Shargie, F., Tang, T. B., Badruddin, N., Kiguchi, M.: Mental stress quantification using EEG signals. In: International Conference for Innovation in Biomedical Engineering and Life Sciences, pp. 15–19, Springer, 31750 Tronoh, Perak, Malaysia (2015)
11. Dietterich, T.G.: Ensemble methods in machine learning. In: Kittler, J., Roli, F. (eds.) MCS 2000. LNCS, vol. 1857, pp. 1–15. Springer, Heidelberg (2000). https://doi.org/10.1007/3-540-45014-9_1

Fatigue Detection Algorithm Based on Discrete Wavelet Transform of EEG Signals

Peixian Wang, Jiawen Li, Yongqi Ren, Leijun Wang, and Rongjun Chen[✉]

School of Computer Science, Guangdong Polytechnic Normal University, Guangzhou 510665, China

crj321@163.com

Abstract. EEG signals are usually used to study brain functions such as cognition, perception, emotion, and sleep, but the collected EEG signals are usually mixed with interference signals such as ocular artifacts. To solve this problem, this paper proposes an adaptive ocular artifacts removal algorithm that combines empirical mode decomposition (EMD), independent component analysis (ICA) and sample entropy (SampEn). In order to detect the fatigue state of the human body, this paper proposes a fatigue detection algorithm based on discrete wavelet transform (DWT). Extract the EEG rhythm wave, reconstruct the Theta wave, Alpha wave and Beta wave, and use the ratio of (Theta + Alpha)/Beta as the fatigue index to measure the fatigue degree of the human body. In 100 independent repeated tests, 91 times of the correct prediction of the state of the subjects, the recognition accuracy of the algorithm reached 91%, and the recognition accuracy of this algorithm reached 91%.

Keywords: EEG collection · EEG analysis · fatigue detection

1 Introduction

With the development of the Internet of Things and intelligent Wearable device technology, Portable smart medical wearable devices such as smart bracelets, and smartwatches, can make people get their health data every time. However, few of them consider combining EEG signals in their systems as an important health monitoring index.

As the most direct bioelectrical signals reflecting brain activities [11], the EEG signals are the signals with low frequency [17] and weak intensity [4], generally maintained at 10–50 μV with a maximum amplitude of no higher than 100 μV. Medical field divide EEG into five rhythm waves: Delta (0.5–4 Hz) [26], Theta (4–8 Hz), Alpha (8–12 Hz), Beta (12–30 Hz) [15] and Gamma (30–50 Hz) in frequency domain. But traditional EEG acquisition equipment is expensive and complex, which is not suitable for wearable devices. NeuroSky Company designs the wearable device named TGAM (ThinkGear AM) Brain Wave Sensor to make it possible to acquire EEG in a low-cost and easy way. EEG signals are often contaminated with various physiological and non-physiological artifacts including, ocular artifacts, ECG artifacts and power frequency interference [1–3]. Power frequency interference can be removed by notch filter. The

J. Ren et al. (Eds.): BICS 2023, LNAI 14374, pp. 291–299, 2024.
https://doi.org/10.1007/978-981-97-1417-9_27

interference of ECG artifacts on the EEG signals are relatively small. As the amplitude of the ocular artifacts are relatively large and the distance from the cerebral cortex is relatively close, it has the greatest impact on the EEG signals [10]. Before using EEG signals for fatigue detection, it is necessary to remove the above-mentioned interference signals.

2 Related Work

2.1 EEG Extraction

As EEG has a low amplitude of vibration, it's easy to be polluted by noise when collected, the main noise source: 50 Hz power frequency interference [18] and ocular artifacts. We can use a notch filter to filter the power frequency. But the frequency of ocular artifacts is 2–13 Hz overlapping with Delta、Theta and Alpha in frequency domain. There are most of the feature extraction methods [13], such as the principal component analysis (PCA)[16], the independent component analysis (ICA) [14]. EEG can be extract by ICA using multichannel collection [8]. Consider TGAM has only one channel, we use a Dimension raising algorithm named EMD [9] to add the channel of EEG to satisfy the condition of ICA. But the order of the independent components output after the ICA algorithm is processed is random, and the research shows that the sample entropy of the EEG signals are larger than the sample entropy of the ocular artifacts. In this paper, the Sample Entropy (SampEn) [12] in nonlinear dynamics is selected as the basis for judging ocular artifacts.

2.2 Fatigue Detection

In the further study of EEG rhythm waves, scholars found that the degree of mental fatigue of subjects can be measured according to the ratio of EEG rhythm waves [7]. During fatigue, Alpha and Theta activities increase, while Beta one decreases [23–25]. The ratio of Alpha wave, Theta wave, and Beta wave (Alpha + Theta)/Beta [22] is significantly different when people are tired and when they are not tired [6]. The ratio (Alpha + Theta)/Beta is called the fatigue index. The fatigue index will increase with the increase of fatigue degree. Considering the time-varying and time-domain continuity of EEG, we use discrete wavelet transform [5, 21]to separate the rhythm waves of EEG. The discrete wavelet transform formula is as follows:

$$c_j[k] = \sum_{n=-\infty}^{\infty} h[n] * 2^{-\frac{j}{2}} x[2n - k], j = J, J - 1, \ldots \ldots, 1 \qquad (1)$$

$$d[k] = \sum_{n=-\infty}^{\infty} g[n] * 2^{-\frac{j}{2}} x[2n - k], j = J, J - 1, \ldots \ldots, 1 \qquad (2)$$

$x[n]$ is the original signal, $h[n]$ and $g[n]$ are low-pass filter and high-pass filter respectively. $k = 0, 1, \ldots \ldots, 2^j - 1$ indicates the sequence index after downsampling.

3 The Proposed Method

This paper presents an adaptive method for removing ocular artifacts by single-channel EEG signals based on empirical mode transformation-independent component analysis (EMD-ICA) based on sample entropy. First, the EEG signals after the notch filter is self-dimensioned by empirical mode decomposition to obtain a limited number of IMF components [19, 20], and each IMF component is regarded as the EEG signals collected by the multi-lead EEG acquisition device. Then, ICA was used to separate the independent components of the obtained IMF weight. Then, the sample entropy is used as the basis to determine the ocular artifacts, and the corresponding independent components are zeroed. Finally, the remaining independent components were transformed by ICA inverse and EMD inverse, to obtain a relatively pure EEG signals with ocular artifacts removed. After getting pure EEG data, we use Wavelet transform to separate Alpha, Beta and Theta and compute the ratio (Alpha + Theta)/Beta. When the ratio is higher than 1.5, it shows the person is tired. Figure 1 is The processing steps of the algorithm.

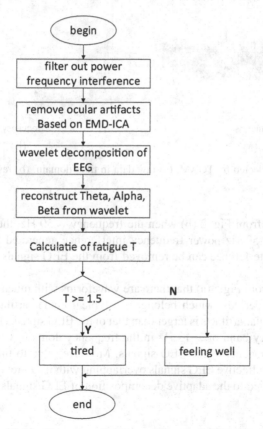

Fig. 1. The processing steps of the whole algorithm.

The calculation formula for classification accuracy is as follows:

$$acc = \frac{rt}{total} * 100\% \tag{3}$$

rt is the number of correct predictions, and *total* is the total number of experiments

4 Experiment

We construct experiment to demonstrate the capabilities of our EEG Extraction Part the Fatigue Detection Part. The time-frequency waveform of EEG signals with power frequency interference and other noise is collected by the single conduction EEG acquisition module TGAM as follow (Fig. 2).

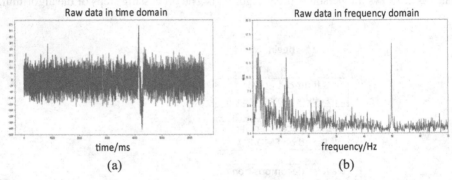

Raw data in time domain
time/ms
(a)

Raw data in frequency domain
frequency/Hz
(b)

Fig. 2. Raw EEG collected by TGAM. (a) raw data in time domain. (b) raw data in frequency domain.

As can be seen from Fig. 2 (b) when the frequency is 50 Hz, there is a very large spike, which is the 50 Hz power frequency interference generated by the radio. The power frequency interference can be removed from the EEG signals by a 50 Hz notch filter.

It can be seen from Fig. 3(a) that there are waveforms with much larger amplitudes than in other time periods, which belong to a typical ocular artifacts. Although the amplitude of the ocular artifacts is larger than that of the EEG signals, the ocular artifacts occupies a frequency band of 2–13 Hz in the frequency domain, which overlaps with the low-band components of the EEG signals. Moreover, due to the existence of the ocular artifacts, the effective EEG signals overlapping with it cannot be detected. In this paper, EMD is applied to the adaptive decomposition of EEG signals containing ocular artifacts.

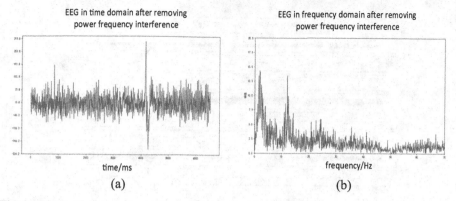

Fig. 3. EEG in time domain and frequency domain after removing power frequency interference. (a) the waveform of EEG in time domain (b)the waveform of EEG in frequency domain.

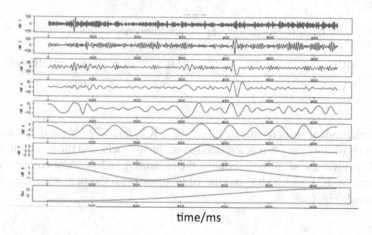

Fig. 4. The EEG signals are decomposed into several IMF and one Res

As shown in Fig. 4 the EEG signals containing the ocular artifacts are decomposed by EMD into 8 IMF components and 1 monotonic component. The frequency of the IMF component acquired decreases in order. These 8 IMF components and monotone components were separately analyzed by ICA for independent component analysis, and 9 independent components were output (Fig. 5 and Table 1).

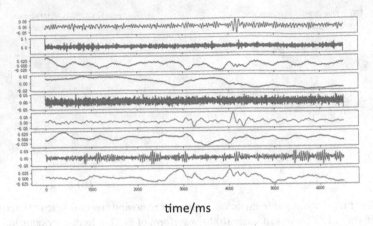

time/ms

Fig. 5. Independent component time domain waveform generated by ICA decomposition

Table 1. Sample Entropy of each independent component

ICs	IC1	IC2	IC3	IC4	IC5	IC6	IC7	IC8	IC9
SampEn	0.36	0.31	0.02	1.65	0.07	1.48	0.04	0.03	0.23

The independent components IC3, IC5, IC7, IC8, and IC9 acquired through EMD-ICA decomposition have been determined as ocular artifacts, and the values of IC3, IC5, IC7, IC8, and IC9 are set to zero. Finally, the remaining independent components were transformed by ICA and EMD, and the EEG signal with the ocular artifacts removed was obtained (Fig. 6).

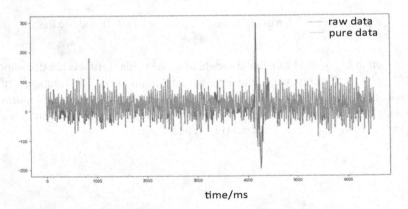

time/ms

Fig. 6. The comparison of raw data and pure data

Perform discrete wavelet transform on EEG to obtain Theta, Alpha, and Beta. Their waveform in the time domain is as follows (Fig. 7) and calculate their ratio.

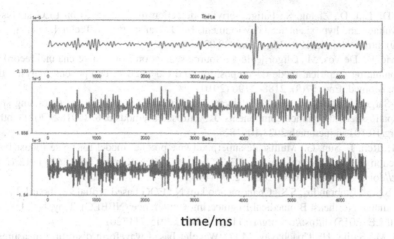

time/ms

Fig. 7. the waveform of Alpha, Theta and Beta in time domain

We conducted one hundred independent repeated experiments. Among them, the fatigue state of the subjects was correctly identified 91 times, and the recognition accuracy was 91%

5 Conclusion

In this article, we propose a novel fatigue detection algorithm, which consists of two parts: EEG signals extraction based on EMD-ICA and fatigue detection based on discrete wavelet transform. The fatigue detection can remove the interference signals in the EEG signals very well, obtain the pure EEG signals, and extract the Alpha, Theta, and Beta rhythm waves through the discrete wavelet for fatigue detection. Experiments show that the recognition accuracy of the fatigue detection algorithm proposed in this paper reaches 91%.

References

1. Mannan, M.M.N., Kamran, M.A., Jeong, M.Y.: Identification and removal of physiological artifacts from electroencephalogram signals: a review. IEEE Access **6**, 30630–30652 (2018). https://doi.org/10.1109/ACCESS.2018.2842082
2. Saini, M., Satija, U.: An effective and robust framework for ocular artifact removal from single-channel EEG signal based on variational mode decomposition. IEEE Sens. J. **20**(1), 369–376 (2019). https://doi.org/10.1109/JSEN.2019.2942153
3. Saini, M., Satija, U., Upadhayay, M.D.: Effective automated method for detection and suppression of muscle artefacts from single-channel EEG signal. Healthc. Technol. Lett. **7**(2), 35–40 (2020). https://doi.org/10.1049/htl.2019.0053
4. Obaid, T., Rashed, H., Ali, A.E.N.: ZigBee based voice controlled wireless smart home system. Int. J. Wirel. Mob. Netw. **6**(1), 47 (2014). https://doi.org/10.5121/ijwmn.2014.6104
5. Li, X., Zhu, C., Xu, C.: VR motion sickness recognition by using EEG rhythm energy ra-tio based on wavelet packet transform. Comput. Methods Programs Biomed. **188**, 105266 (2020). https://doi.org/10.1016/j.cmpb.2019.105266

6. Jing, D., Liu, D., Zhang, S.: Fatigue driving detection method based on EEG analysis in low-voltage and hypoxia plateau environment. Int. J. Transport. Sci. Technol. **9**(4), 366–376 (2020). https://doi.org/10.1016/j.ijtst.2020.03.008

7. Mijović, B., De Vos, M., Gligorijević, I.: Source separation from single-channel record-ings by combining empirical-mode decomposition and independent component analysis. IEEE Trans. Biomed. Eng. **57**(9), 2188–2196 (2010)

8. Abdi-Sargezeh, B., Foodeh, R., Shalchyan, V.: EEG artifact rejection by extracting spatial and spatio-spectral common components. J. Neurosci. Methods **358**, 109182 (2021). https://doi.org/10.1016/j.jneumeth.2021.109182

9. Ma, P., Ren, J., Sun, G.: Multiscale superpixelwise prophet model for noise-robust feature extraction in hyperspectral images. IEEE Trans. Geosci. Remote Sens. **61**, 1–12 (2023). https://doi.org/10.1109/TGRS.2023.3260634

10. Teja, S.S.S., Embrandiri, S.S., Chandrachoodan, N.: EOG based virtual key-board. In: 2015 41st Annual Northeast Biomedical Engineering Conference (NEBEC), Troy, NY, USA, pp: 1–2. IEEE (2015). https://doi.org/10.1109/NEBEC.2015.7117201

11. Saini, M., Satija, U., Upadhayay, M.D.: Wavelet based waveform distortion measures for as-sessment of denoised EEG quality with reference to noise-free EEG signal. IEEE Signal Process. Lett. **27**, 1260–1264 (2020)

12. Cirugeda-Roldán, E.M., Molina-Picó, A., Cuesta-Frau, D.: Comparative study between sample entropy and detrended fluctuation analysis performance on EEG records under data loss. In:2012 Annual International Conference of the IEEE Engineering in Medicine and Biology Society, San Diego, CA, USA, pp. 4233–4236 IEEE, Piscataway (2012). https://doi.org/10.1109/EM-BC.2012.6346901

13. Sun, H., Ren, J., Zhao, H.: Novel gumbel-softmax trick enabled concrete autoencoder with entropy constraints for unsupervised hyperspectral band selection. IEEE Trans. Geosci. Remote Sens. **60**, 1–13 (2021)

14. Joyce, C.A., Gorodnitsky, I.F., Kutas, M.: Automatic removal of eye movement and blink artifacts from EEG data using blind component separation. Psychophysiology **41**(2), 313–325 (2004). https://doi.org/10.1111/j.1469-8986.2003.00141

15. Padfield, N., Ren, J., Qing, C., et al.: Multi-segment majority voting decision fusion for MI EEG brain-computer interfacing. Cogn. Comput. **13**, 1484–1495 (2021). https://doi.org/10.1007/s12559-021-09953-3

16. Park, S.U., Han, J.H., Hong, S.K.: A study on behavioral differentiation EEG data selecting algorithm using LSTM and PCA. In: 2021 24th International Conference on Electrical Ma-chines and Systems (ICEMS), Gyeongju, Republic of Korea, pp. 705–709. IEEE, Piscataway (2021). https://doi.org/10.23919/ICEMS52562.2021.9634648

17. Qin, X., Yang, P., Shen, Y.: Classification of driving fatigue based on EEG signals. In: 2020 International Symposium on Computer, Consumer and Control (IS3C), Taichung City, Taiwan, China, pp. 508–512. IEEE, Piscataway (2020)

18. Yuan, W., Xin. L., Yan, Z.: Driving fatigue detection based on EEG signal In: 2015 Fifth International Conference on Instrumentation and Measurement, Computer, Communication and Control (IMCCC), Qinhuangdao, China, pp. 715–718. IEEE, Piscataway (2015). https://doi.org/10.1109/IMCCC.2015.156

19. Norden, E.H., Shen, Z., Long, S.R.: The empirical mode decomposition and the Hilbert spectrum for nonlinear and non-stationary time series analysis. Proc. Royal Soc. Lond. Ser. A Math. Phys. Eng. Sci. **454**(1971), 903–995 (1998). https://doi.org/10.1098/rspa.1998.0193

20. Kotan, S., Van Schependom, J., Nagels, G.: Comparison of IMF selection methods in class-ification of multiple sclerosis EEG data. In: 2019 Medical Technologies Congress (TIPT-EKNO), Izmir, Turkey, pp. 1–4. IEEE, Piscataway (2019)

21. Padfield, N., Ren, J., Murray, P.: Sparse learning of band power features with genetic channel selection for effective classification of EEG signals. Neurocomputing **463**, 566–579 (2021). https://doi.org/10.1016/j.neucom.2021.08.067
22. Yang, Z., Ren, H.: Feature extraction and simulation of EEG signals during exercise-induced fatigue. IEEE Access **7**, 46389–46398 (2019)
23. Belyavin, A., Wright, N.A.: Changes in electrical activity of the brain with vigilance. Electroencephalogr. Clin. Neurophysiol. **66**(2), 137–144 (1987). https://doi.org/10.1016/0013-4694(87)90183-0
24. Subasi, A.: Automatic recognition of alertness level from EEG by using neural network and wavelet coefficients. Expert Syst. Appl. **28**(4), 701–711 (2005). https://doi.org/10.1016/j.eswa.2004.12.027
25. Rasmussen, P., Stie, H., Nybo, L., et al.: Heat induced fatigue and changes of the EEG is not re-lated to reduced perfusion of the brain during prolonged exercise in humans. J. Therm. Biol. **29**(7–8), 731–737 (2004)
26. Padfield, N., Zabalza, J., Zhao, H., et al.: EEG-based brain-computer interfaces using motor-imagery: techniques and challenges. Sensors **19**(6), 1423 (2019)

Visual Sentiment Analysis with a VR Sentiment Dataset on Omni-Directional Images

Rong Huang[1], Haochun Ou[1], Chunmei Qing[1,3](✉), and Xiangmin Xu[2,3]

[1] School of Electronic and Information Engineering, South China University of Technology, Guangzhou, China
qchm@scut.edu.cn
[2] School of Future Technology, South China University of Technology, Guangzhou, China
xmxu@scut.edu.cn
[3] Pazhou Lab, Guangzhou, China

Abstract. Visual content can affect viewer's emotions, which makes sentiment analysis of visual content more and more concerned. Sentiment analysis on omni-directional images plays an important role in virtual reality (VR) applications such as user behaviour prediction, game scene modelling, psychotherapy, etc. However, due to the serious lack of validated VR emotional datasets, the research progress of sentiment analysis in VR is very slow. In this paper, firstly, we build a VR sentiment dataset containing 1,140 emotion-eliciting omni-directional images. Secondly, a pyramidal dual attention network is proposed to analyse the sentiment task. According to the characteristics of omni-directional images, this network utilizes the dual attention module to capture emotion-eliciting regions and adaptively establish the connection between them. Furthermore, objects of different scales have different contributions to evoke emotions. Therefore, the pyramidal feature hierarchy can analyse objects with different complexity by using multi-layer visual features. Finally, quantitative and qualitative experiments on the self-established dataset illustrate that the proposed network can effectively predict the regions that elicit emotions.

Keywords: Virtual Reality · Dataset · Omni-directional Image · Sentiment Analysis · Emotion Recognition

1 Introduction

Human emotion cognition is considered as an advanced stage of artificial intelligence. Compared with traditional emotion-eliciting stimulus (e.g., pictures, text, voice, video, etc.), virtual reality (VR) is a novel and powerful tool to induce emotion, which adequately stimulates senses to trick users into accepting the virtual environment through real experience and high immersion [1]. Therefore, understanding the emotional impact of VR materials on viewers plays an important role in psychotherapy, user behavior prediction, game scene modelling and so on [2].

The omni-directional image (ODI) in VR is a technology that constructs a virtual environment based on 360° real scenes. It is the projection of all the spatial information

J. Ren et al. (Eds.): BICS 2023, LNAI 14374, pp. 300–309, 2024.
https://doi.org/10.1007/978-981-97-1417-9_28

on the two-dimensional plane. By using the head-mounted display to render it from different perspectives in real time, the observers can achieve all-round visual experience. Therefore, different from traditional 2D images, omni-directional images can obtain more scene information.

In order to analyze the sentiment of VR materials in immersive virtual environments, we established the first VR Sentiment Dataset named as VRSenD, which contains 1,140 emotion-eliciting omni-directional images stored in equi-rectangular projection (ERP) format. Some samples are shown in Fig. 1. It can be seen that the contents of the omni-directional images have strong emotion induction, and VRSenD can be acted as a potential resource of emotion stimulation in VR and served as a good launching pad for researchers. Since human recognition is highly subjective, which is not conducive to explore the emotion induced mechanism in VR, this work focuses on binary sentiment (i.e., positive and negative) analysis problem.

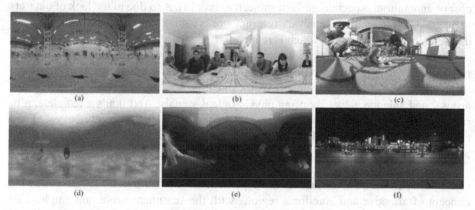

Fig. 1. Samples of omni-directional images from our VRSenD. The first row illustrates the positive samples, while the second row illustrates the negative samples. (a) is a group of happy dancing workers. (b) is a birthday party. (c) is a boy petting a kitten. (d) is a shark chasing the divers. (e) is a scary monster coming at you. (f) is a depressed boy sitting on the stone bench.

The omni-directional image has rich information and all-round perspective. However, some of the content is redundant, which requires the method to have strong object or target location ability. As shown in Fig. 1, all samples illustrate the different scales of objects may have different emotional contributions. Furthermore, due to the ERP format, a whole object may be segmented and projected to both sides like (e) and the content severely distorted in the upper and lower boundaries in (a)–(f). And (c) and (d) illustrate the long-distance dependence caused by large semantic span. According to these characteristics of ODI, the ability to adaptively establish the connections of the emotion-eliciting objects and regions is indispensable. Thus, the traditional sentiment analysis models cannot be directly applied to VR scenes.

Therefore, in this paper a new VR Sentiment Dataset (VRSenD) is established for sentiment analysis. The omni-directional images in this dataset have abundant emotion objects and strong emotion induction. And a novel network (PDANet) is presented for the sentiment analysis on omni-directional images. In this architecture, the dual attention module can adaptively capture the long-distance dependencies between different objects and regions. And the pyramidal feature hierarchy can analyze objects with different complexity. Extensive experiments and visualization results illustrate the effectiveness of the proposed PDANet.

2 Related Works

From the survey [3], most current VR datasets are basically used for saliency detection on 360° video/image to help understand how people perceive and interact with immersive content. In those datasets, there are a lot of scenes with single or simple content, which are not suitable for using as emotion-eliciting materials. Besides, the VR dataset size is relatively small, and most sizes are less than 100. However, CNN-based methods generally rely on the large-scale manually labelled datasets, which are limited by high cost of annotation, especially in such subjective task [4]. And due to the lack of datasets in the field of VR sentiment analysis, relevant works cannot be carried out. Our self-built dataset is over 1,000 in size, which is relatively big enough.

Furthermore, there are very few works to analyze the emotional impact of VR materials. In [2], only four architectural scenes were designed to elicit different emotional responses in subjects, where support vector machine was used to predict the subject's arousal and valence values by using physiological signals. And that's a simple exploration in the field of VR sentiment analysis. As for [5], 73 VR clips are collected and correlations between the standard deviation of head yaw and valence and arousal ratings are analyzed.

There are many works in image sentiment analysis that investigate how images convey rich emotions to influence visual perception [6]. Yang et al. [7] proposed the concept of affective and emotional regions with the sentiment score and emphasized its importance, because different regions may have different effects on the expected emotional expression. And they presented the weakly supervised coupled convolutional network (WSCNet) [8], which considers local regions by integrating both local features and global features in to the final representation for visual sentiment analysis. Ou et al. [9] proposed MCPNet and achieves optimal performance based on the alterable scale and multi-level local regional emotional affinity analysis under the global perspective. Due to the characteristics of ODI, the traditional sentiment analysis method cannot be directly applied, but the design ideas can be used for reference.

3 Dateset

In order to promote the development of sentiment analysis in VR field, we built the VRSenD. The production process mainly includes the following four steps: data collection, data pre-processing, sentiment annotation and dataset statistics.

Data Collection. Omni-directional images (ODIs) on famous photo material websites are not diverse in the content of immersive scenes, so they are not suitable for inducing diverse emotions. Inspired by [10], the relevant emotion-eliciting videos were collected from three major sources: YouTube, Bilibili and existing VR video datasets. Finally, 964

360° VR videos are collected in total. The categories include documentaries, natural scenery, interviews, horror movies, animation, VR games and people's daily activities, which greatly enrich the semantic information of emotions.

Data Pre-processing. The key frames from every 360° video are got by OpenCV automatically. According to the length of VR video and the richness of scene semantics, key frames are intercepted by equal intervals. Then, the high-repetition omni-directional images were deleted by comparing the metric of structural similarity. For VR videos with almost unchanged scenes and single emotional expression, only the most representative frame is reserved, while for other VR videos, no more than 3 frames are reserved. And through manual screening, the ODIs, which have poor quality, incorrect projection format and text occlusion, are removed. Finally, 1,140 ODIs were captured from 964 VR videos.

The projection formats of omni-directional images from different sources are not the same. The most common method for storing omni-directional images is equi-rectangular projection (ERP) and cubemap projection (CMP) [3]. ERP projection makes the content severely distorted in poles, but global contexts is preserved. Instead, CMP projects the content onto the six faces of a cube that is tangent to it, which eliminates the distortion but loses the global information. Because the global contexts can help capture long-distance dependencies and provide a comprehensive understanding of the overall scenario, the ERP storage format is adopted in this work.

Sentiment Annotation. The subjects viewed omni-directional images through an Oculus Rift Head-Mounted Display. The Oculus Rift has a resolution of 2,160 × 1,200 pixels, a 110° field of view and a refresh rate of 90 Hz. The clarity of the ODIs mapping on Unity is within acceptable limits. The sentiment annotation paradigm is as following: First, the subjects sat on rotating chairs in a soundproof room and were first screened for their ability to express emotions normally both physically and mentally. Then they were asked to give arousal and valence values [11] of eight representative images from the IAPS dataset [12] for pre-test. Every eight omni-directional images are divided into a group. The browsing time of each ODI is set to 15 s [13] and the interval between each group is 30 s. The subjects annotated ODIs with binary sentiment labels. In the meantime, we need to prevent careless and deceptive behavior of subjects and annotation inaccuracy due to visual fatigue.

Table 1. Summary of the annotation result of the VR Emotional Dataset. "Five Agree" indicates that all five annotators gave the same sentiment label for the given ODIs.

Sentiment	Five Agree	Four Agree	Three Agree	Sum (%)
Positive	486	199	105	780 (68%)
Negative	179	81	90	360 (32%)
Sum (%)	665 (58%)	280 (25%)	195 (17%)	1,140

Dataset Statistics. In summary, 30 effective subjects participated in this experiment, among which the male and female ratio was 1:1, and each person annotated 300 ODIs on average. Finally, we obtained 1,140 emotion-eliciting omni-directional images. Table 1 shows the statistics of the annotation results. The quantity and proportion of negative labels are 360 and 32%, while positive labels are 780 and 68%. Only a small portion 17% had significant disagreements between the five subjects (3 vs. 2). In addition, the categories of ODIs include but are not limited to the following types: natural landscape, animals, people activities, games, shows, etc.

4 Methodology

In this section, the proposed pyramidal dual attention network (PDANet) will be presented in detail as illustrated in Fig. 2. The ResNet101 [14] is utilized as backbone, since it's a widely accepted effective model in visual tasks.

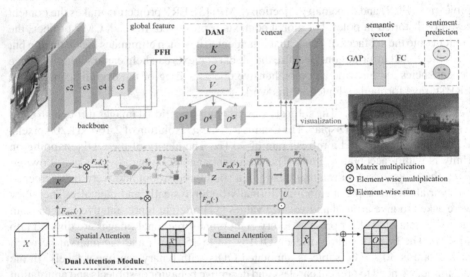

Fig. 2. The architecture of the proposed PDANet. PFH represents the pyramidal feature hierarchy. DAM represents dual attention module. GAP means global average pooling and FC means full connected layer. Below is the framework of dual attention module.

4.1 Network Structure

Given the amount of computation, for the input image, only the features layers c_3, c_4 and c_5 are selected to form the pyramidal feature hierarchy. The extracted features from different CNN layers usually have different properties [15], which make the model have the perception ability of objects of different scales. And they are sampled down to unify the number of channels and the size of feature maps. Then the output features O^l are obtained through the dual attention module. These multi-level contextual features are concatenated as global sentiment representation E which is applied to visualize to verify the effectiveness of the proposed PDANet. Finally, through a global average pooling, it enters the full connection layer for sentiment classification.

$$E = [X_{global}, O^l] = [X^5, O^3, O^4, O^5]. \tag{1}$$

4.2 Dual Attention Module

Dual attention module (DAM) is a core module that adaptively establishes connections between different regions and objects under the guidance of global image representation. As shown in Fig. 2, spatial attention module and channel attention module determine the location and content of attention respectively.

Spatial Attention Module. As illustrated in spatial attention module, firstly, $F_{conv}(\bullet)$ is adopted to the $X \in \mathbb{R}^{c \times h \times w}$ to generate three new feature maps Q, K and V as Query, Key and Value respectively. We take $F_{conv}(\bullet)$ to be a conv1x1 operator to obtain the features of dimension reduction with reduction ratio r. Then three spatial matrix operations include: reshape Q, reshape and transpose K, and reshape V:

$$Q' = \mathbb{R}^{\frac{c}{r} \times hw}, K' = \mathbb{R}^{hw \times \frac{c}{r}}, V' = \mathbb{R}^{c \times hw}. \tag{2}$$

Then the spatial attention map $S \in \mathbb{R}^{hw \times hw}$ is generate by multiplying K' and Q' with a softmax function:

$$S = F_{sa}(Q', K') = \frac{\exp(Q'_i \cdot K'_j)}{\sum\limits_{i=1}^{hw} \exp(Q'_i \cdot K'_j)}. \tag{3}$$

Here, s_{ij} measures the i position's impact on j position in spatial attention map S, which stores affinity coefficients that represent the degree of correlation between the location and key sentiment information. Finally, spatial attention $\overline{X} \in \mathbb{R}^{c \times h \times w}$ is obtained by multiplying $V' \in \mathbb{R}^{c \times hw}$ with $S \in \mathbb{R}^{hw \times hw}$:

$$\overline{X} = S \cdot V' = \alpha \sum\limits_{i=1}^{hw} (s_{ij} V'_i). \tag{4}$$

α is a scale parameter, which is initialized as 0 and gradually learns to assign more weight. It can be inferred from the above equation, $\overline{X}$ selectively aggregates emotional contexts according to the spatial attention map, and similar semantic features can reinforce each other.

Channel Attention Module. Firstly, $F_{sq}(\bullet)$ squeeze global spatial information by simply using global average pooling to generate channel-wise Z:

$$Z = F_{sq}(X) = \frac{1}{hw} \sum\limits_{i=1}^{h} \sum\limits_{j=1}^{w} x_{ij}. \tag{5}$$

The output $Z \in \mathbb{R}^{c \times 1 \times 1}$ of the transformation $X \in \mathbb{R}^{c \times h \times w}$ can be interpreted as a collection of the local descriptors. Then we opt to employ a simple gating mechanism with a sigmoid activation like [16]. Here, δ refers to the ReLU function.

$$U = F_{ex}(Z, W) = sigmoid(W_2 \delta(W_1 Z)). \tag{6}$$

In order to limit model complexity, the reduction ratio r is utilized to form a bottleneck through two fully-connected (FC) layers.

$$W_1 \in \mathbb{R}^{c \times \frac{c}{r}}, W_2 \in \mathbb{R}^{\frac{c}{r} \times c}. \tag{7}$$

Channel attention $\tilde{X}$ is obtained by element-wise multiplying $U \in \mathbb{R}^{c \times 1 \times 1}$ and X:

$$\tilde{X} = U \odot X. \tag{8}$$

Channel attention explicitly establishes channel correlations to enhance the learning of feature maps, so the proposed model can increase its sensitivity to emotion-eliciting features. Specifically, each high-level feature maps can be regarded as the response of a specific sentiment. By mining the inter-dependence between channels, $\tilde{X}$ can capture long-distance dependencies to improve the expression of sentiment. Finally, by the element-wise sum of $\tilde{X}$ and $\overline{X}$, we obtained the output feature maps which can adaptively capture emotion-eliciting regions by focusing on long-distance contexts and establish connections between them.

$$O = \overline{X} \oplus \tilde{X}. \tag{9}$$

5 Experiments

The proposed PDANet uses pre-trained ResNet101 as the backbone. And the input omni-directional images are resized at least 1/8 of the original size. Apply random horizontal flipping and cropping random 512×512 patches as a form of data augmentation. Batch size is 8. And the SGD optimizer with a momentum of 0.9. The learning rate is initialized to 0.001 and drops to 1/10 every 7 epochs of 50 epochs. The VRSenD is randomly divided into 80% for training and 20% for test and prevent ODI of the same video from appearing in both the train and test datasets. All experiments are performed on NVIDIA RTX 2080Ti GPU by PyTorch.

5.1 Comparisons with State-of-the-Art Methods

In this experiment, five-folder cross validation is used to calculate the average accuracy of different models. Here, five classic models are compared: AlexNet [17], Vgg19 [18], Inception-v3 [19], ResNet101 [14], and DenseNet121 [20], respectively. Moreover, in order to clarify the effectiveness of the proposed model, we also add two latest state-of-the-art models (WSCNet [8] and MCPNet [9]) for comparison.

The Imbalance column in Table 2 summarizes the averaged accuracy results on VRSenD. And the accuracy is 1.39% and 1.06% higher comparing with WSCNet and

MCPNet, respectively. Because the number of the positive and negative samples are imbalanced, the accuracy of random prediction could reach about 70%. Therefore, at least the test dataset needs to be balanced to clearly evaluate the proposed model. The result is shown in the Balance column. It can be found that the PDANet still has better performance. And the accuracy is 1.53% and 2.24% higher comparing with WSCNet and MCPNet, respectively.

Table 2. The average accuracy of the five-fold cross validation performance of different models on our VRSenD. The imbalanced column means all samples are used, and the balanced column indicates the same number of positive and negative samples.

Model	Imbalance	Balance
AlexNet [17]	82.83	72.99
Vgg19 [18]	83.46	73.80
Inception-v3 [19]	83.70	75.61
ResNet101 [14]	83.89	75.23
DenseNet121 [20]	83.21	75.48
WSCNet [8]	83.64	76.46
MCPNet [9]	83.97	75.76
Ours	**85.03**	**78.00**

5.2 Visualization

In order to further evaluate our PDANet, a qualitative analysis is utilized by using the class activation mapping [21] to visualize the global emotional feature maps E. The visualization results of ResNet101, WSCNet, MCPNet and PDANet are presented by heat maps as shown in Fig. 3. Although the final sentiment labels are the same, it can be seen that the captured objects or regions have obvious differences.

In Fig. 3(a)–(f), the PDANet can correctly focus on emotion-inducing regions and objects, and give them adaptive attention according to their importance. While the boy sitting on a stone bench in low spirits is too small to be easily ignored in (f), the proposed PDANet still has the ability to capture it. Besides, even ResNet101 can also correctly focus on the same regions like PDANet in (c), the detected emotion-eliciting objects are different. The PDANet can better locate the emotion-inducing objects such as the touch and the cat's response. In (d), the PDANet can locate the dangerous sharks and the divers, which proves the validity of the dual attention module. Because the dual attention module can capture long-distance dependencies, our model can efficiently explore the connections between different objects and regions.

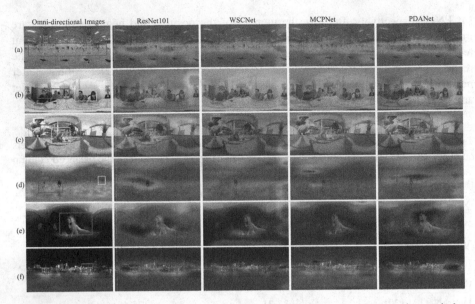

Fig. 3. Visualization comparisons. The bounding box illustrates the significant regions to induce emotion by asking subjects to identify the regions of the image that most influences the evoked emotion.

6 Conclusion

In this work, a new VR Sentiment Dataset (VRSenD) is established, which has abundant emotion objects and strong emotion induction, and can be used to understand the emotional impact of VR materials on viewers. Furthermore, a novel proposed network (PDANet) can accurately capture emotion-eliciting regions and objects, and establish the sentiment correlation of them adaptively. Besides, it has the perceptual ability of multi-scale objects. The code and dataset will available online: https://github.com/HRz zcl99/VRSenD.

Acknowledgement. This work is partially supported by the following grants: National Natural Science Foundation of China (61972163, U1801262), Natural Science Foundation of Guangdong Province (2022A1515011555, 2023A1515012568), Guangdong Provincial Key Laboratory of Human Digital Twin (2022B1212010004) and Pazhou Lab, Guangzhou, 510330, China.

References

1. Zhu, Y., Zhai, G., Min, X., Zhou, J.: Learning a deep agent to predict head movement in 360-degree images. ACM Trans. Multimed. Comput. Commun. Appl. **16**(4), 1–23 (2020)
2. Marín-Morales, J., et al.: Affective computing in virtual reality: emotion recognition from brain and heartbeat dynamics using wearable sensors. Sci. Rep. **8**(1), 1–15 (2018)
3. Xu, M., Li, C., Zhang, S., Le Callet, P.: State-of-the-art in 360 video/image processing: perception, assessment and compression. IEEE J. Sel. Top. Signal Process. **14**(1), 5–26 (2020)

4. She, D., Sun, M., Yang, J.: Learning discriminative sentiment representation from strongly- and weakly supervised cnns. ACM Trans. Multimed. Comput. Commun. Appl. **15**(3s), 1–19 (2019)
5. Li, B.J., Bailenson, J.N., Pines, A., Greenleaf, W.J., Williams, L.M.: A public database of immersive VR videos with corresponding ratings of arousal, valence, and correlations between head movements and self report measures. Front. Psychol. Original Res. **8**(2116) (2017)
6. Zhao, S., et al.: Affective image content analysis: two decades review and new perspectives. IEEE Trans. Pattern Anal. Mach. Intell. (2021)
7. Yang, J., She, D., Sun, M., Cheng, M.-M., Rosin, P.L., Wang, L.: Visual sentiment prediction based on automatic discovery of affective regions. IEEE Trans. Multimed. **20**(9), 2513–2525 (2018)
8. Yang, J., She, D., Lai, Y.-K., Rosin, P.L., Yang, M.-H.: Weakly supervised coupled networks for visual sentiment analysis. In: Proceedings of the IEEE Conference on Computer Vision and Pattern Recognition, pp. 7584–7592 (2018)
9. Ou, H., Qing, C., Xu, X., Jin, J.: Multi-level context pyramid network for visual sentiment analysis. Sensors **21**(6), 2136 (2021)
10. Li, B.J., Bailenson, J.N., Pines, A., Greenleaf, W.J., Williams, L.M.: A public database of immersive VR videos with corresponding ratings of arousal, valence, and correlations between head movements and self report measures. Front. Psychol. **8**, 2116 (2017)
11. Bradley, M.M., Lang, P.J.: Measuring emotion: the self-assessment manikin and the semantic differential. J. Behav. Ther. Exp. Psychiatry **25**(1), 49–59 (1994)
12. Lang, P.J., Bradley, M.M., Cuthbert, B.N.: International affective picture system (IAPS): technical manual and affective ratings. NIMH Cent. Study Emot. Attent. **1**, 39–58 (1997)
13. De Abreu, A., Ozcinar, C., Smolic, A.: Look around you: saliency maps for omnidirectional images in VR applications. In: 2017 Ninth International Conference on Quality of Multimedia Experience, pp. 1–6 (2017)
14. He, K., Zhang, X., Ren, S., Sun, J.: Deep residual learning for image recognition. In: Proceedings of the IEEE Conference on Computer Vision and Pattern Recognition, pp. 770–778 (2016)
15. Zhao, S., Jia, Z., Chen, H., Li, L., Ding, G., Keutzer, K.: PDANet: polarity-consistent deep attention network for fine-grained visual emotion regression. In: Proceedings of the 27th ACM International Conference on Multimedia, pp. 192–201 (2019)
16. Hu, J., Shen, L., Sun, G.: Squeeze-and-excitation networks. In: Proceedings of the IEEE Conference on Computer Vision and Pattern Recognition, pp. 7132–7141 (2018)
17. Krizhevsky, A., Sutskever, I., Hinton, G.E.: ImageNet classification with deep convolutional neural networks. Commun. ACM **60**(6), 84–90 (2017)
18. Simonyan, K., Zisserman, A.: Very deep convolutional networks for large-scale image recognition. arXiv preprint arXiv:1409.1556 (2014)
19. Szegedy, C., Vanhoucke, V., Ioffe, S., Shlens, J., Wojna, Z.: Rethinking the inception architecture for computer vision.. In: Proceedings of the IEEE Conference on Computer Vision and Pattern Recognition, pp. 2818–2826 (2016)
20. Huang, G., Liu, Z., Van Der Maaten, L., Weinberger, K.Q.: Densely connected convolutional networks. In: Proceedings of the IEEE Conference on Computer Vision and Pattern Recognition, pp. 4700–4708 (2017)
21. Zhou, B., Khosla, A., Lapedriza, A., Oliva, A., Torralba, A.: Learning deep features for discriminative localization. In: Proceedings of the IEEE Conference on Computer Vision and Pattern Recognition, pp. 2921–2929 (2016)

Generating Type-Related Instances and Metric Learning to Overcoming Language Priors in VQA

Chongxiang Sun, Ying Yang[✉], Zhengtao Yu, Chenliang Guo, and Jia Zhao

School of Computer and Information, Fuyang Normal University, Fuyang 342001, Anhui, China
yyang@fynu.edu.cn, zhaojia11b@mails.ucas.ac.cn

Abstract. Visual Question Answering (VQA) is a multimodal task that integrates computer vision and natural language processing. It poses a challenge in the field due to language prior, which is influenced by the dataset and the underlying model. Language priori refers to the fact that the model relies on superficial connections between question types and high-frequency answers. In this paper, we propose a joint method of type-related instances and metric learning (TI-ML). This module addresses the language prior associated with the question types. To ensure that the model can learn common features among the instances of different question types, we reduce the distance of instances of the same category in the answer space by metric learning based on the answers. Experimental results show that our method achieves better performance in the ranks of non-pre-trained models on the benchmark dataset VQA-CP v2, meanwhile maintaining high performance in the dataset VQA v2 as well.

Keywords: Visual Question Answering · Language Prior · Self-supervised Learning · Metric Learning

1 Introduction

Visual Question Answering (VQA) has emerged as a prominent cross-modal research task within the field of artificial intelligence. It involves taking an image and its corresponding text question as input and generating an answer as output. Consequently, VQA lies at the intersection of computer vision and natural language processing.

According to recent studies [1, 2] have revealed that the prediction results of VQA models heavily rely on the text modality, primarily due to the strong pseudo-correlation between certain question types and answers. For example, the question type "how many" and the answer "2" often appear together. Previous solutions [3–5] have focused on addressing language priors associated with specific question types, primarily by leveraging differences between answer distributions in the training and test sets. But the models do not really learn the relationship between the feature objects of the instances. In our research, we discovered that the high-frequency answer "dark" in the "what time" question type did not appear in the corresponding question type in the training set,

J. Ren et al. (Eds.): BICS 2023, LNAI 14374, pp. 310–321, 2024.
https://doi.org/10.1007/978-981-97-1417-9_29

but rather appeared frequently in the "is the" question type. This finding highlights the importance of not only addressing language priors for each question type but also establishing connections between instances of different question types in the feature fusion embedding space.

To address this issue, we propose a joint method of type-related instances and metric learning. The method incorporates a metric learning module. By minimizing the distance between instances with the same answers in the cross-modal joint embedding space. The model can better understand the commonalities present in different question type instances that share the same answer. Figure 1 illustrates this concept, where different question type instances with the same answer exhibit certain commonalities (e.g., "bus") in their image and question features. These commonalities play a crucial role in the model's predictive inference process. In addition to the metric learning module, we incorporate a construction type-related instances module into the model, aiming to mitigate the impact of language priors that are strongly associated with specific question types.

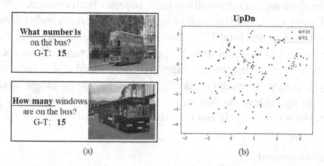

Fig. 1. (a) shows a pair of examples with different question types ("What number is" and "How many"), but with the same answer ("15"). (b) visualizes the projection of each of these instances in the answer space obtained by UpDn [16]. G-T denotes the ground truth.

In summary, our contributions are as follows: i) A self-supervised instance selection strategy is proposed to create image sets in an offline manner, construct a type-related instance for each input instance, and finally utilize such instances through a self-supervised auxiliary task. ii) A metric learning method is introduced to further distinguish different answer instances with the idea of metric learning to establish the connection between different question type instances. This ensures that the model is more likely to produce different answers when predicting the same question type instances rather than answering as high frequency answers.

2 Related Work

Related studies have found that existing VQA models rely excessively on the superficial association between questions and answers, i.e., they give predicted answers without analyzing image information. To address this problem, many scholars have proposed different solutions, which can be classified into the following two categories after comprehensive analysis.

2.1 More Detailed Models

Most of the models in this category are based on integration methods. SAR [6] is a typical model integration approach with a model training into two phases, where the first stage uses a debiasing model to predict the candidate answers. The second stage then reorders the candidate answers by a visual entailment task, and the model eventually performs better. This is since SAR uses LXMERT [7] as the base VQA model for the second stage, which uses the VQA v2 dataset in the pre-training process, So models based on the use of pre-trained parameters generally exhibit superior performance. LPF [8] uses the conditioning factors of the question-only branches as adaptive weights for each training instance, thus reshaping the total VQA loss into a more balanced form. Ramakrishnan [9] pitted the basic VQA model and the question-only branch against each other to reduce the dependence on language. Liang [10] introduced a self-supervised contrast learning scheme to help the model reduce its dependence on language modality. This is achieved by learning the relationship between the counterfactual instances generated by the instance construction module and the original instances.

2.2 Data Enhancement

The overall idea of the approach is to construct a more balanced dataset. For example, CSS [11] balances the original dataset by generating two counterfactual instances to encourage the model to be able to focus more on the key features in the input instances. MMBS [12] reduces the reliance on language by reconstructing the problem and then forming a positive instance with the original image.

In this paper, self-supervised learning approach is used to overcome the language prior. Based on previous research results, two methods exist for constructing self-supervised auxiliary task instances. The first one is Zhu [13], which randomly selects images from the same batch of training data and forms unrelated instance pairs with the original instance's questions. The new instances constructed by this method have low correlation with the original instances in terms of question types. Therefore, it is not ideal in alleviating the language prior closely related to the question type. The second one is Wu [14] who selected the entire instance of the same batch of training data with the same question type as the current instance, but with different ground-truth answers, as the input instance. Due to the limited size of each batch of training data, this method is able to select very limited instances. Therefore, we propose a new method of input instance construction, from which type-related instances are constructed with strong correlation between the question and image aspects and the question type. It shows better performance for solving the language prior problem of the model.

3   Method

The framework of our method is shown in Fig. 2. Next, its working principle will be described in detail.

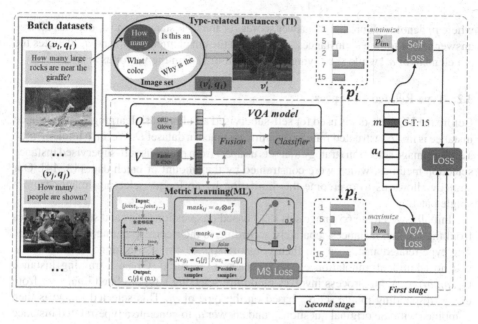

Fig. 2. The framework of our method

3.1   The VQA Paradigm

The goal of the VQA task is to automatically answer textual questions based on a given image, and the commonly used backbone model [15, 16] treats VQA as a multi-label classification task. Specifically, given a VQA dataset $D = \{v_i, q_i, a_i\}_{i=1}^{N}$, each instance i^{th} is a triplet combination of images $v_i \in V$, questions $q_i \in Q$, and answers $a_i \in \mathbb{R}^{|A|}$, and the answers are usually in multi-label form. The basic VQA paradigm aims to learn to map an image-question pair to its answer space $F : V \times Q \to \mathbb{R}^{|A|}$. Given a training instance $(v_i, q_i, a_i) \in D$, the VQA model takes the image-question pair (v_i, q_i) as input, produces a fusion vector $Joint_i$ (Eq. (1)) after the feature fusion function, and finally outputs a predicted probability vector $p_i \in \mathbb{R}^{|A|}$ (Eq. (2)) in the answer space after the classifier.

$$Joint_i = F(v_i, q_i; \phi) \tag{1}$$

$$p_i = softmax(Joint_i) \tag{2}$$

where ϕ denotes the model parameters.

The training objective of the base model is to minimize the multi-label cross-entropy loss. The formula is shown in Eq. (3):

$$L_{vqa} = -\frac{1}{N} \sum_{i=1}^{N} \sum_{c=1}^{|A|} \left[a_{ic} \log p_{ic} + (1 - a_{ic}) \log(1 - p_{ic}) \right] \tag{3}$$

where p_i denotes all predicted answer scores for the ith instances, and a_i denotes all true answer scores for the ith instances. p_{ic} Denotes the cth element of p_i, and a_{ic} denotes the cth element of a_i, while specifying the position of G-T in the vector as m.

3.2 Type-Related Instances

Before the VQA model is used for self-supervised auxiliary task learning, a type-related instance is first constructed for each training instance in dataset D. We have implemented a unique image selection strategy that sets us apart from previous self-supervised instance selection methods, which were constrained by the amount of batch data [13, 14]. Our strategy allows us to overcome this limitation by no longer selecting images from a single batch.

Initially, we create 65 image sets (S_{qt}) based on question types and store the images from the original instances separately according to their corresponding question types. The type-related instances are then constructed, the process is done automatically, and no additional annotation of the instances is required. Given a training instance $(v_i, q_i, a_i) \in D$, the process involves randomly selecting a type-related image $v_{i'}$ from the corresponding image set S_{qt} based on the type of q_i. This selected image is then combined with the original question q_i and answer a_i to generate a type-related instance (v'_i, q_i, a_i). Through this precise image selection method, we ensure that the constructed image-question pairs exhibit a significant correlation between question types.

Inspired by previous work [13], we use a self-supervised module to evaluate the constructed type-related instance. The self-supervised module in our model shares the same network structure as the basic VQA task. The input of the self-supervised module is the constructed type-related instance (v'_i, q_i, a_i) and the output is the predicted value p'_i. Since the type-related instance do not answer the question theoretically correctly, we use the G-T prediction value p'_{im} for them, as an indicator of the evaluation of the model influenced by the language prior. Therefore, the purpose of training the self-supervised module is to minimize p'_{im}, and the self-supervised loss L_{self} is specifically expressed as:

$$L_{self} = \frac{1}{N} \sum_{i}^{N} p'_{im} \tag{4}$$

Using the self-supervised loss L_{self} described above, there is no need to explicitly optimize the model specifically to evaluate the type-related instance of the input data.

3.3 Metric Learning Module

The goal of Metric Learning (ML) is to learn an embedding space between instances in which the vector distance of similar instances is reduced, and the vector distance of dissimilar instances is increased [17]. We perform inter-instance metric learning based on answers for the current training data. For the input instance (v_i, q_i), the correlation between the instance feature fusion vector $Joint_j$ and $Joint_i$ is first expressed by the cosine similarity C_{ij}, as shown in Eq. (5). Then, a mask value $mask_{ij}$ is calculated based on the instance labels a of both, as shown in Eq. (6).

$$C_{ij} = Cosine(Joint_i, Joint_j) \quad i, j \in [1, bs] \tag{5}$$

$$mask_{ij} = a_i \otimes a_j^T \tag{6}$$

where bs denotes the batch size. C_{ij} Denotes the similarity between the current input instance (v_i, q_i) and any instance from the same batch (v_j, q_j). The similarity vector C_i is expressed as $C_i = [C_{i1}, ; \ldots \ldots ; C_{ibs}]$. $mask_{ij}$ is greater than or equal to 0.

Finally, the value $mask_{ij}$ is used as the differentiation condition to classify the instances. The instances with the same answers as (v_i, q_i) are put into the positive instance set, which is denoted as Pos_i. The instances with different answers are put into the negative instance set, which is denoted as Neg_i, as shown in Eq. (7).

$$Pos_i = \begin{cases} C_i[j], \text{ if } mask_{ij} > 0 \\ 0, \text{ otherwise}, \end{cases}$$

$$Neg_i = \begin{cases} 0, \text{ otherwise} \\ C_i[j], \text{ if } mask_{ij} = 0, \end{cases} \tag{7}$$

In this paper, Multi-Similarity Loss (MS loss) [18] is used to learn an embedding space that makes the distance between same-labeled instances closer and the distance between different-labeled instances farther. MS can be expressed in a specific way:

$$L_{ms} = \frac{1}{bs} \sum_{i=1}^{bs} \left\{ \frac{1}{\alpha} \log \left[1 + \sum_{j \in Pos_i} e^{-\alpha(C_{ij}-\mu)} \right] + \frac{1}{\beta} \log \left[1 + \sum_{j \in Neg_i} e^{\beta(C_{ij}-\mu)} \right] \right\} \tag{8}$$

where μ is the similarity margin and is set to 0.5 in our paper. α and β are hyper parameters and is set to 2 and 40.

In summary, our approach can be viewed as an underlying multi-task learning. The total loss formula of the model is obtained as follows:

$$L = L_{vqa} + \lambda_1 L_{self} + \lambda_2 L_{ms} \tag{9}$$

where λ_1 and λ_2 are hyperparameters, and both are greater than 0.

4 Experimental Results and Analysis

4.1 Experimental Setup

We evaluated the performance of the proposed method using the benchmark VQA-CP v2 dataset [19]. The standard VQA accuracy was employed as the evaluation metric for all the following experiments. Our method can be applied to various VQA models.

In this paper, we utilize the UpDn [16] as the base architecture for our method. The training process is divided into two stages. In the first 12 epochs, we only add the metric learning module, and incorporate the self-supervised module in the last 28 epochs. We set the batch size to 512. The parameters are optimized using the Adam optimizer, and the initial learning rate is set to 0.001. However, starting from the 10th epoch of training, the learning rate is halved every 5 epochs.

4.2 Experimental Result

The proposed method is compared with two classical methods, SAN [20] and UpDn [16], as well as other debiasing methods including: (1) single-model methods: AdvReg [9], RUBi [4], HINT [21], SCR [5], DLR [22], SSL [13], CF-VQA [23]; (2) ensemble methods: LMH [3], CSS [11], LMH+MMBS [12], CSS+CL [10], CSS+IntroD [23].

Table 1. Results of the VQA-CP v2 test and VQA v2 val sets using different VQA models

Model	VQA-CP v2 test (%)				VQA v2 val (%)			
	All	Yes/No	Num	Other	All	Yes/No	Num	Other
plain methods								
SAN	24.96	38.35	11.14	21.74	52.02	68.89	34.55	43.80
UpDn	39.74	42.27	11.93	46.05	63.48	81.18	42.14	55.66
ensemble debiasing methods								
LMH	52.45	69.81	44.46	45.54	61.64	77.85	40.03	55.04
CSS	58.95	84.37	49.42	48.21	59.91	73.25	39.77	55.11
CSS+CL	59.18	86.99	49.89	47.16	57.29	67.27	38.40	54.71
LMH+MMBS	56.44	76.00	43.77	49.67	61.87	75.86	40.34	56.95
CSS+IntroD	60.17	89.17	46.91	48.62	62.57	78.57	41.42	56.00
single-model debiasing methods								
AdvReg	41.17	65.49	15.48	35.48	62.75	79.84	42.35	55.16
RUBi	47.11	68.65	20.28	43.18	60.29	76.44	40.22	53.32
HINT	46.73	67.27	10.61	45.88	63.38	81.18	42.99	55.56
SCR	49.45	72.36	10.93	48.02	62.20	78.80	41.60	54.50
DLR	48.87	70.99	18.72	45.57	57.96	76.82	39.33	48.54
SSL	<u>57.59</u>	86.53	<u>29.87</u>	**50.03**	<u>63.80</u>	81.13	43.53	<u>56.00</u>
CF-VQA	53.55	**91.15**	13.03	44.97	63.54	**82.51**	**43.96**	54.30
TI-ML(Ours)	**60.66**	<u>88.53</u>	**51.77**	<u>48.51</u>	**64.56**	<u>81.62</u>	<u>43.64</u>	**57.13**

Performance on the VQA-CP v2 Test Set. As shown in Table 1, we focused on comparing and evaluating the results using the single-model model. Firstly, it can be observed that our approach significantly outperforms the classical method and achieves a 3.07% higher overall accuracy than the second-best single-model method, SSL. Secondly, among all single-model methods, our approach demonstrates the best accuracy performance in the challenging "Num" question category, showing improvements ranging from 21.9% to 41.16% compared to other models. Additionally, our performance in the "Yes/No" and "Other" categories is also commendable, both ranked second in accuracy. Furthermore, despite the ensemble methods employing more complex architectures with higher overall complexity, they still achieve competitive results. For instance, our overall accuracy is 0.49% higher than the best ensemble model, CSS+IntroD.

Performance on the VQA v2 Val Set. The results of the VQA v2 val set [24] are presented in Table 1. It is evident that our method generally outperforms other classical methods, exhibiting an accuracy improvement ranging from 0.76% to 16.32% compared to other models. Particularly noteworthy is its exceptional performance in the "Other" category, ranking first, and its second-best performance in the "Yes/No" and "Num" categories. Our method maintains excellent performance on the VQA v2 dataset and demonstrates significant improvement on the VQA-CP v2 dataset, underscoring the robustness of our approach.

4.3 Ablation Experiment

As shown in Table 2, the "ML" refers to the metric learning module. It can be observed that adding the metric learning module to the UpDn improves accuracy by enabling the module to discover relationships between features of the same answer instances during model training. The "TI" stands for self-supervised learning using type-related instances. The inclusion of the self-supervised learning strategy results in a significant enhancement of the model's performance, confirming the effectiveness of the self-supervised learning module in mitigating the language prior.

Table 2. Ablation studies based on the VQA-CP v2 dataset

Methods	All	Yes/No	Num	Other
UpDn	39.74	42.27	11.93	46.05
+ ML	42.40	43.64	18.31	48.36
+ TI	59.94	88.77	46.96	48.39
+ ML + TI (ours)	**60.66**	**88.53**	**51.77**	**48.51**

4.4 Hyperparameter Analysis

To achieve the optimal performance of the network framework, this section conducts an extensive experimental study on the weights of the self-supervised loss L_{self} and the metric loss L_{ms}. The experimental results are based on the VQA-CP v2 test set, as shown by the red columns depicted in Fig. 3.

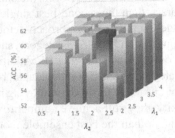

Fig. 3. The accuracy (%) of the VQA-CP v2 test set varies with λ_1 and λ_2 (Color figure online)

4.5 Visualization of Distance in Answer Space

The distance between instances in the answer space was visualized using the t-SNE algorithm, and the results can be seen in Fig. 4. It can be observed that instances of different question types and ground truths exhibit a clearer separation in the answer space after applying the TI-ML. This indicates that our approach enables the VQA model to learn how to accurately distinguish instances with different answers.

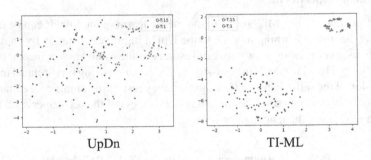

Fig. 4. Visualization of the distance between instances in the answer space using t-SNE

4.6 Qualitative Analysis

This subsection provides a qualitative assessment of the effectiveness of the method proposed in this paper. As depicted in Fig. 5, our method demonstrates the ability to provide correct answers while focusing on the relevant regions. For instance, when responding to the question "How many horses are white?", the UpDn is influenced by the language prior and answers "2" due to its high frequency in the training set. In contrast, our approach accurately focuses on the white horses in the image and answers the question correctly based on visual information. This suggests that our model presented in this paper can reduce the influence of the language prior and provide answers by comprehending the key objects in the image-question pair.

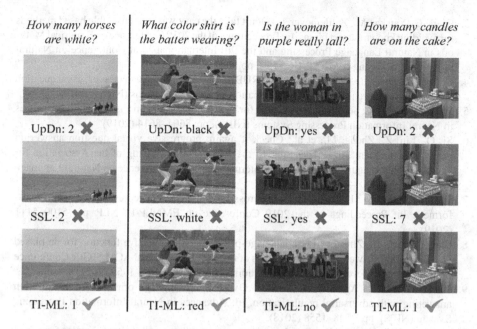

Fig. 5. Qualitative comparison of our method with UpDn [16], SSL [13]

5 Conclusion

In this paper, we propose a new type-related instance and metric learning approach (TI-ML) to further improve the language prior problem faced by existing VQA models. First, a new self-supervised instance selection strategy is used to generate a type-related instance for each instance, which is used to mitigate the language prior that is strongly correlated with the question type. In addition, metric learning enables the model to learn the commonality among instances of the same answer, which further improves the final performance of the method. More advanced results were obtained on the VQA-CP v2 dataset and the VQA v2 dataset, validating the ability of the method to mitigate language priors.

Acknowledgments. This study was funded by the Anhui Provincial Education Department Key Project (2023AH050418, 2022AH051324), the National Natural Science Foundation of China (61906044) and the Key Projects of Natural Science Research in Anhui Colleges and Universities (KJ2020ZD48, KJ2019A0536).

References

1. Antol, S., Agrawal, A., Lu, J., et al.: VQA: visual question answering. In: Proceedings of the IEEE International Conference on Computer Vision (ICCV), pp. 2425–2433 (2015)
2. Agrawal A, Batra D, Parikh D.: Analyzing the behavior of visual question answering models. In: Proceedings of the 2016 Conference on Empirical Methods in Natural Language Processing (EMNLP), pp.1955–1960 (2016)

3. Clark, C., Yatskar, M., Zettlemoyer, L.: Don't take the easy way out: ensemble based methods for avoiding known dataset biases. In: Proceedings of the 2019 Conference on Empirical Methods in Natural Language Processing and the 9th International Joint Conference on Natural Language Processing (EMNLP-IJCNLP), pp. 4069–4082 (2019)
4. Cadene, R., Dancette, C., Cord, M., et al.: RUBi: reducing unimodal biases for visual question answering. In: Advances in Neural Information Processing Systems (NIPS) (2019)
5. Wu, J., Mooney, R.: Self-critical reasoning for robust visual question answering. In: Advances in Neural Information Processing Systems (NIPS), pp. 8604–8614 (2019)
6. Si, Q., Lin, Z., Zheng, M., et al.: Check it again: progressive visual question answering via visual entailment. In: Proceedings of the 59th Annual Meeting of the Association for Computational Linguistics and the 11th International Joint Conference on Natural Language Processing, pp. 4101–4110 (2021)
7. Tan, H., Bansal, M.: LXMERT: learning cross-modality encoder representations from transformers. In: Proceedings of the 2019 Conference on EMNLP-IJCNLP, pp. 5100–5111 (2019)
8. Liang, Z., Hu, H., Zhu, J.: LPF: a language-prior feedback objective function for de-biased visual question answering. In: Proceedings of the 44th International ACM SIGIR Conference on Research and Development in Information Retrieval, pp. 1955–1959 (2021)
9. Ramakrishnan, S., Agrawal, A., Lee, S.: Overcoming language priors in visual question answering with adversarial regularization. In: Advances in Neural Information Processing System (NIPS), pp. 1548–1558 (2018)
10. Liang, Z., Jiang, W., Hu, H., et al.: Learning to contrast the counterfactual samples for robust visual question answering. In: Proceedings of the 2020 Conference on Empirical Methods in Natural Language Processing (EMNLP), pp. 3285–3292 (2020)
11. Chen, L., Yan, X., Xiao, J., et al.: Counterfactual samples synthesizing for robust visual question answering. In: Proceedings of the IEEE/CVF Conference on Computer Vision and Pattern Recognition (CVPR), pp. 10800–10809 (2020)
12. Si, Q., Liu, Y., Meng, F., et al.: Towards robust visual question answering: making the most of biased samples via contrastive learning. In: Proceedings of the 2022 Conference on Empirical Methods in Natural Language Processing (EMNLP), pp. 6650–6662 (2022)
13. Zhu, X., Mao, Z., Liu, C., et al.: Overcoming language priors with self-supervised learning for visual question answering. In: Proceedings of the Twenty-Ninth International Joint Conference on Artificial Intelligence, pp. 1083–1089 (2020)
14. Wu, Y., Zhao, Y., Zhao, S., et al.: Overcoming language priors in visual question answering via distinguishing superficially similar instances. In: Proceedings of the 29th International Conference on Computational Linguistics, pp. 5721–5729 (2022)
15. Mahabadi, R.K., Henderson, J.: Simple but effective techniques to reduce biases. arXiv:1909.06321 (2019)
16. Anderson, P., He, X., Buehler, C., et al.: Bottom-up and top-down attention for image captioning and visual question answering. In: Proceedings of the IEEE Conference on Computer Vision and Pattern Recognition (CVPR), pp. 6077–6086 (2018)
17. Lowe, D.G.: Similarity metric learning for a variable-kernel classifier. Neural Comput. 7(1), 72–85 (1995)
18. Wang, X., Han, X., Huang, W., et al.: Multi-similarity loss with general pair weighting for deep metric learning. In: Proceedings of the IEEE/CVF Conference Computer Vision and Pattern Recognition (CVPR), pp. 5022–5030 (2019)
19. Agrawal, A., Batra, D., Parikh, D., et al.: Don't just assume; look and answer: Overcoming priors for visual question answering. In: Proceedings of the IEEE Conference on Computer Vision and Pattern Recognition (CVPR), pp. 4971–4980 (2018)

20. Yang, Z., He, X., Gao, J., et al.: Stacked attention networks for image question answering. In: Proceedings of the IEEE Conference on Computer Vision and Pattern Recognition (CVPR), pp. 21–29 (2016)
21. Selvaraju, R.R., Lee, S., Shen, Y., et al.: Taking a hint: leveraging explanations to make vision and language models more grounded. In: Proceedings of the IEEE/CVF International Conference on Computer Vision (ICCV), pp. 2591–2600 (2019)
22. Jing, C., Wu, Y., Zhang, X., et al.: Overcoming language priors in VQA via decomposed language representations. In: Proceedings of the AAAI Conference on Artificial Intelligence, pp. 11181–11188 (2020)
23. Niu, Y., Tang, K., Zhang, H., et al.: Counterfactual VQA: a cause-effect look at language bias. In: Proceedings of the IEEE/CVF Conference on Computer Vision and Pattern Recognition (CVPR), pp. 12700–12710 (2021)
24. Goyal, Y., Khot, T., Summers-Stay, D., et al.: Making the V in VQA matter: elevating the role of image understanding in visual question answering. In: Proceedings of the IEEE Conference on Computer Vision and Pattern Recognition (CVPR), pp. 6904–6913 (2017)

Cipher-Prompt: Towards a Safe Diffusion Model via Learning Cryptographic Prompts

Sidong Jiang[1]([✉]), Siyuan Wang[1], Rui Zhang[1], Xi Yang[1], and Kaizhu Huang[2]

[1] Xi'an Jiaotong - Liverpool University, Suzhou, Jiangsu, People's Republic of China
{Sidong.Jiang20,Siyuan.Wang21}@student.xjtlu.edu.cn,
{Rui.Zhang02,Xi.Yang01}@xjtlu.edu.cn
[2] Duke Kunshan University, Kunshan, Jiangsu, People's Republic of China
Kaizhu.Huang@dukekunshan.edu.cn

Abstract. Security and privacy concerns associated with large generative models have recently attracted significant attention. In particular, there is a pressing need to address potential negative issues resulting from the generation of inappropriate images, including explicit, violent, or politically sensitive content. In this work, we propose a lightweight approach to learn cryptographic prompts, named Cipher-prompt, to prevent diffusion models from generating undesirable images that are semantically related to protected prompts. Cipher-prompt utilizes an untargeted attack objective to optimize a black-box model and generate perturbations that maximize the semantic distance between the protected class and the generated images. Therefore, Cipher-prompt does not require retraining or fine-tuning of the generative model or images as the training dataset. To evaluate the effectiveness of our proposed Cipher-prompt, we conduct thorough qualitative and quantitative experiments, measuring the protection failure rate and collateral impact rate. Experimental results show the efficacy of the proposed Cipher-prompt in balancing risk mitigation with the utility of diffusion-based image generation models.

Keywords: Image protection · Diffusion models · Black-box attack

1 Introduction

Recent advancements in Artificial Intelligence for Generative Content (AIGC) have been largely driven by the development of large generative models. The advent of CLIP [7], a multi-modal model maps text with image in hidden space, has expanded frontiers in text-to-image tasks. Afterwards, the field has been revolutionized by powerful image generative models such as DALL-E 2 [8] and Stable Diffusion [9], creating striking images from textual descriptions. Despite the remarkable capabilities of these large generative models, a series of ethical and societal problems arise, such as copyright infringements and image abuse.

Ethical implications of these models are of paramount importance, as their autonomy in generating content may potentially lead to the creation of inappropriate, offensive, or even harmful images. For example, illegal images can be

J. Ren et al. (Eds.): BICS 2023, LNAI 14374, pp. 322–332, 2024.
https://doi.org/10.1007/978-981-97-1417-9_30

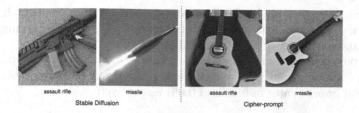

Fig. 1. Introduction of Cipher-prompt: generating cryptographic prompts to mitigate the risk of inappropriate image synthesis by generative models.

generated from violent or sexually explicit prompts. Even worse, in the political arena, the misuse of these technologies could fuel misinformation and propaganda, further polarizing societies. In response to these critical issues, current efforts involve generating adversarial perturbations on the clean samples. This line of methods can mislead the generated results while maintaining the visual similarity between the adversarial sample and the original one [6,10,11]. However, not all generative models are open-source or come with pretrained models; these approaches become inapplicable without the full access to the model. Adversarial Prompting [5] first applies an adversarial attack to prompts, inducing a targeted alternation in the output of the generative model using a black-box optimization approach. Nevertheless, it is limited to attacking individual prompts rather than entire categories.

In this work, we propose Cipher-prompt, a novel cryptographic prompt generation approach designed to prevent diffusion models from generating class-specific risky images. Unlike Adversarial Prompting, our method uses an untargeted attack objective to optimize a black-box model. Additionally, we propose an algorithm that generates perturbations for the protected class by clustering and averaging the CLIP embeddings of augmented prompts. This process aims to amplify the protection range by maximising the semantic boundary of the protected class and increasing the semantic distance between the protected class and the generated images. Importantly, Cipher-prompt is a lightweight model that does not require retraining or fine-tuning the generative model, nor does it rely on images in the training dataset. Combining the base prompt with the generated cryptographic prompt ensures that the generated images do not belong to the protected class while maintaining the essence of the user's original request for prompts outside this class. Furthermore, Cipher-prompt can be utilized to prevent the generation of violent content, such as images depicting firearms or other weapons (as shown in Fig. 1), thereby contributing to the overall safety and ethical standards of image generation.

We summarize our contributions as follows:

- We propose Cipher-prompt, a lightweight framework for generating cryptographic prompts that can prevent diffusion models from generating images related to a protected class.

- We introduce an algorithm for generating a perturbation that maximises the semantic boundary of the protected class. This is achieved by leveraging clustering and averaging of CLIP embeddings of the risky category.
- Our experiments demonstrate the effectiveness of Cipher-prompt in two categories - weapons and people - with low protection failure rates of 0.02 and collateral impact rates of 0.13 and 0.05, respectively.

2   Related Work

Adversarial Attack for Generative Models: Traditionally, the goal of adversarial attacks is to generate adversarial perturbations into the clean samples, which can mislead the classification result while maintaining the visual invariance of the adversarial sample and the clean one [1,4,6]. For our task, we aim to generate adversarial prefix, which we name as "cryptographic prompt", injected before the user inputs to prevent the generation of class-specific images using text-to-image generative models. Adversarial Prompting [5] proposed to generate adversarial prompts as a black-box attack against text-to-image generative models such as Stable Diffusion [9]. It performs targeted attack using Trust Region Bayesian Optimization (TuRBO) [2], a black-box optimization method which efficiently disrupts normal image generation by substituting regular input prompts with generated adversarial ones. However, Adversarial Prompting only enables attacks on individual prompts, not entire categories, as it cannot prevent the generation of semantically similar prompts.

Protection for Generative Models: To prevent misuse of generative models' extensive capabilities, certain models exist that deter these from replicating artistic images or human faces. Anti-DreamBooth [11] aims to obstruct generative models from replicating human faces by introducing perturbations into original images. On the other hand, GLAZE [10] guides generative models towards learning alternative artistic styles, thereby preventing them from copying target artworks. Existing protection methods for generative models focus on perturbing clean images to prevent the learning of new ones but lack the ability to prevent the generation of learned images. To address this, we propose Cipher-prompt for text-to-image generative models to prevent the generation of all risky images related to the protected class.

3   Methodology

We propose Cipher-prompt, a method that generates cryptographic prompts using a black-box optimization model to protect generative models. By adding cryptographic prompts, we can defend against generating images with violence or privacy contents, such as weapons and human images. We first outline the theoretical objective for optimizing a safer text-to-image generative model. Furthermore, we introduce a method for generating average perturbations to enhance model safety. Finally, we employ the TuRBO method as a black-box optimization model to simulate the text-to-image generative process.

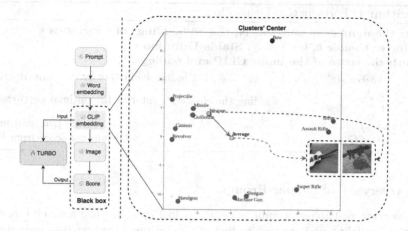

Fig. 2. Illustration of our Cipher-prompt: the method of enhancing the safety of text-to-image generative models by cryptographic prompts. (Left) The black-box for diffusion models. (Right) Finding the center of the risky embeddings.

3.1   Theoretical Objective

In this section, we introduce the theoretical objective for our model. The aim is to deter image generation linked to our protected prompt $\mathcal{P}$ by appending a generated cryptographic prompt to the base prompt, thereby constructing an augmented prompt that efficiently inhibits protected content generation.

Preliminaries: We define the base prompt as p_b and the cryptographic prompt as p_c. The augmented prompt p can then be represented as

$$p = p_b + p_c. \tag{1}$$

The generative model pipeline processes the augmented prompt (see Fig. 2) to score the cryptographic prompt. We employ pretrained models CLIP-ViT large text encoder E_T to embed text, Stable Diffusion as generator ϕ to convert CLIP embeddings into images, and classifier R for scoring.

Optimizing Objective: Given a protected class $\mathcal{P} = \{p_i\}_{i=1}^{N}$ with subclasses p_i, Cipher-prompt estimates a semantic perturbation center e_c that maximizes the semantic boundary of the protected class and simultaneously increases the semantic distance between the protected class and the generated images. We aim to derive a cryptographic prompt p_c through optimal perturbation e_c, maximizing a negative log-likelihood classification loss $\mathcal{L}$ in its untargeted form. This objective is expressed as

$$\arg\max_{e_c} \mathcal{L}(R(\phi(e_c))), \tag{2}$$

where e_c is determined as defined in Algorithm 1.

Algorithm 1. Perturbation Generation

Input: the number of subclass N, the embeddings of subclasses $e^{(i)}$, $0 < i <$
N, **Cluster Center** k, $0 < k < N$, **Stable Diffusion** ϕ.
Output: the score of the input CLIP embeddings.

1: $e_c^{(j)} \leftarrow kmeans(e^{(i)}), j = 1, ..., k$ ▷ Finding k clusters by k-means algorithm

2: $e_c \leftarrow \frac{1}{k} \sum_{j=0}^{k} e_c^{(j)}$ ▷ Finding the clusters' center as the optimal perturbation

3: $I_i \leftarrow \phi(e_c)$ ▷ Generate images by Stable Diffusion model

4: $Score \leftarrow \mathcal{L}(I_i)$ ▷ Obtain the score from Eq. 2

3.2 Overview of Cipher-Prompt

The overview of Cipher-prompt is depicted in Fig. 2. We use a black-box optimization model to emulate text-to-image generation. This text-to-score process, following the pipeline from Adversarial Prompting [5], serves as an objective model to assess the effectiveness of a cryptographic prompt in inhibiting image generation within our protected class $\mathcal{P}$. The aim is to create a perturbation center e_c preventing the generation of all related protected prompts, including subclass p_i or synonyms within $\mathcal{P}$. Input prompts' CLIP embeddings serve as input for TuRBO [2], a black-box optimization technique. It uses the calculated score to locate the optimal embedding e_c, which is then projected into a cryptographic prompt using Adversarial Prompting's token space projection method.

3.3 Perturbation Generation

We design an algorithm for perturbation generation (Algorithm 1), aiming to obtain an optimal CLIP embedding e_c by maximizing the score from the objective function (Eq. 2). Utilizing the Token Space Projection method from Adversarial Prompting [5], the optimal CLIP embedding is projected into a prompt, serving as our final cryptographic prompt. We generate CLIP embeddings for protected prompts $\mathcal{P} = \{p_i\}_{i=1}^{N}$, represented as

$$e^{(i)} = E_T(concat(p_i, p_r)), \tag{3}$$

where p_r represents a random prompt from the vocabulary of CLIP ViT.

A k-means clustering algorithm is applied to classify subclasses into k clusters, from which we compute the centroid as the average embedding e_c. We then apply the Stable Diffusion model to this average embedding to generate an image for the prompt. To score the input for our Cipher-prompt system, a pretrained ResNet model estimates the probability of the image belonging to our protected class, returning the negative logarithm of this probability as the score.

3.4 Trust Region Bayesian Optimization

We deploy a protection mechanism against the generation of images falling into risky categories by utilizing Trust Region Bayesian Optimization (TuRBO) [2] as the black-box optimization model. Our approach incorporates a Gaussian Process (GP) model [3] as the surrogate model. The surrogate model utilizes a CLIP embedding as input and the classification score obtained from the objective model as output, enabling TuRBO to estimate the relationship between perturbations and generated images. Consequently, our method safeguards against generating protected class images while preserving the desired output's integrity.

4 Experiment

4.1 Experiment Setup

For the text-to-image objective module, we utilize various pretrained models via the Hugging Face's transformers and diffusers modules, which include CLIP ViT Large Patch-14, offering an embedding size of 768 and a vocabulary size of 49,408; Stable Diffusion 1.5 for image generation, designed to handle an image size of 512×512; and a ResNet18 model for classifying the generated images.

For the surrogate model, we employ the Gaussian likelihood model from the gpytorch module, initializing it with a hidden layer configuration of 256-128-64 and retaining default settings for the remaining parameters followed by the Adversarial Prompting model [5]. The optimization process continues for up to 10,000 iterations or until the best cryptographic prompt remains unaltered for 3,000 successive iterations. With batch size set to 10, it takes around 24 h to train for total iterations of 10,000 via Nvidia RTX 3090.

4.2 Evaluation Setup

Our evaluation framework utilizes two key metrics: protection failure rate and the collateral impact rate. For each base prompt, we synthesis 100 images with Stable Diffusion 1.5 using the augmented prompt. Both metrics provide the ratio of these failures to total samples.

Protection Failure Rate (PFR) evaluates the effectiveness of Cipher-prompt, quantifying instances where the cryptographic prompt fails prevent the generation of protected class images. It is tested using prompts derived from ImageNet subclass hierarchies and ChatGPT-synthesized protected class prompts.

Collateral Impact Rate (CIR) quantifies unintended effects of our method on unrelated base prompts. If an image from the augmented prompt is not classified to the base prompt's class, it is a "failure". CIR is the ratio of such failures to the total samples.

4.3 Quantitative Result

Table 1 and Table 2 illustrate the PFR and CIR of the protection applied to the "weapon" and "person" classes. For each protected class, we test the cryptographic prompts with base prompts that are both related and unrelated to the protected class. Lower PFR indicates that the model successfully prevents

Table 1. Protection failure rates for the two protected categories. "Default" refers to not using clustering during training; "n-cluster" refers to dividing subclasses of the protected class into n clusters.

	Subclass	Default	4-Cluster	5-Cluster	Synonyms For	Default	4-Cluster	5-Cluster
Weapon	Weapon	0.02	0	0.05	Handgun	0.83	0.01	0.17
	Assault rifle	0.83	0.14	0.6	Heavy machine gun	0.72	0	0.1
	Bow	0	0	0	Shotgun	0.01	0	0.04
	Cannon	0.01	0	0.52	Sniper rifle	0.86	0.04	0.9
	Missile	0.36	0	0.02	A person holding a gun	0.55	0.01	0.73
	Projectile	0.01	0	0.08	A person shooting an arrow	0.68	0.05	0.81
	Revolver	0	0	0.03	Someone fires a gun	0.59	0.01	0.57
	Rifle	0.52	0	0.38	Submachine gun	0.89	0.01	0.15
	Average	0.2	**0.02**	0.21	Average	0.64	**0.02**	0.43
	Subclasss	Default	Average center	5-Cluster	Synonyms For	Default	Average center	5-Cluster
Person	Person	0	0	0	Football player	0.03	0.1	0.01
	Groom	0	0.5	0.04	Astronaut in a spacesuit	0.03	0.16	0.01
	Ballplayer	0	0.09	0	Person in a knight's armor	0	0	0
	Scuba diver	0	0.08	0	Donald Trump	0	0	0
	Average	**0**	0.17	0.01	Average	0.02	0.07	**0.01**

Table 2. Collateral impact rate for the generated cryptographic prompts.

	Cluster center	Dog	Beagle	Chihuahua	Doberman pinscher	Siberian husky	West highland white terrier	Average
Weapon	Default	0.69	0.14	0.94	0.03	0	0	0.3
	4-Cluster	0.95	0.97	0.99	0.46	0.96	0.57	0.82
	5-Cluster	0.4	0.05	0.35	0.01	0	0	**0.13**
People	Default	0.76	0.77	0.95	0.04	0.01	0.01	0.42
	2-Cluster	0	0.01	0.01	0.28	0	0	**0.05**

image generation of the protected class. A Lower CIR for unrelated classes is preferable, as it demonstrates that the image generation for unprotected classes is minimally affected.

4.4   Qualitative Result

In this section, we showcase the effectiveness of our proposed algorithm by implementing protection on the weapon and person classes. We show sample images generated by Stable Diffusion using augmented prompts. The augmentation process involves prepending a cryptographic prompt to the base prompts, with the cryptographic prompts generated by our optimization model. To facilitate the training process, we utilize subclasses of weapon class and person class derived from the ImageNet hierarchy. Our approach utilizes a cluster size of four for the protection of weapon class, and a cluster size of two for the protection of person class. Illustrative examples of prompts, with and without the application of our cryptographic prompt protection, are presented in Fig. 3.

4.5   Ablation Study

In this section, we conduct an ablation study to investigate the effectiveness of different strategies in handling protected classes for image synthesis. The

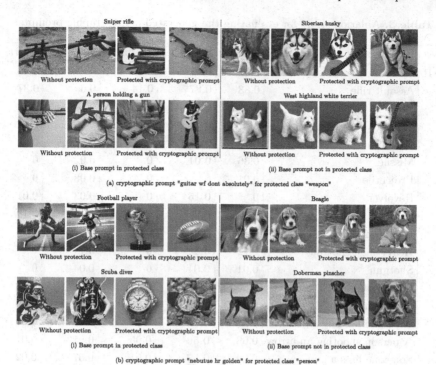

Fig. 3. Sample images generated with augmented prompts. (a) illustrates protection on the class "weapon", while (b) depicts protection on the class "person". Within each sub-figure, (i) demonstrates the protection when the base prompt is within the respective protected class, and (ii) showcases the protection mechanism when the base prompt is unrelated to the protected class.

strategies under scrutiny include: (1) directly using the protected class's name as the base prompt, (2) averaging over the CLIP embedding of the augmented prompt for each subclass of the protected class, and (3) performing clustering over the CLIP embeddings of augmented prompts corresponding to the protected class's subclasses, followed by averaging on the cluster centers.

Direct Use of Protected Class Name as Base Prompt. In the first model variant, we explore the idea of directly using the protected class's name as the base prompt. The assumption here is that the name of the protected class should inherently carry semantic information to guide the protection.

Averaging over CLIP Embeddings of Augmented Prompts. In the second variant, we average over the CLIP embeddings of the augmented prompts for each subclass within the protected class. The aim is to capture a more comprehensive representation of the protected class by considering all its subclasses.

Clustering over CLIP Embeddings and Averaging on Cluster Centers. For the third variant, we perform clustering over the CLIP embeddings of the

Table 3. Ablation study for evaluating the generated cryptographic prompts.

	Base Prompt	Default	3-Cluster	4-Cluster	5-Cluster	Average
PFR	Weapon	0.02	0.08	0	0.05	0
	Assault rifle	0.83	0.87	0.14	0.6	0.37
	Bow	0	0	0	0	0
	Cannon	0.01	0	0	0.52	0
	Guillotine	0.06	0.07	0	0.02	0
	Missile	0.36	0.01	0	0.02	0
	Projectile	0.01	0	0	0.08	0
	Revolver	0	0.15	0	0.03	0.01
	Rifle	0.52	0.29	0	0.38	0.14
	Handgun	0.83	0.81	0.01	0.17	0.23
	Heavy machine gun	0.72	0.48	0	0.1	0
	Shotgun	0.01	0.01	0	0.04	0
	Sniper rifle	0.86	0.58	0.04	0.9	0.03
	Submachine gun	0.89	0.89	0.01	0.15	0.19
	A person holding a gun	0.55	0.77	0.01	0.73	0.01
	A person shooting an arrow	0.68	0.56	0.05	0.81	0.01
	Someone fires a gun	0.59	0.57	0.01	0.57	0.02
	Average	0.41	0.36	**0.02**	0.30	0.06
CIR	Dog	0.69	0.52	0.95	0.4	0.94
	Beagle	0.14	0.74	0.97	0.05	0.71
	Chihuahua	0.94	0.91	0.99	0.35	0.56
	Doberman pinscher	0.03	0.03	0.46	0.01	0.02
	Siberian husky	0	0.04	0.96	0	0.55
	West highland white terrier	0	0.04	0.57	0	0.02
	Average	0.30	0.38	0.82	**0.13**	0.47

augmented prompts for subclasses of the protected class, then compute averages on the cluster centers. This approach aims to identify and leverage semantic centers within the protected class for image synthesis.

The results of these three variants are presented in Table 3. It is observed that averaging on the CLIP embeddings or selecting an appropriate number of clusters can significantly reduce the protection failure rate. This is partially because the name of the protected class does not always constitute the semantic center. Consequently, the methods that explore deeper semantics within the class, such as averaging embeddings or clustering, seem to provide more robust protection. It is also worth noting that the choice of the number of clusters is critical when performing clustering over embeddings. The results show a clear trade-off between reducing the protection failure rate and minimizing the collateral impact rate. Optimal results were achieved with a balance between these two factors. Our ablation study reveals that using deeper semantic information, either through averaging embeddings or through clustering, can enhance the protection mechanism in our image synthesis process. However, careful tuning is required to balance protection efficacy and collateral impact.

5   Conclusion

Given the potency of large generative models, it is crucial to establish mechanisms that inhibit the generation of inappropriate or unacceptable images. This paper presents a novel and efficient method for manipulating the output of diffusion models to avoid the generation of images of a specific class by using a cryptographic prompt. This protective strategy remains effective even with a four-token cryptographic prompt chosen from a randomly selected pool of 100 CLIP tokens. Looking forward, the flexibility of these techniques can be further enhanced to prevent the generation of multiple classes without impinging on the generation of acceptable images. This could be achieved by minimizing the distance between the CLIP embedding of the base prompt and the augmented prompt if the base prompt is unrelated to the protected classes.

Acknowledgements. The work was partially supported by the following: National Natural Science Foundation of China under No. 92370119, No. 62376113, and No. 62206225; Jiangsu Science and Technology Programme (Natural Science Foundation of Jiangsu Province) under no. BE2020006-4; Natural Science Foundation of the Jiangsu Higher Education Institutions of China under no. 22KJB520039; Research Development Fund in XJTLU under No. RDF-19-01-21 and No. RDF-16-01-57.

References

1. Andriushchenko, M., Croce, F., Flammarion, N., Hein, M.: Square attack: a query-efficient black-box adversarial attack via random search. In: Vedaldi, A., Bischof, H., Brox, T., Frahm, J.-M. (eds.) ECCV 2020. LNCS, vol. 12368, pp. 484–501. Springer, Cham (2020). https://doi.org/10.1007/978-3-030-58592-1_29
2. Eriksson, D., Pearce, M., Gardner, J.R., Turner, R., Poloczek, M.: Scalable global optimization via local Bayesian optimization (2019). http://arxiv.org/abs/1910.01739
3. Gardner, J.R., Pleiss, G., Bindel, D., Weinberger, K.Q., Wilson, A.G.: GPyTorch: blackbox matrix-matrix gaussian process inference with GPU acceleration. arXiv preprint arXiv:1809.11165 (2018)
4. Huang, K., Hussain, A., Wang, Q.F., Zhang, R.: Deep Learning: Fundamentals, Theory and Applications, vol. 2. Springer, Heidelberg (2019). https://doi.org/10.1007/978-3-030-06073-2
5. Maus, N., Chao, P., Wong, E., Gardner, J.: Adversarial prompting for black box foundation models. arXiv preprint arXiv:2302.04237 (2023)
6. Qian, Z., Zhang, S., Huang, K., Wang, Q., Zhang, R., Yi, X.: A survey of robust adversarial training in pattern recognition: fundamental, theory, and methodologies. Pattern Recogn. **123**, 108121 (2022)

7. Radford, A., et al.: Learning transferable visual models from natural language supervision. arXiv preprint arXiv:2103.00020 (2021)
8. Ramesh, A., et al.: Zero-shot text-to-image generation. arXiv preprint arXiv:2102.12092 (2021)
9. Rombach, R., Blattmann, A., Lorenz, D., Esser, P., Ommer, B.: High-resolution image synthesis with latent diffusion models. arXiv preprint arXiv:2112.10752 (2021)
10. Shan, S., Cryan, J., Wenger, E., Zheng, H., Hanocka, R., Zhao, B.Y.: GLAZE: protecting artists from style mimicry by text-to-image models (2023). http://arxiv.org/abs/2302.04222
11. Van Le, T., Phung, H., Nguyen, T.H., Dao, Q., Tran, N., Tran, A.: Anti-DreamBooth: protecting users from personalized text-to-image synthesis (2023). http://arxiv.org/abs/2303.15433. Version: 1

Applications

Fusing Multi-scale Attention and Transformer for Detection and Localization of Image Splicing Forgery

Yanzhi Xu[1]([✉]), Jiangbin Zheng[1]([✉]), and Chenyu Shao[2]

[1] School of Software, Northwestern Polytechnical University, Xi'an, China
xyz0926@mail.nwpu.edu.cn, zhengjb@nwpu.edu.cn
[2] School of Computer Science, Northwestern Polytechnical University, Xi'an, China

Abstract. Image forgery detection has attracted widespread attention due to the enormous spread of forged images on the internet and social media. Many existing methods lack global contextual information and ignore the interaction between multi-level representations, which are not conducive to localizing the multi-scale tampered regions. To overcome these limitations, we present a novel method to detect splicing forgery of images by utilizing cross-scale interaction attention and multi-level global information. Firstly, we design a cross-scale interactive attention (CSIA) module for aggregating multi-level convolutional features, focusing selectively on task-relevant information. It allows representations learned at different levels to communicate effectively with each other. Secondly, by introducing the transformer layers with dynamic position embedding, the proposed method can capture the multi-level global contextual correlations of the image in a more flexible way. Extensive experiments have shown that our proposed method outperforms state-of-the-art methods. It can effectively detect, and segment multi-scale tampered regions by aggregating multi-scale local and global feature relevance.

Keywords: Image forgery · Cross-scale interaction attention · Transformer · Global contextual information

1 Introduction

Image forgery is the use of image processing software to manipulate the semantic content of an image to distort the original information conveyed by the image. Splicing and copy-move are two common methods of tampering in real life. Splicing forgery selects random regions from donor images and pastes them into the target image for image composition. In splicing forgery, the forged and real regions are from different source images, as shown in Fig. 1-(a). Due to various post-processing operations masking the traces of manipulation, the naked eye can barely distinguish between the real and tampered areas of the image. Moreover, the number and scale of the forged areas contained in the splicing forged images are different. These factors make the detection of splicing forgery regions a challenging task.

© The Author(s), under exclusive license to Springer Nature Singapore Pte Ltd. 2024
J. Ren et al. (Eds.): BICS 2023, LNAI 14374, pp. 335–344, 2024.
https://doi.org/10.1007/978-981-97-1417-9_31

The tampered images are widely spread on the media and internet. To avoid being tricked by malicious manipulated images, many forgery detection methods have been devised over the past decades. These methods can be broadly classified into two classes: conventional methods and deep learning-based methods. The conventional methods mainly utilize local inconsistencies in specific attributes to locate tampered regions, such as image essence attributes [1], compression artifacts [2], and photo-response non-uniformity (PRNU) [6]. The experimental results of the DCT [11] are shown in Fig. 1.-(c). In recent years, inspired by the success of deep learning in vision tasks, some researchers have exploited CNNs to extract features from image patches and locate tampered regions by classifying patches [7, 8]. However, these methods lack contextual information about the image. To solve this problem, end-to-end models [9, 10] have been proposed. The experimental results of the ManTra [10] are shown in Fig. 1.-(d). These models typically extract features from the entire image without differentiating the importance of the information. Moreover, most of these methods rely only on single-level local features without considering the multi-level local and global information of the image. Such factors hinder these methods from achieving state-of-the-art accuracy when detecting multi-scale splicing forgeries.

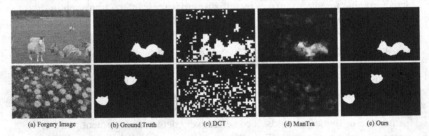

(a) Forgery Image   (b) Ground Truth   (c) DCT   (d) ManTra   (e) Ours

Fig. 1. Two examples of splicing forgery images and corresponding localization results of three methods. (a) The splicing forgery images. (b) Ground truths. (c) The detection results of the traditional method of DCT [11]. (d) The detection result of the deep learning method ManTra [10]. (e) The detection result of the proposed method.

To solve the issues mentioned above, we propose a fusing multi-scale attention and transformer network (FMAT-Net) for the detection of splicing forgery images by aggregating the multi-scale attention and global contextual information. This study makes the following contributions:

- We propose a novel method for the detection and localization of image splicing forgery. It can improve the localization performance of the forged regions by integrating multi-level local attention and global contextual information.
- The designed CSIA module allows adequate interaction of features at different levels and can enhance forgery-related features and suppress irrelevant responses. By introducing the transformer layers with convolution position encoding, the proposed model can obtain multi-level global information. It overcomes the limitation of local features in locating multiple tampered regions at different scales in an image.
- Comprehensive experimental evaluation demonstrates its superior performance in detecting multi-scale tampered areas of splicing forgery images.

2   Related Work

The existing approaches for splicing images forgery detection can be generally divided into two types based on feature extraction methods: traditional ones and deep learning ones.

Traditional Ones.   Traditional feature extraction-based methods mainly include image essence attributes [1], image-compression attributes [2, 11], edge inconsistencies [3, 4], and inherent attribute consistency of imaging devices [5, 6]. These methods detect image forgery based on the inconsistency of specific forgery cues. However, they suffer dramatic performance degradation when the forgery cue is missing or subject to post-processing.

Deep Learning Ones.   Motivated by the tremendous success of deep learning in object detection [12, 15], some researchers have tried to use deep learning techniques to detect forged images [7, 8]. Different from object detection, forgery detection focuses only on the tampered regions. For example, Xiao et al. [7] designed a cascaded CNN network to perform a coarse-to-fine localization of tampered regions at the patch level. However, this method only focuses on local image patches and lacks of relationship between the patches, resulting in incorrect results. Thus, end-to-end-based CNN networks have been proposed for image splicing forgery detection. Bappy et al. [9] proposed to use of LSTM to extract tampering artifacts and fuse them complementarily with spatial domain features to locate manipulated regions. Wu et al. [10] designed a ManTra-Net to detect multiple forgery types by combining self-supervised learning with local anomaly detection methods. Bi et al. [13] exploited residual propagation and residual feedback mechanisms to enhance feature learning of the CNN network and effectively improve splicing image detection accuracy. However, these methods lack global information about the image and are not robust enough. Furthermore, Bi et al. [14] proposed a multi-decoder single-task network to detect tampered regions. A potential limitation of this method is that model extraction features do not distinguish their importance.

3   Proposed Method

As shown in Fig. 2, the proposed FMAT-Net consists of three parts, including an encoder branch, a transition part, and a decoder branch. First, the input images are sent to the encoder, which utilizes ResNet as the backbone to extract multi-level features F_l ($l = 1, 2, 3, 4, 5$) from the images. Second, the CSIA module in the transition part allows the interaction of different levels of information and uses an attention mechanism to enhance local features, thus facilitating the detection of forged regions. Moreover, the depth-wise convolution transformer layer (DTransformer) of the transition part can harvest the multi-level global correlations of the image. This transformer layer uses depth-wise convolution to implicitly encode positional information, allowing flexibility to handle arbitrary input sizes. Then, the multi-scale attention information output by the CSIA module and the global information extracted by the transformer layers are integrated at the corresponding resolution and fed to the decoder branch to complement the information loss. The decoder branch progressively fuses multi-level local attention and global contextual information from the transition layer for forgery region localization.

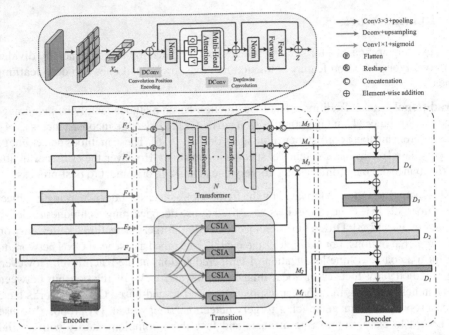

Fig. 2. The proposed model architecture diagram. The CNN encoder branch extracts features and sends them to the transition part. The transition part incorporates multi-level features by the CSIA module and extracts multi-level global contextual information by the Transformer and sends them to the decoder branch. F_l ($l = 1, 2, 3, 4, 5$) represents the l_{th} level outputs of the encoder. M_l ($l = 1, 2, 3, 4, 5$) represents the l_{th} fused features of the transition part. D_l ($l = 1, 2, 3, 4$) represents the l_{th} level outputs of the decoder stage.

3.1 Cross-Scale Interactive Attention (CSIA)

Since single-level features can only represent the information at a specific scale, thus, it is not sufficient to detect multi-scale tampered regions using single-level features. Most feature fusion methods add or concatenate different levels of feature maps directly, ignoring the interaction between different levels of features. In addition, the features extracted by CNN contain a lot of redundant information. Thus, we design a CSIA module to interactively aggregate multi-level information. Therein, the attention mechanism of CSIA adaptively selects forgery-related information from different levels of features and filters irrelevant responses. Since there is a semantic gap between different levels of features, we fuse features from adjacent levels. The architecture of the CSIA is shown in Fig. 3.

First, the features F_a and F_b from different levels of the encoder are passed through the UPConv block to achieve the same resolution and number of channels as the current level features F_c ($H \times W \times C$). $\widehat{F}_a = \mathrm{UPConv}(F_a)$, $\widehat{F}_b = \mathrm{UPConv}(F_b)$. Then, features $\widehat{F}_a$, $\widehat{F}_b$, and F_c are summed element by element.

$$F_{add} = \widehat{F}_a \oplus \widehat{F}_b \oplus F_c \tag{1}$$

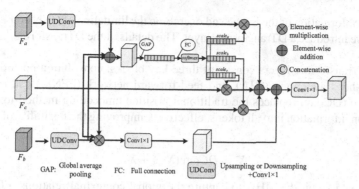

Fig. 3. The architecture of the CSIA module.

Next, the F_{add} uses the global average pooling operation (GAP) to embed spatial information and generate a $1 \times 1 \times C$ feature map, which is passed through the fully connected layers to generate a $3 \times d$ vector so that a corresponding scale vector s_i ($i = a, b, c$) can be obtained for each level features, $d = C/r$, r is a reduction ratio to control the value of the d.

$$s_i = \mathcal{F}(\mathcal{G}(F_{add})), i = a, b, c \tag{2}$$

where $\mathcal{G}(\cdot)$ denotes the global average pooling operation, $\mathcal{F}(\cdot)$ represents the fully connected (FC) layer. Then, a softmax operation is applied on the s_i channel-wise: $scale_i = soft \max(s_i), i = a, b, c$. Next, the features $\widehat{F}_a$, $\widehat{F}_b$, and F_c are multiplied by the corresponding $scale_i$ and element-wise summation of all the results to obtain the final attention information A.

$$A = \sum_{i=a,b,c} scale_i \times [\widehat{F}_a, \widehat{F}_b, F_c] \tag{3}$$

Furthermore, in order to fully exploit the correlations between multi-level features, we perform element-wise multiplication of different levels of features so that different levels of features can interact beneficially and promote each other. It can amplify the differences between real and forged regions and filter the noise interference in the fusion process.

$$F_{mul} = \widehat{F}_a \otimes \widehat{F}_b \otimes F_c \tag{4}$$

Finally, we integrate the attention features A and the interaction information F_{mul} to get the final output of the CSIA module.

3.2 Depth-Wise Convolution Transformer Layer (DTransformer)

Many existing deep learning-based splicing forgery detection methods use CNNs to extract multi-level information from images. However, due to the limited receptive field of the convolution, the features it extracts lack global relevance. To capture the global contextual information of the image and overcome the limitation of CNNs for extracting features, we introduce a DTransformer layer. The details of the DTransformer are shown in Fig. 2.

The DTransformer layers consist of three key blocks: convolution position encoding, multi-head attention (MHA), and feed-forward network (FFN). First, deep-wise convolution (DConv) replaces the traditional position embedding method to integrate the position information into all tokens, effectively improving the flexibility of the input resolution.

$$X = DConv(X_{in}) + X_{in} \tag{5}$$

Then, we exploit the MHA to capture the global contextual relations. Finally, we add FFN to propagate the global contextual correlations.

$$Y = MHA(Norm(X)) + X \tag{6}$$

$$Z = FFN(Norm(Y)) + Y \tag{7}$$

where $Norm(\cdot)$ denotes the normalization function. In our experiments, considering the computation complexity and the performance issues, we set the number of the DTransformer layers N to 6 in all the experiments.

Table 1. Details of the datasets used for the experiments in the study.

Name	CASIA	COLUMB	NIST'16
Plain Splicing	1270	170	234
Plain Training	1220	125	184
Augmented Training	2220	725	784
Testing	50	45	50

4 Experiments

4.1 Data Set

Our experiments were run on four public datasets: Defacto [21], CASIA [16], COLUMB [17], and NIST'16 [18]. Due to we mainly focus on splicing forgery detection, we just select splicing forgery images from four datasets. Defacto is a relatively large dataset containing 105,000 computer-generated splicing images without post-processing operations. To prevent overfitting of the model, we first pre-trained the proposed model on the Defacto dataset and then fine-tuned it on the smaller datasets. For the small data sets, each dataset is randomly separated into a training set and a testing set. Then, we augment the images of each training set by random overturn, random Gaussian noise, and JPEG compression. The experimental images of three small datasets are listed in Table 1. We resize the resolution of images to 256×256.

4.2 Evaluation Metrics

In our experiments, we use pixel-level precision, recall, and F values to evaluate the performance of the proposed FMAT-Net, which are defined as $Precision = TP/(TP + FP)$, $Recall = TP/(TP + FN)$, where TP represents the number of forged pixels predicted correctly, FP denotes the number of real pixels predicted as tampered, and FN indicates the number of altered pixels predicted as real pixels. $F = (2 \times Precision \times Recall)/(Precision + Recall)$, is the weighted harmonic mean of the precision and recall.

4.3 Implementation Details and Methods

We use the PyTorch deep learning framework to implement the proposed method. Before training, we initialize the ResNet using a pre-trained model on the ImageNet and use a stochastic gradient descent optimizer with a momentum of 0.9, and the weight decay is set to 0.0005. The proposed model is optimized by the cross-entropy loss. The batch size of 6 samples learning rate of 0.003 is set. Pre-training and fine-tuning are 100 and 50 epochs, respectively. The proposed model and all the methods used for the experiment run on an NVIDIA GeForce GTX 3090 graphics card.

To evaluate the performance of our proposed method comprehensively, we compare it with the baseline detection methods on the three public datasets. The DCT [11] and CFA [5] are the conventional methods. Their source codes are provided by the authors or reliable third parties [19]. ManTra [10], AttU-Net [20], RRU-Net [13], C2RNet [7], and RTAG [14] are deep learning-based methods. For a fair comparison, the hyperparameters of the comparative methods are adjusted to the values that produce the best performance in the original literature.

Table 2. Results of the different detection methods on the three datasets.

	CASIA			COLUMB			NIST'16		
	P	R	F	P	R	F	P	R	F
DCT	0.346	0.864	0.494	0.387	0.684	0.494	0.124	0.817	0.215
CFA	0.108	**0.916**	0.193	0.546	0.468	0.504	0.164	**0.876**	0.276
ManTra	0.821	0.793	0.807	0.856	0.849	0.852	0.816	0.824	0.82
AttU-Net	0.816	0.787	0.801	0.836	0.796	0.816	0.781	0.776	0.778
RRU-Net	0.848	0.834	0.841	0.923	0.834	0.876	0.783	0.782	0.782
C2RNet	0.581	0.808	0.676	0.804	0.612	0.695	0.468	0.666	0.55
RTAG	–	–	0.815	–	–	0.823	–	–	0.623
Ours	**0.871**	0.843	**0.857**	**0.941**	**0.853**	**0.895**	**0.828**	0.836	**0.832**

4.4 Comparative Experiment and Analysis

All comparison methods are compared on the CASIA, COLUMB, and NIST'16 testing sets based on metrics such as detection precision (P), recall (R), and F-value. The detailed results are given in Table 2, where bold entities indicate the best performance, and '-' indicates that the value of the metric is not provided in the original study.

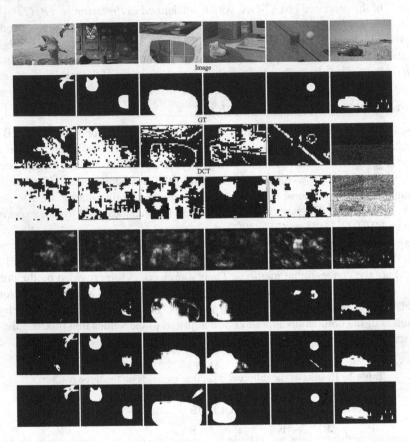

Fig. 4. Comparison of the proposed model with existing splicing forgery detection methods. 1st and 2nd columns are from CASIA, 3rd and 4th columns are from COLUMB, 5th and 6th columns are from NIST'16.

It can be seen from Table 2 that the overall performance of the proposed method performs better than other methods on all three datasets. However, the recall is a little lower than the CFA on the CASIA and NIST'16 datasets. Figure 4 illustrates the visual comparisons of the different methods. We can see from Fig. 4 that the CFA identifies many genuine regions of the image as tampered regions, so the recall is relatively high. Among the deep learning-based methods, our FMAT-Net and RRU-Net can locate most of the tampered areas. However, the proposed model can detect and locate tampered regions and authentic regions more precisely. Furthermore, the detection results of AttU-Net and ManTra are either missing or misjudged. This may be because both methods lack the long-range dependencies of the image. The comprehensive experiments demonstrate the validity of our proposed approach.

5 Conclusion

In this paper, we propose a novel method by aggregating multi-level local attention information and global contextual correlations to detect and locate the tampered regions of the splicing forgery images. Firstly, the designed CSIA module enables different levels of features to interact across scales and guides the model to pay more attention to the critical regions of the image. It enhances the learning of the local discriminative representations, improving the capacity of the model to locate multi-scale tampered regions. Secondly, a global feature mining layer based on the Transformer, dubbed DTransformer, is introduced to extract multi-level global contextual information from the forgery image to improve the detection results. Furthermore, we evaluate the performance of the proposed model and the state-of-the-art detection methods on three benchmark datasets. Extensive experiments indicate that the presented model performs better than the state-of-the-art approaches. It can effectively detect and segment multi-scale tampered regions. However, due to the introduction of the Transformer in the proposed model, its inference time will be longer. Therefore, in the future, we will consider designing more lightweight and effective models to detect splicing forged images.

References

1. Carvalho, T., et al.: Illuminant-based transformed spaces for image forensics. IEEE Trans. Inf. Forensics Secur. **11**(4), 720–733 (2015). https://doi.org/10.1109/TIFS.2015.2506548
2. Iakovidou, C., et al.: Content-aware detection of JPEG grid inconsistencies for intuitive image forensics. J. Vis. Commun. Image Represent. **54**, 155–170 (2018). https://doi.org/10.1016/j.jvcir.2018.05.011
3. Qu, Z., Qiu, G., Huang, J.: Detect digital image splicing with visual cues. In: Katzenbeisser, S., Sadeghi, A.-R. (eds.) IH 2009. LNCS, vol. 5806, pp. 247–261. Springer, Heidelberg (2009). https://doi.org/10.1007/978-3-642-04431-1_18
4. Fang, Z., Wang, S., Zhang, X.: Image splicing detection using color edge inconsistency. In: 2010 International Conference on Multimedia Information Networking and Security (MINES), pp. 923–926 (2010). https://doi.org/10.1109/MINES.2010.196
5. Ferrara, P., Bianchi, T., De Rosa, A., Piva, A.: Image forgery localization via fine-grained analysis of CFA artifacts. IEEE Trans. Inf. Forensics Secur. **7**(5), 1566–1577 (2012). https://doi.org/10.1109/TIFS.2012.2202227
6. Korus, P., Huang, J.: Multi-scale analysis strategies in PRNU-based tampering localization. IEEE Trans. Inf. Forensics Secur. **12**(4), 809–824 (2016). https://doi.org/10.1109/TIFS.2016.2636089
7. Xiao, B., Wei, Y., Bi, X., Li, W., Ma, J.: Image splicing forgery detection combining coarse to refined convolutional neural network and adaptive clustering. Inf. Sci. **511**, 172–191 (2020). https://doi.org/10.1016/j.ins.2019.09.038
8. Liu, B., Pun, C.M.: Exposing splicing forgery in realistic scenes using deep fusion network. Inf. Sci. **526**, 133–150 (2020). https://doi.org/10.1016/j.ins.2020.03.099
9. Bappy, J.H., Simons, C., Nataraj, L., Manjunath, B.S., Roy-Chowdhury, A.K.: Hybrid LSTM and encoder-decoder architecture for detection of image forgeries. IEEE Trans. Image Process. **28**(7), 3286–3300 (2019). https://doi.org/10.1109/TIP.2019.2895466
10. Wu, Y., AbdAlmageed, W., Natarajan, P.: ManTra-Net: manipulation tracing network for detection and localization of image forgeries with anomalous features. In: Proceedings of the IEEE Conference on Computer Vision and Pattern Recognition (CVPR), pp. 9543–9552 (2019). https://doi.org/10.1109/CVPR.2019.00977

11. Ye, S., Sun, Q., Chang, E.C.: Detecting digital image forgeries by measuring inconsistencies of blocking artifact. In: 2007 IEEE International Conference on Multimedia and Expo (ICME), pp. 12–15 (2007). https://doi.org/10.1109/ICME.2007.4284574

12. Yan, Y., Ren, J., Sun, G., et al.: Unsupervised image saliency detection with Gestalt-laws guided optimization and visual attention-based refinement. Pattern Recogn. **79**, 65–78 (2018). https://doi.org/10.1016/j.patcog.2018.02.004

13. Bi, X., Wei, Y., Xiao, B., & Li, W.: RRU-Net: the ringed residual U-Net for image splicing forgery detection. In: Proceedings of the IEEE Conference on Computer Vision and Pattern Recognition Workshops (CVPRW), pp. 30–39 (2019). https://doi.org/10.1109/CVPRW.2019.00010

14. Bi, X., Zhang, Z., Xiao, B.: Reality transform adversarial generators for image splicing forgery detection and localization. In: Proceedings of the IEEE International Conference on Computer Vision (ICCV), pp. 14294–14303 (2021). https://doi.org/10.1109/ICCV48922.2021.01403

15. Ren, J., Wang, Z., Ren, J.: PS-Net: progressive selection network for salient object detection. Cogn. Comput. **14**, 794–804 (2022). https://doi.org/10.1007/s12559-021-09952-4

16. Dong, J., Wang, W., Tan, T.: Casia image tampering detection evaluation database. In: 2013 IEEE China Summit and International Conference on Signal and Information Processing (ChinaSIP), pp. 422–426 (2013). https://doi.org/10.1109/ChinaSIP.2013.6625374

17. Hsu, Y.F., Chang, S.F.: Detecting image splicing using geometry invariants and camera characteristics consistency. In: 2006 IEEE International Conference on Multimedia and Expo (ICME), pp. 549–552 (2006). https://doi.org/10.1109/ICME.2006.262447

18. Guan, H., et al.: MFC datasets: large-scale benchmark datasets for media forensic challenge evaluation. In: 2019 IEEE Winter Applications of Computer Vision Workshops (WACVW), pp. 63–72 (2019). https://doi.org/10.1109/WACVW.2019.00018

19. Zampoglou, M., Papadopoulos, S., Kompatsiaris, Y.: Large-scale evaluation of splicing localization algorithms for web images. Multimed. Tools. Appl. **76**(4), 4801–4834 (2017). https://doi.org/10.1007/s11042-016-3795-2

20. Oktay, O., Schlemper, J., Folgoc, L.: Attention U-net: learning where to look for the pancreas (2018). https://arxiv.org/abs/1804.03999v3

21. Mahfoudi, G., Tajini, B., Retraint, F., Morain-Nicolier, F., Pic, M.: Defacto: image and face manipulation dataset. In: 27th European Signal Processing Conference (EUSIPCO), pp. 1–5 (2019). https://doi.org/10.23919/EUSIPCO.2019.8903181

SAR Incremental Learning via Generative Adversarial Network and Experience Replay

Fei Gao[1,2,4], Chen Fan[1,4], Penghui Chen[1,4($\boxtimes$)], Amir Hussain[2,4], and Huiyu Zhou[3,4]

[1] School of Electronic and Information Engineering, Beihang University, Beijing 100191, China
chenpenehui@buaa.edu.cn
[2] Beihang Hangzhou Innovation Institute Yuhang, Xixi Octagon City, Yuhang District, Hangzhou 310023, China
[3] Edinburgh Napier University, Edinburgh EH11 4BN, UK
[4] University of Leicester, Leicester, UK

Abstract. Incremental learning is a viable approach for addressing the recognition problem of SAR images in data stream scenarios. However, it may suffer from catastrophic forgetting due to insufficient exposure to old category data during training. This paper proposes the integration of a generative adversarial network (GAN) into the iCaRL model to mitigate catastrophic forgetting by generating samples from previously learned categories. Experimental results on our SAR dataset demonstrate that our approach significantly enhances the recognition performance of the iCaRL model.

Keywords: Incremental learning · Generative adversarial network · SAR target recognition

1 Introduction

Synthetic Aperture Radar (SAR) is a high-resolution microwave imaging radar that finds extensive application in fields such as meteorology, geography, and defense. SAR image recognition constitutes an essential component of the SAR system's application, necessitating the model to possess certain new class recognition capabilities [1]. Incremental learning methods are well-suited to meet this requirement.

During the process of learning new tasks, incremental learning models are susceptible to catastrophic forgetting, which refers to the phenomenon of forgetting old tasks while acquiring new ones. Presently, regularization and experience replay are the prevailing techniques utilized to tackle this issue. The regularization reduces the degree to which the model forgets old tasks by designing loss functions. Learning without Forgetting (LwF) [2] uses network distillation and fine-tuning methods to improve the loss function, while Dhar et al. [3] use image attention mechanisms to improve the loss function. The experience replay maintains the recognition ability of the model for old tasks by replaying some old class data during the learning process of new tasks. Incremental Classifier and Representation Learning (iCaRL) [4] is a classic experience replay method, and

J. Ren et al. (Eds.): BICS 2023, LNAI 14374, pp. 345–353, 2024.
https://doi.org/10.1007/978-981-97-1417-9_32

some influential improved algorithms based on the iCaRL algorithm include End-to-End Incremental Learning (EEIL) [5] and Large Scale Incremental Learning (LSIL) [6]. The impact of experience replay on mitigating catastrophic forgetting is particularly significant, and increasing the amount of replayed data can significantly enhance the recognition performance of the model. Therefore, this paper proposes the utilization of old example generation to enhance recognition performance in data-replay-based incremental learning models. The iCaRL algorithm serves as the fundamental model for this study on incremental learning.

Deep generative models use deep neural networks to generate high-quality data by fitting the function relationship between the input distribution and the real data distribution, avoiding explicit feature extraction and greatly reducing the difficulty of model design. Common deep generative models include Restricted Boltzmann Machine (RBM), Deep Belief Network (DBN), Variational Autoencoder (VAE), and Generative Adversarial Networks (GAN) [7, 8], with GAN showing the most prominent performance in image generation. The azimuth attribute has an extremely important influence on SAR image recognition. We introduce the Conditional Generative Adversarial Network (CGAN) [9] with azimuth control into the iCaRL model to generate synthetic samples, thereby enhancing the recognition performance of the model. The main innovations of this paper are as follows:

(1) We have proposed a sample selection algorithm that is based on recognizing old modes and measuring feature distances.
(2) Two sample generation strategies are proposed, including the utilization of a single Generative Adversarial Network (GAN) and multiple GANs.
(3) We incorporate the CGAN model, which employs asymptotic coding control to generate azimuth information for data, into the iCaRL algorithm to enhance its recognition performance.

2 Proposed Method

We propose the generative iCaRL algorithm, which incorporates GAN into the iCaRL algorithm to enhance its performance. Figure 1 illustrates the training process of our proposed method in a data stream scenario. Specifically, only real data is utilized for training the first task, while both selected generated data and real data are employed for subsequent tasks.

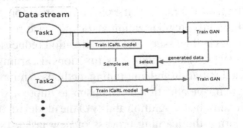

Fig. 1. Training process diagram of generative iCaRL algorithm.

2.1 Generative Adversarial Networks for Sample Generation

The GAN in the generative iCaRL algorithm consists of a generator and a discriminator, and Fig. 2 shows the specific network structure of both. The conditional encoding comprises a 6-dimensional one-hot vector encoding for category control and a 22-dimensional asymptotic encoding for azimuth control. The random noise is a 100-dimensional vector with each element being independent and following a standard normal distribution. The first and last basic blocks of the generator do not contain batch normalization layers, and the last basic block uses a sigmoid activation function, while the other basic blocks use LeakyReLU activation functions. The first and last basic blocks of the discriminator do not contain random dropout layers, and the last basic block does not contain an activation function, while the other basic blocks use LeakyReLU activation functions.

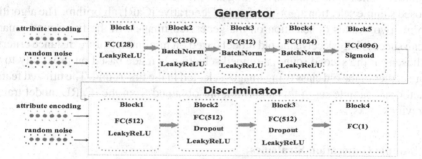

Fig. 2. Structure of GAN with 'FC' Representing Fully Connected Layers and Numbers in Parentheses Indicating Output Neuron Count.

Figure 3 shows a schematic diagram of the asymptotic encoding process. When asymptotic encoding for all integer angles between 0 and 359 degrees is performed, the range is first divided into six intervals, and a 6-dimensional one-hot vector is used to encode the six intervals. Then each interval is further divided into six segments, and a 6-dimensional one-hot vector is used to encode the six segments. Next, a 10-dimensional one-hot vector is used to encode the ten integers within each segment. Finally, the results of the encoding from the previous three steps are concatenated to obtain a 22-dimensional encoding result. For non-integer azimuth angles, the angle is rounded down before applying the rules described above.

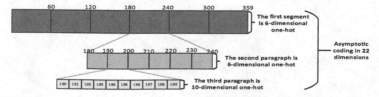

Fig. 3. Schematic diagram of the asymptotic encoding process.

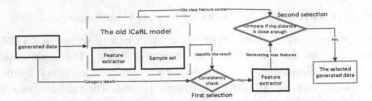

Fig. 4. The Flowchart of the Sample Selection Algorithm.

2.2 Sample Selection Algorithm in the Generative iCaRL Algorithm

To ensure the selection of high-quality generated data, this paper proposes a sample selection algorithm based on old model recognition and feature distance by combining the characteristics of incremental learning and the iCaRL model. Figure 4. Illustrates the proposed sample selection algorithm for the Generative iCaRL algorithm. The algorithm performs two rounds of selection on the generated images. In the first round, images that can successfully deceive the old model based on minimum feature distance criterion are chosen. In the second round, a refined selection is made based on proximity to the feature center of the old class using images selected in the first round. The utilized feature extractor and sample set in this algorithm are obtained from the iCaRL model trained on previous tasks.

2.3 Sample Generation Strategies

We propose two generation strategies, namely one-GAN and Multi-GAN, to address the issue of an increasing number of old categories in the incremental learning process. The flowchart of the two generation strategies employed in the generative iCaRL algorithm is depicted in Fig. 5. The one-GAN generation strategy necessitates retraining the GAN for each task, in order to enable the generator to produce novel category data. The training data of GAN is the same as that of the iCaRL model. The Multi-GAN generation strategy employs distinct generators to produce data for previous categories in each task. Each task is associated with an exclusive generator, which is then added to the generator set after being trained.

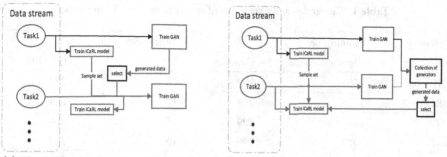

(a) The Generation Strategy of one-GAN   (b) The Generation Strategy of Multi-GAN

Fig. 5. The Flowchart of Generation Strategy.

3 Experiment

3.1 Experiment Data

This article selects data from the Moving and Stationary Target Acquisition and Recognition (MSTAR) dataset with the same collection conditions and scenarios as the experimental data. The selected data includes six types of targets: 2S1, BRDM_2, D7, T62, ZIL_131, and ZSU_23_4. The optical and SAR images of each target are shown in Fig. 6. Each task includes imaging pictures covering a full 360-° range, with azimuth intervals typically falling between 1 and 2°. The training dataset is characterized by an elevation angle of 17°, whereas the test dataset exhibits an elevation angle of 15°. Table 1 presents a breakdown of image categories and their corresponding quantities in both datasets.

Fig. 6. MSTAR Dataset Example.

Table 1. Categories and Quantities of the Training and Testing Sets.

Category	Training Set		Testing Set	
	Elevation Angle	Quantity	Elevation Angle	Quantity
2S1	17°	299	15°	274
BRDM_2	17°	298	15°	274
D7	17°	299	15°	274
T62	17°	299	15°	273
ZIL_131	17°	299	15°	274
ZSU_23_4	17°	299	15°	274

3.2 Experiment Settings

Three experiments were conducted to compare the recognition performance of the traditional iCaRL algorithm, the generative iCaRL algorithm with the one-GAN generation strategy, and the generative iCaRL algorithm with the Multi-GAN generation strategy.

The specific scenario for incremental learning established in this paper is illustrated in Fig. 7. The six categories of data mentioned above are distributed across five tasks, with the first task encompassing two target categories and each subsequent task containing one target category. The data categories included in each task differ from one another.

Fig. 7. Experimental Incremental Learning Scenario Illustration.

The training process of the generative iCaRL model comprises two stages: GAN training and iCaRL model training.The model parameters are updated using gradient descent, while the Adam adaptive learning rate algorithm dynamically adjusts the learning rate. In the training process of GAN, the mean squared error is utilized as the loss function, while employing an initial learning rate of 0.0002, a first-order momentum decay rate of 0.5, a second-order momentum decay rate of 0.999, and setting the total number of training epochs to 2000. The traditional iCaRL model has the same parameter settings as the iCaRL model in the generative iCaRL model. The cross-entropy function is employed as the loss function, with an initial learning rate of 0.001, a first-order momentum decay rate of 0.9, a second-order momentum decay rate of 0.999, and a total training epoch set to 400. The maximum number of expanded samples for each target category in the image selection algorithm is 250, and the number of generated samples per expansion for each target category is 365.

3.3 Experiment Result

In this section, we will conduct a quantitative comparison of the recognition performance among the traditional iCaRL algorithm, the generative iCaRL algorithm with one-GAN generation strategy, and the generative iCaRL algorithm with Multi-GAN generation strategy.

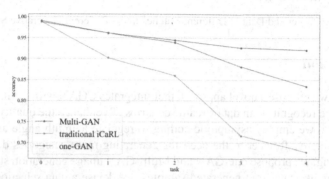

Fig. 8. Comparison of Recognition Accuracy between Different Generative Strategies of iCaRL Model and Traditional iCaRL Model.

The recognition accuracy curves of the generative iCaRL model with different generation strategies and the traditional iCaRL model as the task changes are presented in Fig. 8. The results demonstrate that both the one-GAN and Multi-GAN generation strategies are effective in mitigating the catastrophic forgetting problem encountered by the iCaRL model during incremental learning. Notably, the Multi-GAN generation strategy outperforms its counterpart in this regard.

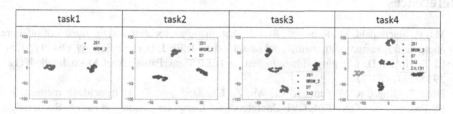

Fig. 9. T-SNE Visualized Feature Maps of Generated and Training Data Selected in Different Batches under Multi-GAN Generative Strategy.

Figure 9 illustrates the feature reduction visualization results of both Multi-GAN generated data and training data, indicating that the generated data possess favorable category attributes. Consequently, the employment of Multi-GAN generation strategy significantly enhances model recognition accuracy.

Figure 10 displays the data generated by 2S1 using different generation strategies for each task. The Multi-GAN generated data exhibit clearer backgrounds and contours, as well as more comprehensive azimuth information, making it a more effective strategy for generating high-quality data.

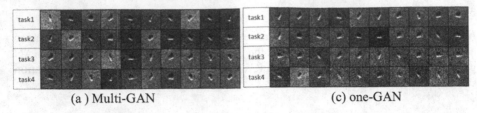

(a) Multi-GAN   (c) one-GAN

Fig. 10. 2S1 Generated Data of Different Batches under Different Generative Strategies.

4   Conclusion

In this paper, we propose a novel approach that integrates CGAN with the iCaRL model for SAR image recognition in data stream scenarios. To enhance the quality of CGAN-generated data, we employ asymptotic coding to regulate azimuth angle and augment azimuthal diversity. To cater to the need for generating more categories during incremental learning, we propose one-GAN and Multi-GAN image generation strategies. To select high-quality data from generated samples, we devise a data selection algorithm based on old model recognition and feature distance. The experimental results demonstrate that both generative strategies can enhance the recognition performance of the model, with Multi-GAN being more effective in improving the recognition performance of the iCaRL model.

Acknowledgments. This research was funded by the National Natural Science Foundation of China under Grant 62371022. The authors would like to thank them for their support in this work.

References

1. Ma, P., Macdonald, M., Rouse, S., Ren, J.: Automatic geolocation and measuring of offshore energy infrastructure with multimodal satellite data. IEEE J. Oceanic Eng. **49**, 66–79 (2023)
2. Li, Z., Hoiem, D.: Learning without forgetting. IEEE Trans. Pattern Anal. Mach. Intell. **40**(12), 2935–2947 (2017)
3. Dhar, P., Singh, R. V., Peng, K. C., Wu, Z., Chellappa, R.: Learning without memorizing. In: Proceedings of the IEEE/CVF Conference on Computer Vision and Pattern Recognition, pp. 5138–5146 (2019)
4. Rebuffi, S.A., Kolesnikov, A., Sperl, G., Lampert, C.H.: ICARL: incremental classifier and representation learning. In Proceedings of the IEEE Conference on Computer Vision and Pattern Recognition, pp. 2001–2010 (2017)
5. Castro, F.M., Marín-Jiménez, M.J., Guil, N., Schmid, C., Alahari, K.: End-to-end incremental learning. In: Proceedings of the European Conference on Computer Vision (ECCV), pp. 233–248 (2018)
6. Wu, Y., et al.: Large scale incremental learning. In: Proceedings of the IEEE/CVF Conference on Computer Vision and Pattern Recognition, pp. 374–382 (2019)
7. Kammoun, A., Slama, R., Tabia, H., Ouni, T., Abid, M.: Generative adversarial networks for face generation: a survey. ACM Comput. Surv. **55**(5), 1–37 (2022)

8. Xi, Y., et al.: DRL-GAN: dual-stream representation learning GAN for low-resolution image classification in UAV applications. IEEE J. Select. Top. Appl. Earth Observat. Remote Sens. **14**, 1705–1716 (2020)
9. Mirza, M., Osindero, S.: Conditional generative adversarial nets. arXiv pre-print arXiv:1411. 1784 (2014)

Face Reenactment Based on Motion Field Representation

Si Zheng, Junbin Chen, Zhijing Yang$^{(\boxtimes)}$, Tianshui Chen, and Yongyi Lu

School of Information Engineering, Guangdong University of Technology,
Guangzhou 510006, China
{3117002554,2112003148}@mail2.gdut.edu.cn,
{yzhj,chentianshui,yylu}@gdut.edu.cn

Abstract. Face reenactment is a challenging problem that aims to transfer facial and head motions from a source actor to a target actor. However, existing GAN-based methods often struggle to adequately capture the diverse content within face reenactment videos, resulting in unrealistic transitions between frames. In this paper, we propose a novel facial reenactment framework based on motion field representation. Our approach effectively combines the target actor's identity with the source actor's motion, enabling separate modeling and learning of the portrait and background regions in video frames. As a result, we are able to generate highly realistic portrait images. Extensive experimental evaluations demonstrate that our algorithm outperforms many state-of-the-art facial reenactment methods, highlighting its superiority in this domain.

Keywords: Face reenactment · Conditional GAN · Flow warping · Full head reenactment · Generative adversarial network

1 Introduction

Face reenactment aims to transfer the facial expressions, pose, and gaze of a source actor to a target video. In recent years, brain-inspired learning models have gained significant research interest and have been widely adopted. Among these models, face reenactment technology has garnered considerable attention due to its vast potential applications in media, entertainment, and virtual reality. One of its primary benefits is improving the efficiency of audio and video production. Existing face reenactment methods can be categorized into subject-specific methods [1,2] and subject-agnostic methods [3,4].

Existing subject-specific face reenactment methods effectively utilize the video information of the target actor. However, most of these methods rely on parameterized face models or facial landmarks to transmit motions, resulting

This work is supported in part by the Guangdong Basic and Applied Basic Research Foundation (nos. 2021A1515011341, 2023A1515012561), and Guangdong Provincial Key Laboratory of Human Digital Twin (2022B1212010004).

in the incomplete transfer of upper body movements and potentially unstable synthesized portrait videos. Traditional 3D-based subject-specific methods [5–8] focus on constructing 3D models for the facial area [9] of the target actor, primarily emphasizing the transfer of expressions without considering the head motion. Nevertheless, limitations arise as only the facial area is reconstructed, leading to a mismatch between the head and shoulders in the synthesized images due to the inability to transfer motion in areas such as the shoulders. The introduction of GANs [2, 10–12] has brought advancements in face reenactment techniques. Notably, *RecycleGAN* [11] successfully transfers the entire head motion from the source actor to the target actor. The *DVP* [13] parameterized face model reconstructs 3D faces for both the target and driving videos and GANs are employed to synthesize realistic images. The *Head2Head* [1] series methods utilize previous frames in the GAN generator for image synthesis. However, since these methods do not consider shoulder or other background conditions as inputs, the accumulation of errors can affect the quality of synthesized videos.

On the other hand, subject-agnostic face reenactment methods face limitations as they rely on only one or a few images of the target person, making it challenging to synthesize high-fidelity videos. However, by leveraging unsupervised learning methods to train and decouple motion and appearance, it becomes possible to achieve complete transfer of upper body motions.

Subject-agnostic methods in face reenactment can generate portrait videos of a target person performing various actions using just one image of the target person [14–17]. While 3D model-based methods [17–20] can generate realistic facial interiors, they are not applicable to new poses. Another approach that involves 3D models combines *StyleGAN* [21, 22] and 3D Morphable Face Models (3DMMs) [9, 23] to synthesize the entire head region. *Vid2Vid* [12] incorporates information injection into its generator by dynamically determining parameters in the SPADE [24] module. *FOMM* [25] learns a set of unsupervised keypoints and their local affine transformations to model more complex motions. These methods are learned through unsupervised training [4, 25, 26], and although they may still exhibit issues with image quality, they are capable of more comprehensively transferring motions of the entire upper body.

In this paper, we propose a face reenactment framework based on Motion Field Representation to tackle the aforementioned challenges. By combining the strengths of subject-specific and subject-agnostic face reenactment methods, our approach offers a comprehensive solution. We begin by partitioning the portrait image into distinct portrait and background regions. Subsequently, we employ an unsupervised approach to learn the keypoint motion within the portrait region, enabling the transfer of motion from the source actor to the target actor. For the background region, we fuse the left and right reference backgrounds extracted from the target video to create a synthesized background for the target frame. Finally, we utilize a neural network that takes both the portrait and background conditions as inputs to train and synthesize realistic images.

2   Methodology

Our method aims to generate a photo-realistic portrait video of the given target actor that imitates the actions of the source actor while preserving the target's identity, appearance, and background variation. The overall workflow of our proposed method is illustrated in Fig. 1. We first present a brief overview of how the portrait condition is captured in Sect. 2.1. Next, we provide a detailed description of the network architecture proposed for background extraction in Sect. 2.2. Finally, we introduce the video synthesis process in Sect. 2.3.

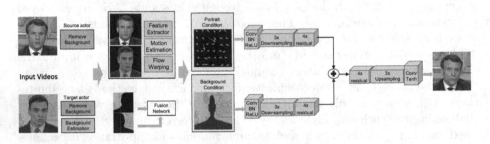

Fig. 1. Pipeline of our approach.

2.1   Portrait Condition

Given the target video $V_t = \{I_t^1, ..., I_t^M\}$ and any source video $V_s = \{I_s^1, ..., I_s^N\}$, we select a frame image from V_t as the target reference image I_t. Here, M and N represent the number of frames in the video. The objective at this stage is to transfer the motion from V_s to V_t, aligning the actions of the source actor with the target actor. We model the movement of the portrait area based on the classical unsupervised motion transfer model *FOMM* [25], model the movement as a series of key points and their nearby affine changes, and then obtain an optical flow field representing the movement of the person.

First, an encoder ε is employed to extract appearance-related features F_t from the target reference image I_t. These features capture the visual characteristics and details of the target person's appearance.

$$F_t = \varepsilon(I_t) \tag{1}$$

Next, the motion estimation module is utilized to calculate the motion between the target reference image I_t and the driving video frame I_d^i. This module computes the optical flow p, which represents the displacement of pixels between two consecutive frames. Additionally, the motion estimation module also estimates an occlusion map O, which identifies areas where occlusions occur due to objects or body parts obstructing the view. The estimated motion is then applied to the appearance features of the target reference image F_t to obtain the distorted appearance feature $\widetilde{F}_t$. This process accounts for the motion information and aligns the appearance features with the corresponding regions in the driving video frame, considering the observed optical flow and occlusion map.

$$\widetilde{F}_t = O \bigodot W(p, F_t) \tag{2}$$

where $W(,)$ represents the warping operation, and $\bigodot$ represents the Hadamard product.

2.2 Background Condition

In many instances, the background extracted from a single image may be incomplete or contain artifacts. To overcome the issues of ghosting and instability in the background region of the synthesized video, we introduce a background estimation module. This module utilizes the background of the current target frame I_t^i as a reference. By leveraging the available background information, the module generates a background that is consistent with the target frame and ensures stability throughout the synthesized video.

To ensure the completeness of background information, we adopt a two-step approach to extract the left and right backgrounds from the target video V_t. Firstly, we utilize object detection techniques [27] to select the leftmost portrait image P_{left} from the target video. This image provides a clear view of the background on the right side. Then, using image segmentation techniques, we generate a portrait mask that covers the portrait area in P_{left}. By applying the inverse of the portrait mask, we can extract the right background image, denoted as BR. Finally, we extract the correct portion from BR to obtain the appropriate reference background, denoted as BR_{ref}, for the video sequence. Similarly, we follow the same procedure to obtain the left reference background, denoted as BL_{ref}.

Our background estimation module consists of two encoder networks, denoted as e_L and e_R, along with a background fusion network $\mathcal{H}$. Each encoder network (e_X, where X can be either L or R) takes the reference background BX_{ref} and the target frame I_t^i as inputs. The encoder network predicts six real values, denoted as $VX_1^i, ..., VX_6^i$, which represents the affine background transformation between BX_{ref} and I_t^i, such that $A_X^i = \begin{bmatrix} VX_1^i & VX_2^i & VX_3^i \\ VX_4^i & VX_5^i & VX_6^i \end{bmatrix}$.

Then BX_{ref} is warped according to the A_X^i and the warped background $\widetilde{\mathcal{B}}_{left}^i$ and $\widetilde{\mathcal{B}}_{right}^i$ can be obtained as background conditions. $\widetilde{\mathcal{B}}_{left}^i$ and $\widetilde{\mathcal{B}}_{right}^i$ should match each other and correspond to the backgrounds in I_t^i. Thus:

$$\widetilde{\mathcal{B}}_{left}^i = \mathcal{W}(e_L(I_t^i, BR_{ref}), BR_{ref}) \tag{3}$$

$$\widetilde{\mathcal{B}}_{right}^i = \mathcal{W}(e_R(I_t^i, BL_{ref}), BL_{ref}) \tag{4}$$

Afterward, a background fusion network $\mathcal{H}$ consisting of a series of 2D convolution blocks without any upsampling or downsampling operation is designed to fuse $\tilde{\mathcal{B}}^i_{left}$ and $\tilde{\mathcal{B}}^i_{right}$. $\mathcal{H}$ takes $\tilde{\mathcal{B}}^i_{left}$ and $\tilde{\mathcal{B}}^i_{right}$ as inputs and output a background image $\tilde{I}^i_B$ size of $H \times W \times 3$, where H and W is the width and height of the video frame:

$$\tilde{I}^i_B = \mathcal{H}(\tilde{\mathcal{B}}^i_{left}, \tilde{\mathcal{B}}^i_{right}) \tag{5}$$

In this way, a background video $\tilde{V}_B = \{\tilde{I}^1_B, ..., \tilde{I}^M_B\}$ can be obtain as background condition (Fig. 2).

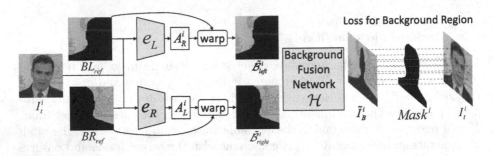

Fig. 2. The framework of our background estimation module. Given a target frame, our background estimation module can synthesize the corresponding background.

2.3   Video Synthesis

Our video synthesis module receives two inputs, namely: portrait condition $\widetilde{F}^{i-1:i+1}_t$, and background condition $\tilde{I}^{i-1:i+1}_B$. These conditions are downsampled through different branches and obtained high-level features. The resulting high-level features are fused and upsampled to a realistic image y^i. We label the module as $\mathcal{G}$. Thus:

$$y^i = \mathcal{G}(\widetilde{F}^{i-1:i+1}_t, \tilde{I}^{i-1:i+1}_B) \tag{6}$$

While training, the source and target videos are the same footage. That is, in the synthetic video, each frame y^i should be a replication of the corresponding frame I^i_t in the original target video which serves as ground truth. At test time, The motion of the generated target in the portrait part, including the shoulders, will be based on the motion of I^i_t. For the background, we can select the background video $\tilde{V}_B$ generated by the training set as the background condition input. In this way, the background of the generated video will be consistent with $\tilde{V}_B$. Optionally, select one of the frames as the background condition input, the generated video background will be static.

3   Experiments

3.1   Dataset and Evaluation Metrics

We evaluated our method on the dataset provided by *Head2head++* [2]. This dataset consists of eight original target videos representing different identities, with each video having a minimum duration of ten minutes. Prior to conducting the experiments, we followed the same preprocessing steps as *Head2head++* [2] by cropping and resizing each video to a resolution of 256 pixels in height and width. To assess the quality of the predicted portrait images, we employed four commonly used metrics: Average Pixel Distance (APD), Masked Average Pixel Distance for the Portrait Region (APD-P), Masked Average Pixel Distance for the Background Region (APD-B), and Fréchet Inception Distance (FID). These metrics provided quantitative measurements to evaluate the accuracy and similarity between the generated portraits images and the ground truth images.

3.2   Ablation Study

In order to assess the performance of our proposed background extraction method on dynamic background videos, we have developed a quantitative evaluation approach. This evaluation method allows us to objectively measure the quality and effectiveness of our background extraction technique. we conducted a comparison between the full model, which includes the background condition input, and a variant of the model without the background condition input. By comparing the performance of these two models, we can evaluate the impact of the background condition input on the overall synthesis quality of the videos.

Table 1. The effect of background condition. All metrics are averages of all eight target videos in the dataset. For all metrics, the lower the better.

Model	FID ↓	APD ↓	APD-P ↓	APD-B ↓
w/o BGC	6.70	13.74	17.07	9.02
Full Model	**6.27**	**12.6**	**16.91**	**6.58**

The impact of background conditions is demonstrated in Table 1. Our observations indicate that the model without background conditions exhibits inferior performance compared to the full model. The full model, which incorporates background conditions, demonstrates superior synthesis quality and generates more realistic images. All evaluation metrics show improvements when background conditions are included, resulting in smoother and more visually appealing videos.

As can be seen in the second row of Fig. 3, when removing the background condition input and only conditioning on portrait information (w/o BGC), the background region in the generated video has noticeable artifacts. As shown in the second row, without conditioning on the background, the change of background in the video is abrupt (blinking from left to right). In the third row, This problem is solved by conditioning on the background.

This observation highlights the effectiveness of our approach in leveraging background information to generate more accurate and realistic backgrounds in the synthesized videos. By incorporating background conditions, our network

temporal axis

Fig. 3. Significance of conditioning on background in video perspective.

can better capture and reproduce the visual characteristics and details of the real background, leading to a more faithful representation of the scene.

3.3 Comparison Study

We conducted a comparative analysis of our method with the current state-of-the-art face reenactment methods on the aforementioned dataset. The comparison included two subject-agnostic methods, namely *X2Face* [4] and *FOMM* [25]), as well as two subject-specific methods, *Head2Head++* [2] and *Vid2Vid* [12]. The results of the experiments are presented in Table 2.

We performed self-reconstruction experiments to quantitatively evaluate the synthesis quality of our method in comparison to *X2Face* [4] and *FOMM* [25]. The experimental results are presented in Fig. 4. It was observed that subject-agnostic methods, which primarily rely on a single image for face reenactment, tend to lack contextual information about the target actor, leading to subpar results. In contrast, our method excels in generating videos that closely resemble real ones, encompassing both the portrait characters and background information. This showcases the superiority of our approach in achieving more realistic and comprehensive face reenactment results.

Table 2. Quantitative comparison with X2Face, FOMM, Vid2Vid and Head2Head++. All metrics are averages of all eight target videos in the dataset. For all metrics, the lower the better.

Method	FID ↓	APD ↓	APD-P ↓	APD-B ↓
X2Face	74.8	33.34	33.51	33.06
FOMM	21.43	20.21	20.3	19.74
Vid2Vid	8.19	16.49	17.37	14.86
Head2Head++	7.8	14.93	**16.87**	12.07
Ours	**6.27**	**12.6**	16.91	**6.58**

GroundTruth   X2Face   FOMM   Ours

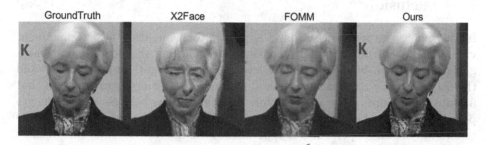

Fig. 4. Self-reconstruction experiments compared with X2Fcae and FOMM.

We also conducted cross-actor reenactment experiments with *Vid2Vid* [12], comparing the results with *Head2Head++* [2] and *Vid2Vid* [12]. These models are regarded as excellent face reenactment methods capable of synthesizing realistic videos. The comparison results, depicted in Fig. 5, reveal that our method outperforms *Head2Head++* [2] and *Vid2Vid* [12] across most metrics, which aligns with the findings from the qualitative evaluation. This can be attributed to the fact that our method models various components in portrait images, including the background, shoulders, and head, enabling more comprehensive control over video generation. As a result, our method achieves superior performance in terms of realism and fidelity compared to the other methods.

Source   Ours   Vid2Vid   Target Identity

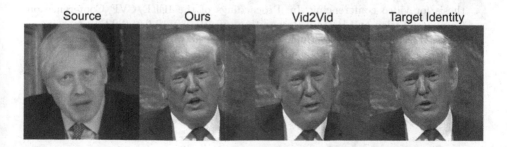

Fig. 5. Cross-actor reenactment compared with Vid2Vid.

In the quantitative comparison presented in Table 2, our method exhibits lower values in terms of Average Pixel Distance (APD) and Average Pixel Distance for Background Region (APD-B) compared to the comparison methods. This indicates that our method effectively fuses the background information and generates more realistic images. Moreover, our method also outperforms the comparison methods in terms of the Fréchet Inception Distance (FID) index, which measures the realism of the synthesized images. These results further validate the effectiveness of our approach in achieving superior synthesis quality and realism compared to the comparison methods.

4   Conclusion

In this paper, we introduce a novel approach for generating face reenactment videos that focus on separately modeling and learning the portrait and background regions in video frames. By leveraging the target actor's identity and the source actor's motion, our method generates realistic portrait images that imitate the actions of the source actor. Additionally, we address the challenges posed by dynamic backgrounds by developing a background generation module tailored to the target person. Our experimental results showcase the effectiveness of our method, surpassing the performance of current advanced face reenactment video generation methods. Our proposed method provides a promising solution for generating high-quality and realistic face reenactment videos, opening up new possibilities for enhanced audio and video production and immersive experiences.

References

1. Koujan, M.R., Doukas, M.C., Roussos, A., Zafeiriou, S.: Head2head: video-based neural head synthesis. In: 2020 15th IEEE International Conference on Automatic Face and Gesture Recognition (FG 2020), pp. 16–23. IEEE (2020)
2. Doukas, M.C., Koujan, M.R., Sharmanska, V., Roussos, A., Zafeiriou, S.: Head2head++: deep facial attributes re-targeting. IEEE Trans. Biometr. Behav. Ident. 3(1), 31–43 (2021)
3. Wang, T.-C., Mallya, A., Liu, M.-Y.: One-shot free-view neural talking-head synthesis for video conferencing. In: Proceedings of the IEEE/CVF Conference on Computer Vision and Pattern Recognition, pp. 10 039–10 049 (2021)
4. Wiles, O., Koepke, A., Zisserman, A.: X2face: A network for controlling face generation using images, audio, and pose codes. In: Proceedings of the European Conference on Computer Vision (ECCV), pp. 670–686 (2018)
5. Suwajanakorn, S., Seitz, S.M., Kemelmacher-Shlizerman, I.: Synthesizing obama: learning lip sync from audio. ACM Trans. Graph. (ToG) 36(4), 1–13 (2017)

6. Thies, J., Zollhöfer, M., Nießner, M.: Deferred neural rendering: image synthesis using neural textures. ACM Trans. Graph. (TOG) **38**(4), 1–12 (2019)

7. Thies, J., Zollhöfer, M., Nießner, M., Valgaerts, L., Stamminger, M., Theobalt, C.: Real-time expression transfer for facial reenactment. ACM Trans. Graph. **34**(6), 183–1 (2015)

8. Thies, J., Zollhofer, M., Stamminger, M., Theobalt, C., Nießner, M.: Face2face: real-time face capture and reenactment of RGB videos. In: Proceedings of the IEEE Conference on Computer Vision and Pattern Recognition, pp. 2387–2395 (2016)

9. Blanz, V., Vetter, T.: A morphable model for the synthesis of 3D faces. In: Proceedings of the 26th Annual Conference on Computer Graphics and Interactive Techniques, pp. 187–194 (1999)

10. Wu, W., Zhang, Y., Li, C., Qian, C., Loy, C.C.: Reenactgan: learning to reenact faces via boundary transfer. In: Proceedings of the European Conference on Computer Vision (ECCV), pp. 603–619 (2018)

11. Bansal, A., Ma, S., Ramanan, D., Sheikh, Y.: Recycle-gan: unsupervised video retargeting. In: Proceedings of the European Conference on Computer Vision (ECCV), pp. 119–135 (2018)

12. Wang, T.-C., Liu, M.-Y., Zhu, J.-Y., Liu, G., Tao, A., Kautz, J., Catanzaro, B.: Video-to-video synthesis. Adv. Neural. Inf. Process. Syst. **31**, 1144–1156 (2018)

13. Kim, H., et al.: Deep video portraits. ACM Trans. Graph. (TOG) **37**(4), 1–14 (2018)

14. Averbuch-Elor, H., Cohen-Or, D., Kopf, J., Cohen, M.F.: Bringing portraits to life. ACM Trans. Graph. (TOG) **36**(6), 1–13 (2017)

15. Burkov, E., Pasechnik, I., Grigorev, A., Lempitsky, V.: Neural head reenactment with latent pose descriptors. In: Proceedings of the IEEE/CVF Conference on Computer Vision and Pattern Recognition, pp. 13 786–13 795 (2020)

16. Chen, L., Maddox, R.K., Duan, Z., Xu, C.: Hierarchical cross-modal talking face generation with dynamic pixel-wise loss. In: Proceedings of the IEEE/CVF Conference on Computer Vision and Pattern Recognition, pp. 7832–7841 (2019)

17. Geng, J., Shao, T., Zheng, Y., Weng, Y., Zhou, K.: Warp-guided gans for single-photo facial animation. ACM Trans. Graph. (ToG) **37**(6), 1–12 (2018)

18. Fried, O., et al.: Text-based editing of talking-head video. ACM Trans. Graph. (ToG) **38**(4), 1–14 (2019)

19. Nagano, K., et al.: pagan: real-time avatars using dynamic textures. ACM Trans. Graph. (ToG) **37**(6), 1–12 (2018)

20. Olszewski, K., et al.: Realistic dynamic facial textures from a single image using gans. In: Proceedings of the IEEE International Conference on Computer Vision, pp. 5429–5438 (2017)

21. Karras, T., Laine, S., Aila, T.: A style-based generator architecture for generative adversarial networks. In: Proceedings of the IEEE/CVF Conference on Computer Vision and Pattern Recognition, pp. 4401–4410 (2019)

22. Karras, T., Laine, S., Aittala, M., Hellsten, J., Lehtinen, J., Aila, T.: Analyzing and improving the image quality of stylegan. In: Proceedings of the IEEE/CVF Conference on Computer Vision and Pattern Recognition, pp. 8110–8119 (2020)

23. Egger, B., et al.: 3d morphable face models-past, present, and future. ACM Trans. Graph. (TOG) **39**(5), 1–38 (2020)

24. Park, T., Liu, M.-Y., Wang, T.-C., Zhu, J.-Y.: Semantic image synthesis with spatially-adaptive normalization. In: Proceedings of the IEEE/CVF Conference on Computer Vision and Pattern Recognition, pp. 2337–2346 (2019)

25. Siarohin, A., Lathuilière, S., Tulyakov, S., Ricci, E., Sebe, N.: First order motion model for image animation. Adv. Neural. Inf. Process. Syst. **32**, 7137–7147 (2019)
26. Hwang, G., Hong, S., Lee, S., Park, S., Chae, G.: Discohead: audio-and-video-driven talking head generation by disentangled control of head pose and facial expressions. arXiv preprint arXiv:2303.07697 (2023)
27. Bochkovskiy, A., Wang, C.-Y., Liao, H.-Y.M.: Yolov4: optimal speed and accuracy of object detection. arXiv preprint arXiv:2004.10934 (2020)

LWGSS: Light-Weight Green Spill Suppression for Green Screen Matting

Anhui Bai[1], Zhijing Yang[1(✉)], Jinghui Qin[1], Yukai Shi[1], and Kai Li[2]

[1] School of Information Engineering, Guangdong University of Technology,
Guangzhou 510006, China
2112103151@mail2.gdut.edu.cn, {yzhj,ykshi}@gdut.edu.cn
[2] ZEGO, Shenzhen 518000, China

Abstract. Green spill is a significant challenge in green screen matting and it affects the overall visual quality of the images. Due to the limitations of user interaction and the lack of specific datasets containing green spill information, it is difficult to generate high-quality foreground images automatically in green screen matting. In this paper, we propose a light-weight green spill suppression method for green screen matting, named as LWGSS, to generate high-quality foreground images without green spill in a concise way. Specifically, we adopt an anomaly detector to detect the anomalous pixels with green spill, and normalize these pixels through the spill removal module. With the help of the green spill suppression stage, our method can break the limitations of user interaction and specific datasets. Additionally, we switch the learning objectives from predicting the alpha mattes of foregrounds to predict the alpha mattes of background by using label inversion, which enables us to address all kinds of objects. Furthermore, we present a light-weight network with the smallest model size and parameters to further improve the inference speed. Extensive experiments on Composition-1k and Distinctions-646 demonstrate the superiority of our LWGSS.

Keywords: Green screen matting · Green spill suppression · Label inversion · Lightweight

1 Introduction

Green screen matting is the process of accurately extracting alpha mattes and visually appealing foreground images. With the development of brain-inspired and deep learning [1–5], green screen matting is becoming increasingly important in scenarios that require background replacement, such as video conferencing tools and movie special effects.

This work is supported in part by the Guangdong Basic and Applied Basic Research Foundation (nos. 2021A1515011341, 2023A1515012561), and Guangdong Provincial Key Laboratory of Human Digital Twin (2022B1212010004).

J. Ren et al. (Eds.): BICS 2023, LNAI 14374, pp. 365–374, 2024.
https://doi.org/10.1007/978-981-97-1417-9_34

The green spill phenomenon affects the overall quality of the images and can be considered a huge challenge in green screen matting. When light reflections cause the transmission of green color from the background onto foreground, it gives rise to abnormal green areas in the extracted foreground images, commonly known as green spill phenomenon. To address this issue, Smith et al. [6] proposed a conceptual model study where a separate color spill image with its alpha channel was used for modeling. Aksoy et al. [7] proposed a color unmixing algorithm to obtain high-quality images without green spill. Luo et al. [8] employed polarization and disparity-based optical flow algorithms to suppress color spill. However, these methods require complex algorithms and user interaction, which are time-consuming and not conducive to real-world applications. As deep learning has been proven effective in natural image matting [1,9–16], many researchers have turned their attention towards employing deep learning methods to tackle green screen matting problems [17,18]. Maul et al. [17] created a dataset for suppressing green spill and trained a deep learning model to predict images with good visual effects. Jin et al. [18] constructed a dataset with foreground objects containing green spill, and removed it by learning the features of green spill from the foreground objects. However, these methods require datasets that include green spill information, which are labor-intensive and costly.

In addition, the ability of matting methods to handle specific object categories is primarily determined by the types of objects included in their training datasets. However, existing datasets can not cover all of them. Besides, deep learning methods prefer to use large models with complex model structures and a large number of parameters [13,14] to obtain better performance, which significantly increase the inference time. In conclusion, the above problems make previous methods unsuitable for applications in the real world.

To address the aforementioned limitations, we propose a Light-Weight framework with Green Spill Suppression stage for green screen matting (LWGSS). Unlike previous methods which require a significant amount of user interaction or specific datasets containing green spill information to solve green spill problems, we propose an anomaly detector to detect the anomalous pixels with the green spill problem, and normalize these pixels through the spill removal module. Besides, we propose label inversion to enable our LWGSS to handle various types of objects. Furthermore, we propose a light-weight network, named as LW-Net, to improve the inference speed. Extensive experiments on two pubilc datasets demonstrate our method performs better than previous methods.

In summary, the contributions are summarized as follows:

- We propose a novel LWGSS to obtain high-quality foreground images without green spill for green screen matting. Our approach can solve the green spill problem without user interaction and specific datasets containing green spill information by utilizing anomaly detector and spill removal module in the green spill suppression stage.
- To break the constraints of data diversity and improve the inference speed, we propose label inversion strategy and a light-weight network LW-Net to realize a better performance with lower cost.

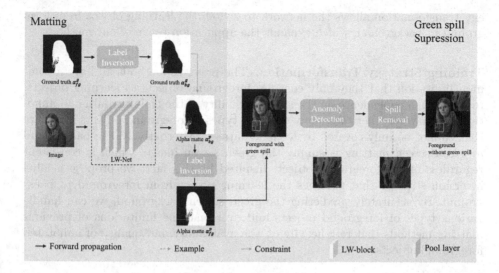

Fig. 1. The overview of our LWGSS. Left: the matting stage consists of LW-Net and label inversion to predict the alpha matte. Right: the green spill suppression stage consists of anomaly detector and spill removal module to generate high-quality images without green spill. (Color figure online)

– Extensive experiments on Composition-1k and Distinctions-646 demonstrate the superiority of our LWGSS for green screen matting.

2 Method

In this section, we propose a light-weight framework for green screen matting that includes a green spill suppression stage to enhance the quality of the foreground image. The overview of our matting framework is shown in Fig. 1, which illustrates the two main stages: the matting stage and the green spill suppression stage. The matting stage is primarily responsible for predicting the corresponding alpha matte based on the input image. It comprises two main parts: the label inversion strategy and the LW-Net. The label inversion strategy converts the learning object from foreground to background, enabling us to handle all kinds of objects without corresponding types of training data. Subsequently, the LW-Net predicts the alpha matte of the background using a green screen image as input. In the green spill suppression stage, we detect anomalous areas containing pixels with the green spill phenomenon in the foreground image using an anomaly detector. After suppressing the green spill phenomenon through the spill removal module, we can obtain a visually pleasing image.

2.1 Label Inversion

Our proposed label inversion strategy includes two branches: training strategy transformation and foreground-background transformation. The training strategy transformation allows the network to switch the learning object from foreground to background, which expands the application range of our method.

Training Strategy Transformation. The previous methods all have a common limitation that they only consider foreground objects as learning objects, resulting in a dependence on the size and quality of the training data and annotations. However, there are many different types of foreground objects in nature that cannot be fully covered by a single dataset. Alternatively, the background in green screen matting is simply the green screen image, which is consistent regardless of the foreground object. Inspired by this fact, we propose a label inversion strategy that switches the learning object from foreground to background. By accurately predicting the green screen background, we can handle various types of foreground objects and overcome the limitations of previous matting methods that rely heavily on the availability and quality of annotated foreground objects.

Foreground-Background Transformation. The pipeline for label inversion is shown in Fig. 1. The alpha matte of foreground and background can be converted into each other according to Eq. 1.

$$\alpha_{fg} = 1 - \alpha_{bg} \tag{1}$$

Here, $\alpha_{fg} \in [0,1]$ represents the alpha matte of foreground and $\alpha_{bg} \in [0,1]$ stands for the alpha matte of background, respectively.

2.2   LW-Net

Our proposed matting network stands apart from previous methods by using a simpler model structure with fewer parameters, which enables faster inference speed. By combining with anti-alpha, we can learn the green screen features directly instead of learning the foreground objects, which reduces the difficulty of training the network, thus allowing us to have better performance while being lightweight. As illustrated in Fig. 1, our LW-Net consists of 4 LW-blocks and 1 pooling layers. To reduce model complexity and the number of parameters, each LW-block consists of 3×3 convolution and 1×1 convolution, instead of convolution layers with larger convolution kernels. We use the green screen image as the input of the LW-Net and predict the alpha matte of background α_{bg}^{p}. In addition, the output of the LW-Net is supervised by the ground truth α_{bg}^{g}.

2.3   Green Spill Suppression

Traditional methods for green screen matting often require significant user interaction, while deep learning methods rely on datasets that include green spill information. To overcome these limitations, we propose a green spill suppression stage that grays out anomalous pixels to eliminate green spill without manual interaction or specialized datasets. This approach enables our method to achieve better results with greater flexibility and efficiency. Our green spill suppression stage is composed of an anomaly detector and a spill removal module. By detecting pixels containing green spill using an anomaly detector, we can prevent extracting relevant features from foreground objects containing green spill, thus reducing the cost of creating specific datasets containing foregrounds with green spill. Moreover, our green spill suppression is automatic and efficient since we can remove the anomalous pixels without user interaction.

Anomaly Detector. Images with green spill often contain pixels with abnormally high green channel values, causing these regions to appear green. To address this issue, our approach views the detection of these anomalous areas as the first key step in suppressing green spill. The details are presented as follows:

- Step 1: We perform channel separation on the foreground image to obtain the values of the R, G, and B channels separately.
- Step 2: Green spill results from a pixel's green channel value exceeding the weighted average of the other two channels, which denotes anomalous green color. To determine the threshold of the green channel for detecting anomalous pixels, we need to consider the corresponding values of the red and blue channel for each image. Green spill phenomenon will happen if a pixel's green channel value is higher than the threshold. Therefore, the threshold is critical and can be described using the following formula:

$$G_{max} = \frac{2b + r}{3} \qquad (2)$$

  where $G_{max} \in [0,255]$ denotes the threshold of the green channel, and b and $r \in [0,255]$ represents the value of the blue and red channel for the same pixel. The choice of coefficients for b and r is based on a comparison of the performance of experiments using different coefficients.
- Step 3: We can determine whether a pixel is an anomalous pixel by comparing its G_{max} with green channel value. If the green channel value of a pixel is higher than G_{max}, then the pixel can be identified as an anomalous pixel.

Spill Removal Module. Once the anomalous areas in the foreground image have been detected, we can utilize the spill removal module to suppress the green spill to obtain visually pleasing images. The details are presented as follows:

- Step 1: If the pixel is an anomalous pixel containing green spill, its green channel value will be replaced with G_{max}, achieving the removal of the green spill. It can be described as follows:

$$G_i = \begin{cases} G_i^{max} & G_i \geq G_i^{max} \\ G_i & others \end{cases} \qquad (3)$$

Table 1. The quantitative comparison on Composition-1k testing dataset and Distinctions-646 testing dataset. Upper part: auxiliary-based approaches. Lower part: trimap-free approaches.

Dataset	Composition-1k					Distinctions-646				
Metrics	SAD↓	MSE↓	MAD↓	GRAD↓	CONN↓	SAD↓	MSE↓	MAD↓	GRAD↓	CONN↓
BGMV2	230.37	0.0767	0.1019	121.19	236.80	163.04	0.0686	0.0874	209.10	162.78
KNN	160.06	0.0480	0.0713	128.32	62.35	82.76	0.0303	0.0519	**79.20**	**55.59**
IndexNet	**60.23**	**0.0076**	**0.0272**	**71.49**	**58.73**	**79.19**	**0.0172**	0.0398	123.91	79.00
GFM	226.69	0.0757	0.0978	146.59	151.91	233.44	0.0855	0.1023	229.81	186.02
MODNet	212.23	0.0708	0.0956	132.73	168.22	177.37	0.0698	0.0883	216.50	145.72
AIM	93.95	0.0243	0.0472	**71.89**	91.36	96.68	0.0274	0.0506	103.83	96.54
Ours	**90.76**	**0.0162**	**0.0392**	141.98	**89.65**	**60.86**	**0.0201**	**0.0398**	**86.70**	**58.14**

where G_i represents the green channel value of the i-th pixel, and G_i^{max} denotes the threshold of the green channel of the i-th pixel.
- Step 2: We need to merge the adjusted R, G, and B channel values to obtain the visually pleasing image.

3   Experiment

3.1   Quantitative Result

To validate the effectiveness of our proposed method, we conduct experiments on the test set of two public datasets: Distinctions-646 [19] and Adobe Composition-1k [1]. As our model takes green screen images as input, we synthesize new green screen images by compositing the foregrounds from the common matting dataset with green screen backgrounds. Similar to previous methods, we use five common metrics to evaluate the quality of the predicted alpha matte, including the sum of absolute differences (SAD), mean squared error (MSE), mean absolute difference (MAD), gradient (GRAD), and connectivity (CONN). To show the superiority of our LWGSS, we compare LWGSS with the following methods: auxiliary-based methods(KNN [20], IndexNet [21], BGMV2 [22]), and trimap-free methods(MODNet [15], AIM [13], GFM [14]).

Composition-1k Test Set: The quantitative results on composition-1k test set are shown in the middle part of Table 1. Compared to auxiliary-based methods, our method achieves superior performance over KNN and BGMV2. However, IndexNet outperforms our method, due to its utilization of additional trimap and large-scale model architecture. In contrast, our approach only requires a single green screen image as input and employs a light-weight model structure. Among trimap-free methods, our LWGSS shows superior performance over MODNet and GFM across all metrics, except for a slightly lower score on GRAD when compared to AIM. Despite the light-weight model design, our method still delivers competitive results, which shows the superiority of our approach.

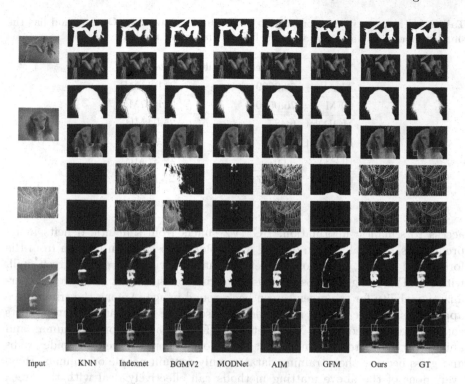

Fig. 2. Qualitative comparison with other methods. Our method performs better and achieves great visual effects for all kinds of objects without auxiliary inputs.

Distinctions-646 Test Set: For the Distinctions-646 dataset, the comparative results are summarized in the right part of Table 1. For those auxiliary-based methods, our method shows the superiority on all metrics compared to BGMV2, while slightly worse than IndexNet in SAD, MSE and KNN in GRAD, CONN. It is worth noting that both IndexNet and KNN demand trimaps as additional input to provide comprehensive semantic information, which is helpful to refine the transition area and effectively improve their performance. Besides, we can observe that our method achieves the best performance in all metrics among the trimap-free methods with the help of our proposed label inversion strategy.

3.2 Qualitative Results

To verify the superiority of our method, we conducte a qualitative comparison with the other 6 matting methods by predicting the alpha matte and the corresponding foreground image. As illustrated in Fig. 2, KNN demonstrates good performance in humans and animals, but encounters challenges when dealing with intricate details such as spider webs. IndexNet demonstrates superior capability in capturing intricate details across various object categories, but also produces graffiti around the outlines of foreground objects. Due to insufficient

Table 2. The quantitative comparison on Parameters and Size. Our method has the smallest parameters and size, which is lighter than other methods.

Method	Parameters (Million)	Size
AIM	76.0699	211.26 MB
GFM	76.0580	211.26 MB
MODNet	6.487	25.04 MB
IndexNet	5.9535	22.97 MB
BGMV2	5.007	19.37 MB
Ours	**0.0017**	**25 KB**

access to semantic information, BGMV2 suffers from semantic deficits in its predictions, especially when dealing with spider webs. As can be seen from the foreground images, the above 3 methods have varying degrees of green spill, with BGMV2 being the most serious. This may be due to its insufficient generalization ability for the green screen background image. Among the trimap-free approaches, AIM captures more adequate semantic information, but predicts coarse details. In contrast, MODNet and GFM are able to predict human and animal details more accurately, but they do not perform well on spider webs and glass because their training datasets only contain people or animals. However, none of the above matting methods can effectively deal with the green spill problem. In comparison, our LWGSS predicts a more accurate alpha matte for handling all kinds of objects with the help of label inversion strategy. By incorporating a green spill suppression stage, we effectively solve the problem of green spill, resulting in a visually pleasing foreground image.

3.3 Model Parameters, Size

Table 2 presents a comprehensive comparison of our proposed LWGSS with other matting approaches in terms of model size and parameters. The complex model architectures results in much bigger model sizes and parameters for AIM and GFM. In contrast, our approach stands out as being extremely light-weight, showing the smallest model size and parameters. It also means that we can infer with fewer computer resources and enhance the inference speed, making it simpler to apply our method to real-world settings.

3.4 Ablation Study

The core component of our proposed method is the green spill suppression stage, which utilizes an anomaly detector to identify abnormal regions containing green spill pixels, and subsequently employs a spill removal module to eliminate the green spill phenomenon. To verify its effectiveness, we conduct ablation experiments on green spill suppression. As illustrated in Fig. 3, we show the results for two different types of objects. Specifically, each group of results includes the

Fig. 3. For each group, Left: green screen images. Middle: method without green spill suppression. Right: method with green spill suppression. (Color figure online)

test image and two foreground images predicted based on whether or not they pass green spill suppression stage. We can observe that the method without green spill suppression stage still exhibits green spill phenomenon around the foreground object as shown in the middle column of (a) and (b). In contrast, the method with green spill suppression stage normalizes the area with green spill pixels to generate a more visually appealing foreground image, which illustrates the advantages of green spill suppression stage.

4   Conclusion

In this paper, we propose a light-weight green spill suppression method for green screen matting named LWGSS. By taking only a green screen image as input, our method can predict an accurate alpha matte and high-quality foreground images without green spill. While the proposed label inversion enables us to address all kinds of objects, the proposed green spill suppression stage can detect the anomalous pixels with green spill by anomaly detector and normalize these pixels by spill removal module to get rid of the constraints of user interaction and specific datasets. With the proposed LW-Net, our LWGSS contains the smallest size and parameters, making it light-weight. Extensive experiments on two popular public benchmarks demonstrate our robust performance.

References

1. Xu, N., Price, B., Cohen, S., Huang, T.: Deep image matting. In: Proceedings of the IEEE Conference on Computer Vision and Pattern Recognition, pp. 2970–2979 (2017)
2. Ma, P., et al.: Multiscale superpixelwise prophet model for noise-robust feature extraction in hyperspectral images. IEEE Trans. Geosci. Remote Sens. **61**, 1–12 (2023)
3. Li, Y., et al.: Cbanet: an end-to-end cross-band 2-d attention network for hyperspectral change detection in remote sensing. IEEE Trans. Geosci. Remote Sens. **61**, 1–11 (2023)
4. Ma, P., Ren, J., Zhao, H., Sun, G., Murray, P., Zheng, J.: Multiscale 2-d singular spectrum analysis and principal component analysis for spatial-spectral noise-robust feature extraction and classification of hyperspectral images. IEEE J. Sel. Topics Appl. Earth Observat. Remote Sens. **14**, 1233–1245 (2021)

5. Xie, G., Ren, J., Marshall, S., Zhao, H., Li, R., Chen, R.: Self-attention enhanced deep residual network for spatial image steganalysis. Dig. Signal Process. **139**, 104063 (2023)
6. Smith, A.R., Blinn, J.F.: Blue screen matting. In: Proceedings of the 23rd Annual Conference on Computer Graphics and Interactive Techniques, pp. 259–268 (1996)
7. Aksoy, Y., Aydin, T.O., Pollefeys, M., Smolić, A.: Interactive high-quality green-screen keying via color unmixing. ACM Trans. Graph. (TOG) **36**(4), 1 (2016)
8. Luo, Y.: Color spill suppression in chroma keying. Ph.D. dissertation, Université d'Ottawa/University of Ottawa (2020)
9. Li, Y., Lu, H.: Natural image matting via guided contextual attention. In: Proceedings of the AAAI Conference on Artificial Intelligence, vol. 34, no. 07, pp. 11 450–11 457 (2020)
10. Sun, Y., Tang, C.-K., Tai, Y.-W.: Semantic image matting. In: Proceedings of the IEEE/CVF Conference on Computer Vision and Pattern Recognition, pp. 11 120–11 129 (2021)
11. Fang, X., Zhang, S.-H., Chen, T., Wu, X., Shamir, A., Hu, S.-M.: User-guided deep human image matting using arbitrary trimaps. IEEE Trans. Image Process. **31**, 2040–2052 (2022)
12. Dai, Y., Lu, H., Shen, C.: Learning affinity-aware upsampling for deep image matting. In: Proceedings of the IEEE/CVF Conference on Computer Vision and Pattern Recognition, pp. 6841–6850 (2021)
13. Li, J., Zhang, J., Tao, D.: Deep automatic natural image matting. arXiv preprint arXiv:2107.07235 (2021)
14. Li, J., Zhang, J., Maybank, S.J., Tao, D.: Bridging composite and real: towards end-to-end deep image matting. Int. J. Comput. Vision **130**(2), 246–266 (2022)
15. Ke, Z., Sun, J., Li, K., Yan, Q., Lau, R.W.: Modnet: real-time trimap-free portrait matting via objective decomposition. In: Proceedings of the AAAI Conference on Artificial Intelligence, vol. 36, no. 1, pp. 1140–1147 (2022)
16. Chen, G., et al.: Pp-matting: High-accuracy natural image matting. arXiv preprint arXiv:2204.09433 (2022)
17. Maul, P.: A deep learning approach to color spill suppression in chroma keying (2019)
18. Jin, Y., Li, Z., Zhu, D., Shi, M., Wang, Z.: Automatic and real-time green screen keying. Vis. Comput. **38**(9–10), 3135–3147 (2022)
19. Qiao, Y., et al.: Attention-guided hierarchical structure aggregation for image matting. In: Proceedings of the IEEE/CVF Conference on Computer Vision and Pattern Recognition, pp. 13 676–13 685 (2020)
20. Chen, Q., Li, D., Tang, C.-K.: KNN matting. IEEE Trans. Pattern Anal. Mach. Intell. **35**(9), 2175–2188 (2013)
21. Lu, H., Dai, Y., Shen, C., Xu, S.: Indices matter: learning to index for deep image matting. In: Proceedings of the IEEE/CVF International Conference on Computer Vision, pp. 3266–3275 (2019)
22. Lin, S., Ryabtsev, A., Sengupta, S., Curless, B.L., Seitz, S.M., Kemelmacher-Shlizerman, I.: Real-time high-resolution background matting. In: Proceedings of the IEEE/CVF Conference on Computer Vision and Pattern Recognition, pp. 8762–8771 (2021)

Visual-Textual Attention for Tree-Based Handwritten Mathematical Expression Recognition

Wei Liao[1], Jiayi Liu[1], Jianghan Chen[1], Qiu-Feng Wang[1(✉)],
and Kaizhu Huang[2]

[1] School of Advanced Technology, Xi'an Jiaotong-Liverpool University, Suzhou,
China
Qiufeng.Wang@xjtlu.edu.cn
[2] Data Science Research Center, Duke Kunshan University, Suzhou, China

Abstract. Handwritten mathematical expression recognition (HMER) has attracted much attention and achieved remarkable progress under the encoder-decoder framework. However, it is still challenging due to complex structures and illegible handwriting. In this paper, we propose to refine the encoder-decoder framework for HMER. Firstly, we propose a multi-scale vision and textual attention fusion mechanism to enhance the contexts from both spatial and semantic information. Next, most of HMER works simply regard the HMER as a sequence-to-sequence problem (i.e., Latex string), ignoring the structure information in the mathematical expressions. To overcome this issue, we utilize a tree decoder to capture such structure contexts. Furthermore, we propose a parent-children mutual learning method to enhance the learning of our encoder-decoder model. Extensive experiments on the HMER benchmark datasets of CROHME 2014, 2016 and 2019 demonstrate the effectiveness of the proposed method.

Keywords: Handwritten mathematical expression recognition · Tree decoder · Visual-textual attention · Mutual learning

1 Introduction

Handwritten mathematical expression recognition (HMER) aims to convert handwritten mathematical expressions into a digital format (e.g., Latex string), which is beneficial for various applications like document recognition, intelligent education, and academic paper writing tools, etc. However, HMER is a challenging problem [1,2] such as recognizing diverse mathematical symbols, handling complex spatial relationships between symbols, and dealing with variations in handwriting styles. Traditional approaches [3–5] commonly utilize custom-designed rules and obtain limited recognition performance on complex mathematical expressions, because these rules heavily rely on prior knowledge to define expression structure and positional relationships among symbols, which is challenging.

Recent advancements of deep neural networks [6] have led to the development of encoder-decoder architectures for HMER [7–11]. These models

J. Ren et al. (Eds.): BICS 2023, LNAI 14374, pp. 375–384, 2024.
https://doi.org/10.1007/978-981-97-1417-9_35

incorporate attention mechanisms and regard HMER as a sequence-to-sequence prediction task, achieving significant performance improvements. However, these approaches usually overlook complex structures in mathematical expressions. To overcome this issue, tree-based models [12–15] convert images into tree structures to exploit spatial relationships explicitly, while these methods lack syntactic constraints and may encounter structural prediction errors. In addition, the semantic contexts from predicted characters at different time intervals are usually under-explored in most of current HMER methods.

In this paper, we aim to improve the HMER performance under the encoder-decoder framework. To this end, we firstly propose to fuse both visual and textual attention, enabling multi-scale spatial contexts and improved semantic perception for the decoder. Secondly, we adopt a tree reasoning algorithm in the decoder to capture spatial structure constrains in the mathematical expression tree. Finally, we propose a parent-children mutual learning strategy to enhance the learning process via complementary attention positions.

2   Related Works

Similar to most of computer vision tasks, handwritten mathematical expression recognition (HMER) can also be divided into traditional machine learning based and deep learning based methods. In this section, we will briefly review related works according to such two categories, together with some specific tree-based related works.

Traditional machine learning methods [3] segment input images into math symbols for classification, where it is challenging to obtain correct symbol segmentation. In addition, some works [4,5] analyze symbols and structure with global information, but they usually struggle with complex expressions due to specific grammars, prior knowledge, and parsing algorithms.

Deep learning has achieved promising progress in HMER. The WYGIWYS model [7] utilizes a typical attention-based encoder-decoder neural network, which combines convolutional and recurrent neural networks with an attention mechanism for LaTeX string generation. To address issues like coverage and attention drift, models such as WAP [9], BTTR [16], and ABM [11] incorporate coverage attention, bidirectional training, and attention aggregation, respectively. Furthermore, multi-scale attention, as demonstrated by [11,17], captures both local and global information. However, such sequence-to-sequence based models commonly overlook the complicated structures in mathematical expressions, resulting in the unsatisfied recognition performance.

Tree-structured approaches have been proposed to capture structural relationships. Zhang et al. [14] introduce a tree-based model with a Bidirectional-LSTM for online HMER, while Zhang et al. [13] utilize an Sequential Relationship Decoder (SRD) for tree structures. In the offline HEMR, Zhang et al. [12] develop DWAP-TD, which decompose the target structure into subtrees with parent-child relationships. Wu et al. [15] propose a similar tree structure with separate decoders. However, these methods lack rule constraints and often over-look syntactic relationships, leading to structure prediction errors in practical applications.

3   Methodology

In this paper, we adopt an encoder-decoder framework for handwritten math-ematical expression recognition (HMER), and Fig. 1 shows an overview of our model including three components. The encoder extracts a sequence of features from the handwritten images. Subsequently, a visual-textual attention fusion block is employed to capture both spatial and semantic contexts and align the information. Specifically, the visual attention incorporates a multi-scale cover-age attention mechanism, which utilizes two branches of different sizes to cap-ture both local and global information from the features. Meanwhile, the textual attention captures information about word dependencies from LaTeX sequences. In the decoder, we propose a tree reasoning algorithm for the construction of tree node sequences. In the following sections, we will provide detailed descriptions of these three components.

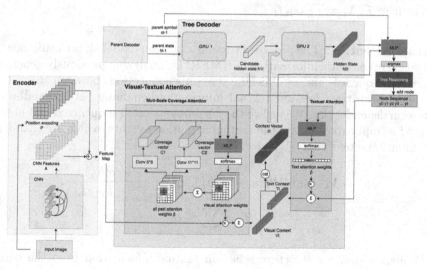

Fig. 1. Overview of the proposed encoder-decoder model with visual-textual attention and tree-based decoder.

3.1   Encoder

We use DenseNet [18] as the backbone in the encoder to extract a sequence of features from input image as shown by $\mathcal{A} = [a_1, a_2, ..., a_M]$. To integrate the position information, we consider positional information by a sinusoidal position embedding sequence $\mathcal{P}$ [16]. Finally, we add such two sequence to form a final feature sequence $\mathcal{F}$ by $\mathcal{F} = \mathcal{A} + \mathcal{P}$.

3.2   Visual-Textual Attention

Multi-scale Coverage Attention. In this work, the visual attention mechanism help the encoder focus related regions of the input image where the coverage attention [9] is commonly employed to address the issues of over-parsing and under-parsing in HMER. It keeps track of past alignment information β_t by assigning higher attention weights α_t to unrecognized regions and preventing repeated attention, which can be calculated by $\beta_t = \sum_{l=1}^{t-1} \alpha_l$. In addition, we utilize a multi-scale coverage attention with two branches to capture both local and global information (C_s and C_l) by applying a 5×5 convolutional kernel U_s and a 11×11 convolutional kernel U_l, which is similar to [11] as shown by

$$C_s = U_s\beta_t, \ \ C_l = U_l\beta_t, \tag{1}$$

$$e_{ti} = V_a^T \tanh(W_a h_t^1 + U_a f_i + W_s C_s + W_l C_l),$$
$$\alpha_{ti} = \frac{exp(e_{ti})}{\sum_{k=1}^{L} exp(e_{tk})}. \tag{2}$$

Finally, the visual context vector V_t is obtained by attention weights α_t and the feature map $\mathcal{F}$: $V_t = \sum_i \alpha_{ti} f_i$.

Textual Attention. The proposed textual attention model aims to enhance the utilization of semantic information in HMER. By analyzing previously generated expression sequences and establishing semantic connections between characters, the model prioritizes relevant textual information, enabling better handling of long-term dependencies. We utilize the current generated sequence o and hidden state h_t^1 as input to calculate the textual attention weights, which are then used to compute the text context vector T_t:

$$b_{ti} = V_b^T \tanh(W_o o_i + W_b h_t^1), \ \ \gamma_{ti} = \frac{exp(b_{ti})}{\sum_{k=1}^{L} exp(b_{tk})}, \tag{3}$$
$$T_t = \sum_i \gamma_{ti} o_i.$$

While the decoder structure relies on textual information from the parent node for prediction, the textual attention model considers global textual information and dependencies beyond the parent node, improving semantic understanding. Finally, the fused context vector c_t is given by $c_t = [V_t : T_t]$.

3.3   Tree Decoder

In an expression tree, each node represents a mathematical expression symbol while branches connecting the nodes represent the structural relationships between the symbols. The node symbols contains all possible mathematical expression symbols, while the structural relationships include right, above, below, superscript, subscript, left-subscript, and inside in this work.

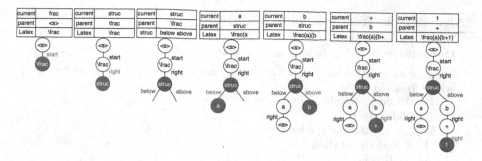

Fig. 2. One example $\frac{b+1}{a}$ of tree reasoning process. The table shows predicted symbols for the current and parent nodes in each step, and the expression tree below highlights the current node in red. (Color figure online)

In contrast to previous tree models [12,15] that predicted both symbolic and structural relationships simultaneously, we incorporate a rule-based approach as described in Algorithm 1 to guide the tree construction process. A special structure node is included to handle complex relationships, while simple expressions require no additional relationship prediction. An example is given in Fig. 2, for each node, the trained model determines if it has a complex structure by structure symbol. If it does, the model predicts its relationships and then predicts the children of each structural relationship branch separately. If the node represents a mathematical expression (ME) symbol, it is directly added to the result sequence without determining the relationship. The algorithmic rules aid in determining symbols with special or multiple relationships, while simplify complex relation construction, and the structure node also facilitates attention distribution across distant nodes in each branch.

We employs the decoder module consists of two GRU structures similarly to [9,10], aiming to predict sequences of node symbols in the tree decoding. The first GRU takes the parent node's hidden state $\mathbf{h}_p^2$ and predicted symbol embedding $\mathbf{e}_p$ as input, with its output serving as a candidate hidden state $\mathbf{h}_t^1$ to capture short-term contextual information:

$$\mathbf{h}_t^1 = GRU_1(\mathbf{h}_p^2, \mathbf{e}_p). \tag{4}$$

The second GRU takes the context vector $\mathbf{c}_t$ from the attention module, producing the final hidden state $\mathbf{h}_t^2$, which represents the node's historical state in symbol generation.

$$\mathbf{h}_t^2 = GRU_2(\mathbf{c}_t, \mathbf{h}_t^1) \tag{5}$$

Following the Algorithm 1, the decoder firstly predicts the current symbol $\mathbf{o}_t$ under the parent node Y_p and input image X:

$$P_{symbol}(\hat{\mathbf{o}}_t | Y_p, X) = softmax(\mathbf{W}_s(\mathbf{W}_e\mathbf{e}_p + \mathbf{W}_h\mathbf{h}_t^2 + \mathbf{W}_c\mathbf{c}_t)),$$
$$o_t = argmax(\hat{\mathbf{o}}_t) \tag{6}$$

Algorithm 1: Tree Reasoning Algorithm

Input: Image's feature map: $\mathcal{F}$
Output: List of node labels in expression tree $\mathcal{L}$.

Initialize an empty relation stack $\mathcal{S}$; A start node as the current parent node;
while *True* **do**
  Predict current node $\mathcal{Y}_t$ by parent node $\mathcal{Y}_p$;
  Append $\mathcal{Y}_t$ in $\mathcal{L}$;
  if $\mathcal{Y}_t$ *is not the end symbol* **then**
    if $\mathcal{S}$ *is empty* **then**
      ⌊ Break;
    else
      ⌊ Pop a node from $\mathcal{S}$ as $\mathcal{Y}_p$;

  else if $\mathcal{Y}_t$ *is the structural symbol* **then**
    Predict the relationships and push $\mathcal{Y}_p$ into $\mathcal{S}$;
    Pop a node from $\mathcal{S}$ as $\mathcal{Y}_p$
  else
    ⌊ Use $\mathcal{Y}_t$ as $\mathcal{Y}_p$;

return $\mathcal{L}$

If the predicted symbol is a structure symbol, the structure relationship under the symbol is predicted by

$$P_{structure}(\mathbf{r}_t|Y_p, X) = sigmod(\mathbf{W}_r(\mathbf{W}_p\mathbf{e}_p + \mathbf{W}_h\mathbf{h}_t^2 + \mathbf{W}_c\mathbf{c}_t)). \qquad (7)$$

Then, the decoder determines the parent node for the next step.

3.4 Parent-Children Mutual Learning

As visual attention is important for spatial contexts in the decoder, we adopt a mutual learning method [11] to align the attention. In this work, we perform bidirectional learning on the attention probabilities as illustrated in Fig. 3. The model predicts nodes in parent-to-children (P2C) order and the inverse order (C2P), facilitating prediction from parent to child nodes and vice versa. The node attention probability sequence in both directions are

$$\alpha_{p2c} = [\alpha_1, \alpha_2, ..., \alpha_t], \ \alpha_{c2p} = [\alpha'_t, \alpha'_{t-1}, ..., \alpha'_1]. \qquad (8)$$

Then we employ KL divergence to quantify the disparity between the attention distributions of predicted parent and child nodes

$$L_{kl} = -\sum_i \alpha_{c2p} \log \frac{\alpha_{c2p}}{\alpha_{p2c}} \qquad (9)$$

In addition, we use cross-entropy loss function in this work to evaluate the prediction of both symbols L_{symbol} by Eq. (6) and structures L_{struc} by Eq. (7). Similarly, we also consider a cross-entropy loss L_{symbol}^{rev} for symbol prediction from the reverse order (i.e., C2P). In summary, the final loss function can be represented by

$$L = L_{symbol} + L_{struc} + L_{symbol}^{rev} + L_{kl} \qquad (10)$$

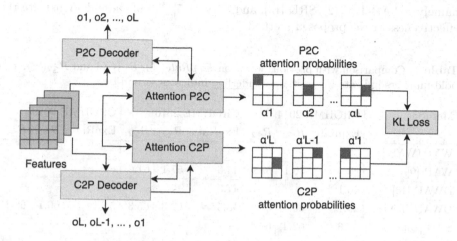

Fig. 3. Parent-Children Mutual Learning. P2C and C2P means parent-to-children and reserve order (children-to-parent), respectively.

4   Experiments

Datasets and Experimental Settings. The proposed model is trained on the CROHME 2014 training dataset, which consists of 8,836 handwritten mathematical expressions with 111 classes of mathematical symbols. We evaluate the model on the HMER benchmark testsets of CROHME 2014, CROHME 2016, and CROHME 2019 with 986, 1,147, and 1,199 expressions, respectively.

All experiments are implemented by PyTorch on a single Nvidia Tesla V100-SXM4. The batch size is 8. The feature dimension was set to 684, GRU hidden state sizes were set to 256, and word embedding and relation embedding dimensions were set to 256. The Adadelta optimizer with $\rho = 0.95$ and $\epsilon = 10^{-6}$ was employed during training. The final model was trained by 200 rounds of epochs, and there was no data augmentation applied during training. We evaluate the performance by Expression Level Rate ($ExpRate$) [9] to indicate the correctness of whole predicted string. Moreover, we also report two relaxation metrics including $R_{\leq 1}$, $R_{\leq 2}$ that allow maximum of 1 and 2 character errors, respectively.

Results. We compare our model with previous works including both string and tree decoder based methods on CROHME 2014, 2016 and 2019 datasets as shown in Table 1. We can see our proposed method achieves the best ExpRate on all CROHME test datasets, outperform the second-best counterpart 1.5%, 1.2%, and 0.8% on CROHME2014, 2016 and 2019, respectively. The performance of $R_{\leq 1}$, $R_{\leq 2}$ are definitely higher than $ExpRate$ as they allow one and two character errors. In addition, our tree approach combined with the reasoning algorithm, yields superior results compared to other methods utilizing tree decoders, namely DWAP-TD [12], SRD [13], and TVv2 [15]. All of these demonstrate the effectiveness of the proposed method.

Table 1. Comparison with previous works on $ExpRate(\%)$, $R_{\leq 1}(\%)$ and $R_{\leq 2}(\%)$. The bold numbers indicate the best, and underline indicates second-best.

Method	CROHME 2014			CROHME 2016			CROHME 2019		
	ExpRate	$R_{\leq 1}$	$R_{\leq 2}$	ExpRate	$R_{\leq 1}$	$R_{\leq 2}$	ExpRate	$R_{\leq 1}$	$R_{\leq 2}$
WYGIWYS [7]	36.4	–	–	–	–	–	–	–	–
WAP [9]	40.4	56.1	59.9	44.6	57.1	61.6	–	–	–
DWAP [10]	50.1	–	–	47.5	–	–	–	–	–
DWAP–TD [12]	49.1	64.2	67.8	48.5	62.3	65.3	51.4	66.1	69.1
SRD [13]	53.6	62.4	66.4	–	–	–	–	–	–
TDv2 [15]	53.6	–	–	55.2	–	–	58.7	–	–
BTTR [16]	54.0	66.0	70.3	52.3	63.9	68.6	53.0	66.0	69.1
Li et al. [19]	56.6	69.1	75.3	54.6	69.3	73.8	–	–	–
ABM [11]	56.9	_73.7_	**81.2**	52.9	69.6	_78.7_	53.9	71.1	78.7
CAN [20]	57.0	**74.2**	_80.6_	56.1	_71.5_	**79.5**	54.9	71.9	_79.4_
Ding et al. [8]	_58.7_	–	–	_57.7_	70.0	76.4	_61.4_	**75.2**	**80.2**
Ours	**59.2**	71.9	74.9	**58.5**	**73.0**	77.7	**62.2**	_72.7_	76.7

Ablation Studies. In this work, we refine the encoder-decoder framework by three components. For the decoder, we adopt a tree reasoning algorithm. To better align the contexts in the decoder, we propose a visual-textual attention fusion, where a multi-scale coverage attention is utilized. To better train the model, we propose a parent-children mutual learning method. The Table 2 shows the results of the combination of different components, and we can see that using TD, MS, VT, and ML can bring an average performance improvement of 7.4%, 4.4%, 2.8%, 1.2% on the three datasets, respectively, indicating the effectiveness of different components in HMER.

Furthermore, we visualize the attention positions during one expression tree construction, as shown by one example in Fig. 4. We can see that the attention focuses on the center area of symbol nodes (e.g., the first sub-picture) while large parts of structure nodes (e.g., the second sub-picture). For the structure node (e.g., the second picture), the attention will iteratively focus on the symbols (e.g., the sub-figure 3 to sub-figure 9) within such structure.

Table 2. Ablation Study on *ExpRate*(%). "Baseline" means the ordinary sequence model without our proposed methods, "+TD", "+MS", "+VT", "+ML" mean the tree reasoning decoder, multi-scale coverage attention, visual-textual attention fusion, and parent-child mutual learning, respectively.

Method	CROHME 2014	CROHME 2016	CROHME 2019
Baseline (B)	42.0	45.4	45.2
B+TD	52.6	50.1	52.0
B+TD+MS	55.3	56.6	56.1
B+TD+VT	58.4	57.6	60.4
B+TD+MS+VT+ML (ours)	59.2	58.5	62.2

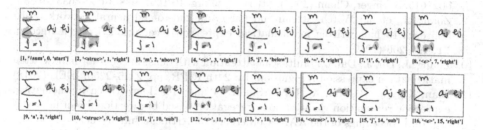

Fig. 4. One attention visualization sample, and the attention areas are highlighted.

5 Conclusion

In this paper, we propose a refined encoder-decoder model to improve handwritten mathematical expression recognition (HMER). Specifically, we propose a tree reasoning algorithm in the tree decoder, and utilize a visual-textual attention fusion block to better capture contexts during the decoding process. Furthermore, we propose a parent-children mutual learning method to better optimize the model. Finally, our proposed method has improved the recognition performance on the benchmark HMER datasets, indicating its effectiveness.

Acknowledgements. This research was funded by National Natural Science Foundation of China (NSFC) no. 62276258, Jiangsu Science and Technology Programme (Natural Science Foundation of Jiangsu Province) no. BE2020006-4, Xi'an Jiaotong-Liverpool University's Key Program Special Fund no. KSF-E-43.

References

1. Chan, K.F., Yeung, D.Y.: Mathematical expression recognition: a survey. Int. J. Doc. Anal. Recogn. **3**, 3–15 (2000)
2. Zanibbi, R., Blostein, D.: Recognition and retrieval of mathematical expressions. Int. J. Doc. Anal. Recogn. **15**, 331–357 (2012)
3. Zanibbi, R., Blostein, D., Cordy, J.R.: Recognizing mathematical expressions using tree transformation. IEEE T-PAMI **24**(11), 1455–1467 (2002)

4. Álvaro, F., Sánchez, J.A., Benedí, J.M.: An integrated grammar-based approach for mathematical expression recognition. Pattern Recogn. **51**, 135–147 (2016)

5. MacLean, S., Labahn, G.: A new approach for recognizing handwritten mathematics using relational grammars and fuzzy sets. Int. J. Doc. Anal. Recogn. (IJDAR) **16**, 139–163 (2013)

6. Huang, K., Hussain, A., Wang, Q.F., Zhang, R.: Deep Learning: Fundamentals, Theory and Applications, vol. 2 (2019)

7. Deng, Y., Kanervisto, A., Ling, J., Rush, A.M.: Image-to-markup generation with coarse-to-fine attention. In: International Conference on Machine Learning, pp. 980–989. PMLR (2017)

8. Ding, H., Chen, K., Huo, Q.: An encoder-decoder approach to handwritten mathematical expression recognition with multi-head attention and stacked decoder. In: Lladós, J., Lopresti, D., Uchida, S. (eds.) ICDAR 2021. LNCS, vol. 12822, pp. 602–616. Springer, Cham (2021). https://doi.org/10.1007/978-3-030-86331-9_39

9. Zhang, J., et al.: Watch, attend and parse: an end-to-end neural network based approach to handwritten mathematical expression recognition. Pattern Recogn. **71**, 196–206 (2017)

10. Zhang, J., Du, J., Dai, L.: Multi-scale attention with dense encoder for handwritten mathematical expression recognition. In: 24th ICPR, pp. 2245–2250. IEEE (2018)

11. Bian, X., Qin, B., Xin, X., Li, J., Su, X., Wang, Y.: Handwritten mathematical expression recognition via attention aggregation based bi-directional mutual learning. In: Proceedings of the AAAI Conference on Artificial Intelligence, vol. 36, pp. 113–121 (2022)

12. Zhang, J., Du, J., Yang, Y., Song, Y.Z., Wei, S., Dai, L.: A tree-structured decoder for image-to-markup generation. In: International Conference on Machine Learning, pp. 11076–11085. PMLR (2020)

13. Zhang, J., Du, J., Yang, Y., Song, Y.Z., Dai, L.: SRD: a tree structure based decoder for online handwritten mathematical expression recognition. IEEE Trans. Multimedia **23**, 2471–2480 (2020)

14. Zhang, T., Mouchère, H., Viard-Gaudin, C.: A tree-BLSTM-based recognition system for online handwritten mathematical expressions. Neural Comput. Appl. **32**, 4689–4708 (2020)

15. Wu, C., et al.: TDv2: a novel tree-structured decoder for offline mathematical expression recognition. In: Proceedings of the AAAI Conference on Artificial Intelligence, vol. 36, pp. 2694–2702 (2022)

16. Zhao, W., Gao, L., Yan, Z., Peng, S., Du, L., Zhang, Z.: Handwritten mathematical expression recognition with bidirectionally trained transformer. In: Lladós, J., Lopresti, D., Uchida, S. (eds.) ICDAR 2021. LNCS, vol. 12822, pp. 570–584. Springer, Cham (2021). https://doi.org/10.1007/978-3-030-86331-9_37

17. Fu, J., et al.: Dual attention network for scene segmentation. In: CVPR, pp. 3146–3154 (2019)

18. Huang, G., Liu, Z., Van Der Maaten, L., Weinberger, K.Q.: Densely connected convolutional networks. In: CVPR, pp. 4700–4708 (2017)

19. Li, Z., Jin, L., Lai, S., Zhu, Y.: Improving attention-based handwritten mathematical expression recognition with scale augmentation and drop attention. In: 17th ICFHR, pp. 175–180 (2020)

20. Li, B., et al.: When counting meets HMER: counting-aware network for handwritten mathematical expression recognition. In: Avidan, S., Brostow, G., Cissé, M., Farinella, G.M., Hassner, T. (eds.) ECCV 2022. LNCS, vol. 13688, pp. 197–214. Springer, Cham (2022). https://doi.org/10.1007/978-3-031-19815-1_12

WildTechAlert: Deep Learning Models for Real-Time Detection of Elephant Presence Using Bioacoustics in an Early Warning System to Support Human-Elephant Coexistence

Yen Yi Loo[1,2] (iD), Naufal Rahman Avicena[1,3,4] (iD), Noah Thong[1] (iD),
Abdullah Marghoobul Haque[1,5] (iD), Yenziwe Temawelase Nhlabatsi[1,5] (iD),
Safa Yousif Abdalla Abakar[1,5] (iD), Kher Hui Ng[5] (iD), and Ee Phin Wong[1(✉)] (iD)

[1] Management and Ecology of Malaysian Elephants (MEME), School of Environmental and
Geographical Sciences, University of Nottingham Malaysia, Jalan Broga, 43500 Semenyih,
Selangor, Malaysia
EePhin.Wong@nottingham.edu.my
[2] Sunway Centre for Planetary Health, Sunway University, Jalan Universiti, Bandar Sunway,
Petaling Jaya, Selangor, Malaysia
[3] Southeast Asia Biodiversity Research Institute, Chinese Academy of Sciences & Center for
Integrative Conservation, Xishuangbanna Tropical Botanical Garden, Chinese Academy of
Sciences, Mengla 666303, Yunnan, China
[4] Yunnan International Joint Laboratory of Southeast Asia Biodiversity Conservation & Yunnan
Key Laboratory for Conservation of Tropical Rainforests and Asian Elephants, Menglun,
Mengla 666303, Yunnan, China
[5] School of Computer Science, University of Nottingham Malaysia, Jalan Broga, 43500
Semenyih, Selangor, Malaysia

Abstract. Human-elephant conflict (HEC) in Malaysia occurs in areas where
humans and Asian elephants (*Elephas maximus*) share landscapes, especially at
forest edges and agricultural lands. Due to increased anthropogenic activities, such
as logging, agriculture, and infrastructure developments, a large portion of natural habitat has been converted and fragmented, resulting in increased interactions
between people and wild elephants. This presents a unique challenge of conserving
elephants, safeguarding human livelihoods and ensuring habitat connectivity for
endangered wildlife. The ability to detect and predict the presence of elephants
in human-dominated landscapes can help increase safety for communities and
provide insights for decision-makers to better manage conflict situations and promote a harmonious human-elephant coexistence. In the past decade, technologies
used, such as camera traps, drone imaging, and GPS tracking, have allowed us
to understand activity patterns, habitat preferences, and behaviour of elephants,
but they often have limitations in terms of cost, coverage effort, and manpower
needed. The increasing application of artificial intelligence (AI) in conservation
offers an opportunity to develop real-time solutions to detect elephant presence in
areas with a high risk of human-elephant encounters and HEC. In this study, we
present a prototype of an early warning system, WildTechAlert, which combines
bioacoustics and AI technology. Bioacoustics is the study of biological sounds;
and the prototype device, being able to record sounds travelling from all angles,

J. Ren et al. (Eds.): BICS 2023, LNAI 14374, pp. 385–399, 2024.
https://doi.org/10.1007/978-981-97-1417-9_36

may potentially have higher detection rates from its surroundings (omnidirectional coverage) than sensors with fixed angles, such as camera traps. A sound stored in the raw waveform format can be converted into a spectrogram, which is a visual representation of the sound signature in the frequency, time, and amplitude domains. The spectrograms of elephant sounds were used as images for training a deep learning model using convolutional neural networks. WildTechAlert includes a device equipped with bioacoustic sensors which connect to a cloud-based deep learning algorithm, a user-friendly web interface, and mobile notification alert functions. The alerts serve as a tool to increase the safety of communities and plantation workers and promote preparedness for mitigation actions. We trained binary and multiclass classifiers to detect elephant sounds and achieved up to a 94% accuracy in detecting elephant sounds with the binary classifier. The performance of this system has yet to be tested in field conditions but shows potential in HEC management and to promote coexistence between humans and elephants.

Keywords: Bioacoustics · Deep Learning · Elephants · Human-Elephant Coexistence · Human-Elephant Conflict · Malaysia

1 Introduction

Machine learning has gained popularity worldwide in many fields due to its potential in mitigating real-world issues [1]. In conservation, machine learning (ML) algorithms, or ML-assisted approaches, can speed up large-scale problem-solving processes by automating labour-intensive tasks such as recognizing and labelling information or data [2–6]. Like other countries with elephants, Malaysia experiences human-elephant conflict (HEC) [7–9], with challenges to establish a harmonious coexistence [10], and with growing concerns over incidents where wild elephants ventured into housing areas. Often, HEC occurs in areas where humans and elephants share landscapes, especially at forest edges and agricultural lands [11–13]. This overlapping presence increases interactions that may cause negative impacts to humans and elephants, such as injuries, fear, loss of life, and damage to crops [9, 14–16]. Rapid development has reduced the habitat range of elephants and other wildlife, and elephants are attracted to forest edges and plantation areas due to the abundance of food [11, 17]. Predicting HEC requires understanding of elephant behaviour and ecology using long-term datasets from landscape-wide studies, which can benefit from machine learning algorithms.

Elephants provide important ecological functions. As a wide-ranging species, with home ranges reaching several hundreds of square kilometres, they help ensure seed dispersal and genetic diversity of large-seeded plants [18]. Protecting this endangered megafauna and its natural habitats is crucial for maintaining healthy ecosystems [18]. However, in landscapes where humans and elephants co-occur, elephants may forage in croplands and cause damage to crops and infrastructures, which may drive negative perceptions towards elephants in local communities [11, 19]. Comprehensive strategies and interventions for human-wildlife conflict management have been proposed [20–23]. Present studies using camera trap data have shown that elephants are active mainly during dawn and dusk, possibly to avoid encountering humans (MEME, unpubl. Data) [24, 25]. These tools have been important to understand the activity patterns and movement dynamics of elephants in the landscape, enabling prediction of the seasonal presence of elephants in a given area. In the case of elephants, early warning systems can help reduce human mortality and injuries at a landscape level [26]. Real-time notifications of elephant presence can increase safety for communities and provide insights for decision-makers to better manage conflict situations and promote a more harmonious human-elephant coexistence [10, 26].

The increasing application of artificial intelligence (AI) in conservation offers an opportunity to develop real-time solutions to detect elephant presence in areas with a high risk of human-elephant encounters. Elephant presence can be detected using their vocalisations as elephants communicate over short and long distances using different types of calls [27, 28]. Research in Sri Lanka had classified up to 14 call types for Asian elephants (*Elephas maximus maximus*). Furthermore, as Asian elephants are vocal learners [29], learned calls may vary beyond these categories [27]. Elephant calls can be detected from the infrasound range (<20 Hz) [28] to the human-audible range 5 kHz or 6 kHz, such as trumpets or roars [30]. The calls of elephants have distinctive features that can be used to differentiate them from other species and background noise. The invention of spectrographic representation of sound in the 1950s [31], enabled the conversion of raw waveforms into a spectrographic image with three domains, frequency (kHz) in the y-axis, time (seconds, s) in the x-axis, and amplitude (decibels, dB) as the intensity of a colour gradient in the x- and y-axes (Fig. 1) which effectively advances the study of animal sounds and communication. These spectrograms are equivalent to regular images with RGB values, which can be used to train a model to automatically detect distinctive features of target objects, or in this study, using deep learning methods [33–35].

Convolutional neural network (CNN) is a deep learning architecture that is composed of multiple building blocks: convolution, pooling, and fully connected layers designed to recognise grid-like patterns such as image or sound data following inspiration from the sensory perception of the brain, which exhibits patterns of connectivity [36, 37]. The CNN architecture has been successfully applied for wildlife conservation including automated wildlife monitoring [38] and species identification to aid data processing [39]. Automated detections of Asian elephant sounds using CNN have recently been developed by Avicena [40] with training datasets of annotated elephant sounds from Sri Lanka [27] and Malaysia (Avicena et al., in review). This advancement enables us to close the gap between fundamental and applied deep learning of elephant bioacoustics.

In light of this development, we present a prototype of an early warning system in this study, 'WildTechAlert', which combines bioacoustics and AI technology to notify users of elephant presence in near real-time. This paper tests the performance of the model with new datasets from focal recordings in Johor, Malaysia, and outlines the methodological framework of the early warning system, which includes a cloud-based deep learning algorithm, a user-friendly web interface, and mobile notification alert functions. We then discuss the potential of this tool to increase the safety of communities and plantation workers and to promote preparedness for mitigation actions. Testing the performance of this system in field conditions has novel and important implications for managing HEC and moving towards a more harmonious coexistence between humans and elephants.

2 Methods

2.1 Model Development

Model Architecture. We trained three types of classifiers to extract elephant sounds from raw recordings of the elephant's natural environment (Table 1). The models were based on a similar model built for the classification of elephant and non-elephant sounds [40]. The overall architecture of the model by Avicena [40] was maintained. However, a few extra layers were added and the output shapes of the model's predictions were changed according to their purpose. The models were built using PyTorch, which is a Python module for specialised neural network development [41]. Firstly, we trained a binary classifier to differentiate between elephant and non-elephant sounds (dataset: MEME-DS). Secondly, we trained a multiclass classifier to differentiate between elephant and environmental sounds (dataset: ENV-DS). Thirdly, we trained another multiclass classifier to detect elephant sounds as well as anthropogenic sounds related to poaching or logging (dataset: ANTH-DS). Lastly, we used an unseen dataset for inference (dataset: INF-DS). The sample size of each dataset is outlined below.

Table 1. Sample sizes of datasets used for model development.

Dataset	Description	Dataset sample size
MEME-DS	Elephants and non-elephants	1040 (520 elephants, 520 non-elephants)
ENV-DS	Elephants and other forest sounds	2545 (520 elephants, 2025 others)
ANTH-DS	Elephants, non-elephants and forest threats	1265 (520 elephants, 745 others)
INF-DS	Elephants and non-elephants, used for inference	49 (26 elephants, 23 non-elephants)

Binary Classification. The dataset used to train the binary classifier consisted of Asian elephant (*Elephas maximus indicus*) sounds that were obtained by Management and Ecology of Malaysian Elephants (MEME) from Belum-Temengor Forest Complex, Perak (passive acoustic recordings, Frontier Labs BAR-LT and Wildlife Acoustics Song Meter 4TS) and Padang Hijau, Kluang district in Johor (focal recordings by YYL, Zoom H5 Handy recorder with Sennheiser ME66/K6 directional microphone module), as well as non-elephant sounds from publicly available datasets such as warblr10k and freefield1010. Warblr10k consists of smartphone recordings of random sounds including weather noise, human sounds, and traffic noise and freefield1010 consists of field recordings from around the world. This dataset was labelled MEME-DS. There were 520 audio files on elephant sounds and 520 on non-elephant sounds. The binary classifier consists of six layers which include a convolution layer, a Rectified Linear Unit (ReLU) activation function and batch normalisation (Fig. 1). The CNN is a common deep learning technique for image recognition, classification, and object detection [42]. The convolutional layers are a fundamental component in the model in performing feature extraction. This is ideal for audio data that is being converted to spectrogram images before passing into the model.

Multiclass Classification. We also trained two different multiclass classifiers using ENV-DS and ANTH-DS. Firstly, the Forest Environmental Sound Classification (FSC22) dataset consisted of 27 classes of forest sounds, such as environmental sounds (e.g. fire, rain), vehicles, humans, and others [43]. This dataset was referred to as ENV-DS and consisted of 520 elephant sounds from MEME-DS, and 75 sounds each from the other 27 classes from FSC22, making a total of 28 distinct classes. The multiclass model consisted of a similar architecture as the binary classifier; however, the multiclass classifier upsampled the inputs to a higher amount compared to the binary classifier (Fig. 2).

The ANTH-DS model was trained using a subset of ENV-DS and combined with MEME-DS which consisted of elephant, non-elephant, as well as sounds of anthropogenic forest threats, such as gunshots, chainsaw and axe classes. This dataset was referred to as ANTH-DS and consisted of 520 elephant sounds, 520 non-elephant sounds and 75 sounds each from the other three classes. The multiclass model produces outputs of different shapes from the linear layer, which depended upon the number of classes in the dataset used for training, 28 for ENV-DS and 10 for ANTH-DS (Fig. 2).

For the purpose of inference, another unseen dataset was created which consisted of 26 elephant sounds as well as 23 non-elephant sounds, referred to as INF-DS.

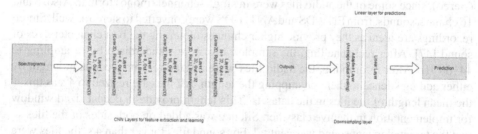

Fig. 1. Architecture of the binary classifier, MEME-DS.

Transfer Learning Classification. A transfer learning technique of training a ResNet-50 model on the datasets was also implemented (Fig. 3). ResNet-50 is a Residual Network and consists of 50 parameter layers in the architecture of the network [44]. The ResNet-50 model's input and out-put layers were fine tuned to fit the spectrogram data and were trained using the ENV-DS and ANTH-DS datasets. The cross-entropy loss function was used for the calculation of loss. This is a very common loss function used with large capacity deep neural networks in state-of-the-art supervised learning problems [45].

Data Preprocessing. To train the CNN, the sound files were converted into an image tensor by normalising, augmenting, and converting to a Mel spectrogram. The data normalisation phase ensures that the audio files have the same dimensions for the model

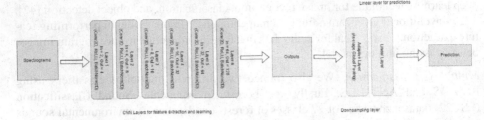

Fig. 2. Architecture of the multiclass classifiers, ENV-DS and ANTH-DS.

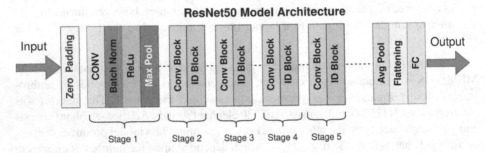

Fig. 3. Architecture of the transfer learning classifier, ResNet-50 [46].

to read consistent input features. Firstly, all sound files were converted to two channels (stereo), since some of the audio files were in single-channel (mono) format. Also, 8 and 16 channel sounds, from ENV-DS and ANTH-DS were converted to stereo as well. Stereo recordings are ideal as they provide higher chances of getting the best characteristics of sound [47]. After rechannelling, the sample rates were standardised by resampling to 44100 Hz. After resampling, the audio files were resized to four seconds in length by either adding silence to them or snipping the length. All files are resized to 4 s as this is the mean length of the files in the datasets. This length provides a decent sized window for implementation in the live classifier. Silence was added on both sides of the files so that the target elephant sound was centred. For sound files longer than 4 s, the files were snipped at the start and end, while also ensuring the centering of the elephant sound.

Data augmentation is a suite of techniques which enhance the size and variability of the training data, allowing better deep learning models to be built [48]. It has shown to produce better accuracies for classification and is very useful in combating overfitting [49]. There were two data augmentation techniques applied after normalising the audio files. The first one was a time-shift technique that was applied by randomly shifting the audio in the temporal domain, allowing the creation of the first round of augmented data.

After the first round of augmentation, the audio files are converted into a Mel spectrogram. Studies have demonstrated that Mel spectrograms—typically used for human speech processing—perform well in the recognition of acoustic events [50]. The spectrograms are generated with a fastfourier transform of size 1024 and 128 mel filterbanks. The spectrogram is then converted from the power/amplitude scale to the decibel scale, with a minimum negative cutoff of 80 units. Then, the generated spectrogram is further augmented by masking random frequencies and random time steps using SpecAugment [51, 52]. This is done by adding horizontal and vertical bars on top of the spectrograms, respectively.

Batch normalisation helps in stabilising the output of each layer through rescaling and recentering as well as increases the speed of the model's training process [53]. The ReLU activation function has good sparsity, fast convergence speed and solves the gradient dispersion problem which is caused by some of the other activation functions [54]. After feeding into the CNN layers, the inputs were downsampled in the adaptive layer by reducing the sampling rate to half of the original rate. The function of downsampling is to reduce input features whilst retaining most of the important information, essentially to reduce the computational complexity of the model and increase efficiency. The inputs are then passed to the linear layer where the model predicts and performs binary classification. Finally, the outputs of the linear layer are passed through a sigmoid function which generates a score for each class. This was used to generate the confidence rating for each detection.

Training and Validation. All models were validated with the k-fold cross validation where the datasets were split into 10 folds and for each of the 10 iterations, one distinct fold was abstained, and the models were trained with the other k-1 folds ($n = 9$). For each of the folds, the number of epochs was set to 20 and the batch size was set to 32.

Testing and Inference. Inference was carried out on the models with the INF-DS and the performance of models was evaluated with accuracy, precision, recall and F1 score. The models were also deployed into a real-time detector and its performance was manually monitored. This real-time detector recorded 4 s of clips from the system's microphone and passed it to the model after preprocessing and converting into a Mel spectrogram. The models then predicted what class the audio file belonged to and this was manually checked by the user.

Evaluation. The performance of the models was determined by their accuracy, precision, recall, and F1 score. The accuracy metric calculates the percentage of predictions correctly made by the model. The precision of a model is the ratio between true positives and the sum of true positives and false positives, while recall is the ratio between true positives and the sum of true positives and false negatives. F1 score is the harmonic mean of precision and recall and reflects the precision and robustness of the classifier [55].

2.2 Early Warning System

The trained models enable near real-time detection of elephant sounds. The detection of elephants will immediately send an early warning notification to the users living within a certain radius. A website is used to support the users with differentiated permission access to increase security.

Notification and Alert System. Social messaging platforms like Telegram and Whatsapp serve as notification platforms for WildTechAlert, leveraging its messaging capabilities and popularity. A dedicated group link can be retrieved on plantation/village account creation. The group contains a bot which will notify whenever an elephant sound has been detected. The system employs an intelligent detection algorithm that analyses incoming data over a 5-min period. When the algorithm identifies a series of consistent elephant presence signals during this timeframe, a confirmed detection is registered. This signals the bot to notify in the respective group chat. Additionally, the notification process also sends a notification on the website. This ensures that even if a detection is missed, the time is logged and available for the user to refer back to. As the system sends periodic updates and alerts to the users, the bandwidth used is a moderate few megabits per second.

User Interface and Functionality. Due to the sensitive information of endangered wildlife such as Asian elephants, the website will feature security policies. As such, the site will be designed to have differentiated access levels. Only registered and verified users will have the ability to access and view the heatmap of conflict hotspots. This information can be filtered by date and location. There will be features to allow WildTechAlert researchers to manually validate the detections that are below the confidence threshold, before they are passed on as notifications. The team management page allows users to manage their team and provide access to the system.

3 Results

3.1 Model Performance

There were three main model iterations, with the first one being a binary classifier and the other two being multi-class classifiers. In this section, each model's success will be measured according to their detection accuracy of the samples used for training, testing, and inference.

Binary Classifier. The binary classifier achieved an average k-fold test accuracy of 95.6%. The average training and test losses were 0.35 and 0.27, respectively. For inference, the model was tested with the inference dataset and achieved an overall accuracy of 93.88%, and in detecting elephants, a precision of 96.00%, recall of 92.30% and F1-score of 94.11%. Figure 4 shows the confusion matrix for the inference dataset.

The model was also run as a live detector—looping continuously to analyse a continuous audio stream—with the scores being displayed. The results yielded were comparable to the initial testing accuracy, although the probabilities for prediction were found to be in a low threshold (0.55–0.70).

Multiclass Classifier. The multiclass classifier model, which aimed to detect and further sub-classify sound types, delivered an average test accuracy of 7.44% on ENV-DS. This implied that the model was not able to differentiate sounds between 28 sound categories for various forest and elephant sounds. The model had an average test loss of 8.77, along with 2.00% precision, 5.00% recall and 2.86% F1-score.

On ANTH-DS with four sound classes, it was able to deliver an average accuracy of 33.04% which implied that learning of patterns was minimal and predictions made were slightly above random. With ANTH-DS, the model had a test loss of 1.89, with a 10.00% precision, 12.00% recall and 10.90% F1-score.

CNN Model Confusion Matrix

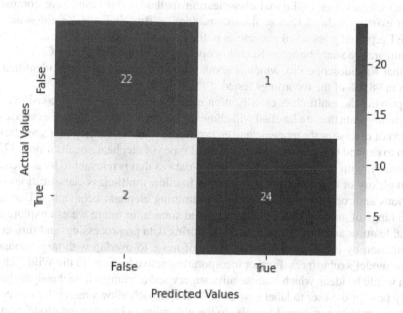

Fig. 4. Confusion matrix for the binary classifier. True refers to elephant audio and false refers to non-elephant audio.

ResNet-50 Classifier. The ResNet-50 classifier had an average accuracy of 32.41% with a loss of 1.87, almost similar to ANTH-DS.

4 Discussion

WildTechAlert is among the first automated elephant early warning system prototype using bioacoustics and AI, and the prototype device, being able to record sounds travelling from all angles, may potentially have higher detection rates from its surroundings (circular coverage) than sensors with fixed angles, such as camera traps. The current binary CNN model used for WildTechAlert is expected to detect elephant sounds with 94% accuracy and 96% precision, thus, shows strong potential to be used as an alert system for improving preparedness and safety of people. The performance of the multiclass were unsatisfactory (just slightly above random), possibly due to the higher number of sound classes and limited number of sound samples for the models to learn from. Nevertheless, the binary CNN model built with local Asian elephant sounds (*Elephas maximus indicus*) performed well when inferred with unseen elephant and non-elephant sounds (Fig. 4). In previous research, we found CNN model built with Asian elephant sounds from Sri Lanka (*Elephas maximus maximus*) did not perform well when inferred with Asian elephant sounds from Malaysia (Avicena et al., in review). The results here confirmed that CNN model built with local data can perform better.

Several machine learning methods exist for photos or videos recognition system, such as those utilising CCTV live feed or camera trap data [56]. Neural networks have provided some advantages over traditional classification methods. Our results are comparable to other existing audio detection algorithms, such as BirdNET and Kaleidoscope Pro. BirdNET reportedly achieved accuracies in the range of 80 to 90% in focal recordings. In a comparative study between Kaleidoscope Pro and BirdNET, Pérez-Granados [57] found that Kaleidoscope Pro, which is another audio analyser, correctly identified bird species in 98.4% of the recordings tested.

Improving the multi-class classification model will include several considerations. More training data that are labelled will allow for higher accuracy and more precise classification of diverse anthropogenic and environment sounds. For elephant sounds, there is room to expand the classification to different types of elephant vocalisations [27, 30]. Additionally, with the availability of a curated dataset that is relevant to local scenarios, the complexity of the model can be increased to allow multi-class classifications based on various anthropogenic sounds and for examining elephant behaviours. Due to the diverse range of anthropogenic and background sounds, in future we can explore noise removal features and enhance target sounds during data preprocessing, and further data augmentation by introducing different types of noise to overlap with target sounds to increase model's robustness. Further incorporating active learning in the WildTechAlert system would be ideal, which is an iterative supervised learning where the algorithm can actively prompt the user to label new data. This approach allows researchers to continuously incorporate new sound samples to the algorithm to improve the model accuracy and readings.

The next stage is to test the prototype in the field to complete the proof of concept. The early warning system significantly enhances safety for humans in areas prone to human-elephant conflict by providing timely alerts so communities can avoid venturing into those areas when elephants are present. There are examples of such approaches as carried out by the Nature Conservation Foundation in Annamalai Hills, Karnataka, India that were successful in reducing fatalities for humans (NCF, unpubl. Data) [26]. Our system differs as it utilises bioacoustic sensors in remote areas to detect, identify, and confirm the presence of elephants. The detection of elephants then triggers the system to notify users of this information. This allows the stakeholders to take precautionary measures to keep a safe distance, or stay safely in their houses, or avoid working in the specific fields where the elephants are present. The website and alerts can help communities prepare for a potential encounter and may reduce the risk of sudden, unexpected encounters with elephants which can be dangerous. The members of public or untrained staff are not encouraged to approach the elephants on their own for safety reasons. Thus, when the system detects elephant sounds in the monitored entrances and in areas where elephants occasionally or unexpectedly appear, such as in housing areas or staff quarters, it is able to send automatic alerts to wildlife rangers or specially trained personnel to safely help direct the elephants to the nearest forest and they can carry out crowd control to minimise danger to people and elephants. The system's management features can facilitate collaborations between different plantations, local communities and governmental agencies. By sharing data and information on elephant movement and behaviour, plantation owners in HEC landscapes can increase the safety of their staff and families while allowing the elephants to move safely to the forest or designated elephant food bank areas. Since elephants are endangered species and there are concerns over poaching, alerts and information on the location of the elephants will need to be scrambled and presented over a wider area.

Conservation and artificial intelligence can work hand in hand to close the gap between data that require intensive effort to process and actions that require rapid feedback for action. The sharing of elephant presence data will support further research and collectively configure strategies that allow win-win scenarios for humans and elephants through evidence-based conservation efforts. New developments on internet connectivity, such as Starlink [58], would also be useful to improve the real-time data collection, early warning system, or monitoring system in forest habitats. This integration is ripe for multi-disciplinary collaboration between biologists, conservationists, computer scientists, and systems engineers, to realise the potential of human-elephant coexistence.

Acknowledgments. We would like to thank Dr. Shermin de Silva (Trunks and Leaves, Inc) and Assoc. Prof. Dr. Tomas Maul for prior collaboration in model development (Avicena et al., in review). We give our appreciation to Yayasan Sime Darby (grant number: NVHH0007) for funding field operations of this project under Management and Ecology of Malaysian Elephants (MEME). We also acknowledge MEME's field assistants, Param, Cherang, and Hussein, MEME's staff Aina Amyrah Ahmad Husam, Lim Jia Cherng, Chan Yik Khan, and MEME's co-investigator, Dr. Cedric Tan Kai Wei for their support. Finally, we give our appreciation to the plantation partners, stakeholders of Achieving Coexistence with Elephants (ACE), Forestry Department of Malaysia, Forestry Department of Perak, Forestry Department of Johor, and the Department of Wildlife and National Parks Peninsular Malaysia (PERHILITAN) who have provided permits (JH/100 Jld. 31(7) and JPHL&TN (IP): 100-34/1.24 Jld. 17(29) respectively), and access to field sites for recording of elephant sounds.

Disclosure of Interests. Noah Thong's and Loo Yen Yi's salary is from Yayasan Sime Darby during research. Abdullah Marghoobul Haque, Yenziwe Temawelase Nhlabatsi and Safa Yousif Abdalla Abakar report receiving stipends from Yayasan Sime Darby. Wong Ee Phin reports receiving funding grant from Sime Darby Foundation for research but not salary. As WildTechAlert is still in prototype development, we do not receive any commercial benefit from the publication of this paper, not the authors nor the organizations they are affiliated, or the funders involved. There is no conflict of interest as there are no consultancies nor commercialization involved during research or publication of this manuscript.

References

1. Sarker, I.H.: Machine learning: algorithms, real-world applications and research directions. SN Comput. Sci. **2**, 160 (2021). https://doi.org/10.1007/s42979-021-00592-x
2. Whytock, R., et al.: High performance machine learning models can fully automate labeling of camera trap images for ecological analyses. bioRxiv (2020). https://doi.org/10.1101/2020.09.12.294538
3. Kulkarni, R., Minin, E.D.: Automated retrieval of information on threatened species from online sources using machine learning. Methods Ecol. Evol. **12**(7), 1226–1239 (2021). https://doi.org/10.1111/2041-210X.13608

4. Ditria, E.M., Buelow, C.A., Gonzalez-Rivero, M., Connolly, R.M.: Artificial intelligence and automated monitoring for assisting conservation of marine ecosystems: a perspective. Front. Mar. Sci. **9**, 918104 (2022). https://doi.org/10.3389/fmars.2022.918104(2022)
5. Marrable, D., et al.: Accelerating species recognition and labelling of fish from underwater video with machine-assisted deep learning. Front. Mar. Sci. **9**, 944582 (2022). https://doi.org/10.3389/fmars.2022.944582
6. Nundloll, V., Smail, R., Stevens, C., Blair, G.: Automating the extraction of information from a historical text and building a linked data model for the domain of ecology and conservation science. Heliyon **8**, e10710 (2022). https://doi.org/10.1016/j.heliyon.2022.e10710(2022)
7. Ahmad Zafir, A.W., Magintan, D.: Historical review of human-elephant conflict in Peninsular Malaysia. J. Wildlife Parks **31**, 1–19 (2016). https://wildlife.gov.my/images/stories/penerb itan/jurnal/2016/Jilid_31/01_HISTORICAL_REVIEW_OF_HUMAN-ELEPHANT.pdf
8. Lim, T., et al.: Mapping the distribution of people, elephants, and human-elephant conflict in temengor forest complex, Peninsular Malaysia. Malay. Nat. J. Spec. Ed. **2017**, 25–43 (2017)
9. Saaban, S., Zamahsasri, A.I., Wan Nordin, S.N., Gopalakrishnan, L., Elagupillay, S.T.: On the trail of our elephants in the Central Forest Spine. Department of Wildlife and National Parks Peninsular Malaysia, Kuala Lumpur (2021)
10. Wong, E.P., et al.: Living with elephants: evidence-based planning to conserve wild elephants in a megadiverse South East Asian country. Front. Conserv. Sci. **2**, 682590 (2021). https://doi.org/10.3389/fcosc.2021.682590
11. de la Torre, J.A., et al.: Sundaic elephants prefer habitats on the periphery of protected areas. J. Appl. Ecol. **59**(12), 2947–2958 (2022). https://doi.org/10.1111/1365-2664.14286
12. Fernando, P., et al.: Ranging behavior of the Asian elephant in Sri Lanka. Mamm. Biol. **73**(1), 2–13 (2008). https://doi.org/10.1016/j.mambio.2007.07.007
13. Calabrese, A., et al.: Conservation status of Asian elephants: the influence of habitat and governance. Biodivers. Conserv. **26**(9), 2067–2081 (2017). https://doi.org/10.1007/s10531-017-1345-5
14. Tan, A.S.L., de la Torre, J.A., Wong, E.P., Thuppil, V., Campos-Arceiz, A.: Factors affecting urban and rural tolerance towards conflict-prone endangered megafauna in Peninsular Malaysia. Glob. Ecol. Conserv. **23**, e01179 (2020). https://doi.org/10.1016/j.gecco.2020.e01179
15. Fernando, P., Pastorini, J.: Range-wide status of Asian elephants. Gajah **35**, 15–20 (2011). https://doi.org/10.5167/uzh-59036
16. Williams, C., et al.: Elephas maximus. The IUCN red list of threatened species 2020, e.T7140A45818198 (2020). https://doi.org/10.2305/IUCN.UK.2020-3.RLTS.T7140A45818198.en
17. Yamamoto-Ebina, S., Saaban, S., Campos-Arceiz, A., Takatsuki, S.: Food habits of Asian elephants Elephas maximus in a rainforest of northern Peninsular Malaysia. Mammal Study **41**(3), 155–161 (2016). https://doi.org/10.3106/041.041.0306
18. Campos-Arceiz, A., Blake, S.: Megagardeners of the forest – the role of elephants in seed dispersal. Acta Oecologica **37**(6), 542–553 (2011). https://doi.org/10.1016/j.actao.2011.01.014
19. de la Torre, J.A., et al.: There will be conflict – agricultural landscapes are prime, rather than marginal, habitats for Asian elephants. Anim. Conserv. **24**(5), 720–732 (2021). https://doi.org/10.1111/acv.12668
20. Hill, C., Osborn, F., Plumptre, A.J.: Human-wildlife conflict: identifying the problem and possible solutions. Albertine Rift Technical Report Series, vol. 1. Wildlife Conservation Society (2002)
21. Leslie, S., Brooks, A., Jayasinghe, N., Koopmans, F.: Human Wildlife Conflict mitigation: Lessons learned from global compensation and insurance schemes. Annex Report,

HWC SAFE Series. WWF Tigers Alive (2019). https://wwfeu.awsassets.panda.org/downlo ads/wwf_human_wildlife_conflict_mitigation_annex.pdf

22. IUCN SSC Position Statement on the Management of Human-Wildlife Conflict. IUCN Species Survival Commission (SSC) Human-Wildlife Conflict Task Force (2020). https://www.iucn.org/sites/default/files/2022-11/2021-position-statement-man agement-hwc_en.pdf

23. Mohammadi, F., Mahmoudi, H., Ranjbaran, Y., Ahmadzadeh, F.: Compilation and prioritizing human-wildlife conflict management strategies using the WASPAS method in Iran. Environ. Challenges **7**, 100482 (2022). https://doi.org/10.1016/j.envc.2022.100482

24. Adams, T.S.F., Leggett, K.E.A., Chase, M.J., Tucker, M.A.: Who is adjusting to whom?: Differences in elephant diel activity in wildlife corridors across different human-modified landscapes. Front. Conserv. Sci. **3**, 872472 (2022). https://doi.org/10.3389/fcosc.2022.872472

25. Gaynor, K.M., Branco, P.S., Long, R.A., Gonçalves, D.D., Granli, P.K., Poole, J.H.: Effects of human settlement and roads on diel activity patterns of elephants (*Loxodonta africana*). Afr. J. Ecol. **56**(4), 872–881 (2018). https://doi.org/10.1111/aje.12552

26. Kumar, A.M., Raghunathan, G.: Fostering human-elephant coexistence in the Valparai landscape, Annamalai Tiger Reserve, Tamil Nadu. In: South Asian Association for Regional Cooperation (ed.) Human-Wildlife Conflict in the Mountains of SAARC Region: Compilation of Successful Management Strategies and Practices. SAARC Forestry Centre, Thimphu, Bhutan (2014)

27. de Silva, S.: Acoustic communication in the Asian elephant, Elephas maximus maximus. Behaviour **147**(7), 825–852 (2010). https://doi.org/10.1163/000579510X495762

28. Stoeger, A.S., de Silva, S.: African and Asian elephant vocal communication: a cross-species comparison. In: Witzany, G. (ed.) Biocommunication of Animals, pp. 21–39. Springer, Dordrecht (2014). https://doi.org/10.1007/978-94-007-7414-8_3

29. Stoeger, A.S., Manger, P.: Vocal learning in elephants: neural bases and adaptive context. Curr. Opin. Neurobiol. **28**, 101–107 (2014). https://doi.org/10.1016/j.conb.2014.07.001

30. Nair, S., Balakrishnan, R., Seelamantula, C.S., Sukumar, R.: Vocalizations of wild Asian elephants (Elephas maximus): structural classification and social context. J. Acoust. Soc. Am. **126**(5), 2768–2778 (2009). https://doi.org/10.1121/1.3224717

31. Marler, P.: Science and birdsong: the good old days. In: Marler, P.R., Slabbekoom, H. (eds.) Nature's Music: The Science of Birdsong, pp. 1–38. Elsevier (2004)

32. Ma, P., Ren, J., Zhao, H., Sun, G., Murray, P., Zhang, J.: Multiscale 2-D singular spectrum analysis and principal component analysis for spatial–spectral noise-robust feature extraction and classification of hyperspectral images. IEEE J. Sel. Top. Appl. Earth Observ. Remote Sens. **14**, 1233–1245 (2021). https://doi.org/10.1109/JSTARS.2020.3040699

33. Chaturvedi, A., Yadav, S.A., Salman, H.M., Goyal, H.R., Gebregziabher H., Rao, A.K.: Classification of sound using convolutional neural networks. In: 5th International Conference on Contemporary Computing and Informatics (IC3I), Uttar Pradesh, India, pp. 1015–1019 (2022). https://doi.org/10.1109/IC3I56241.2022.10072823

34. Ma, P., et al.: Multiscale superpixelwise prophet model for noise-robust feature extraction in hyperspectral images. IEEE Trans. Geosci. Remote Sens. **61**, 1–12 (2023). https://doi.org/10. 1109/TGRS.2023.3260634

35. Xie, G., Ren, J., Marshall, S., Zhao, H., Li, R., Chen, R.: Self-attention enhanced deep residual network for spatial image steganalysis. Digit. Sig. Process. **139**, 104063 (2023). https://doi. org/10.1016/j.dsp.2023.104063

36. Yamashita, R., Nishio, M., Do, R.K.G., Togashi, K.: Convolutional neural networks: an overview and application in radiology. Insights Imaging **9**, 611–629 (2018). https://doi.org/ 10.1007/s13244-018-0639-9

37. Christin, S., Hervet, É., Lecomte, N.: Applications for deep learning in ecology. Methods Ecol. Evol. **10**(10), 1632–1644 (2019). https://doi.org/10.1111/2041-210X.13256

38. Nguyen, H., et al.: Animal recognition and identification with deep convolutional neural networks for automated wildlife monitoring. In: IEEE International Conference on Data Science and Advanced Analytics (DSAA), Tokyo, Japan, pp. 40–49 (2017). https://doi.org/10.1109/DSAA.2017.31

39. de Silva, E.M.K., et al.: Feasibility of using convolutional neural networks for individual-identification of wild Asian elephants. Mamm. Biol. **102**, 931–941 (2022). https://doi.org/10.1007/s42991-021-00206-2

40. Avicena, N.R.: CNN-based bioacoustics classification of elephant and non-elephant sounds (2020). https://github.com/aalavicena/FYP-Project-Repository-Audio-Classification-of-Ele phants-and-Non-Elephants-Sounds

41. Paszke, A., et al.: PyTorch: an imperative style, high-performance deep learning library (2019). https://doi.org/10.48550/arXiv.1912.01703. arXiv:1912.01703

42. Premarathna, K.S.P., Rathnayaka, R.M.K.T., Charles, J.: An elephant detection system to prevent human-elephant conflict and tracking of elephant using deep learning. In: Proceedings of ICITR 2020 – 5th International Conference on Information Technology Research: Towards the New Digital Enlightenment (2020). https://doi.org/10.1109/ICITR51448.2020.9310798

43. Bandara, M., Jayasundara, R., Ariyarathne, I., Meedeniya, D., Perera, C.: FSC22 dataset. IEEE Dataport (2022). https://doi.org/10.21227/40ds-0z76

44. Jaju, S., Chandak, M.: A transfer learning model based on ResNet-50 for flower detection. In: 2022 International Conference on Applied Artificial Intelligence and Computing (ICAAIC), Salem, India, pp. 307–311 (2022). https://doi.org/10.1109/ICAAIC53929.2022.9792697

45. Akata, Z., Perronnin, F., Harchaoui, Z., Schmid, C.: Good practice in large-scale learning for image classification. IEEE Trans. Pattern Anal. Mach. Intell. **36**(3), 507–520 (2014). https://doi.org/10.1109/TPAMI.2013.146

46. Wikimedia Commons. https://commons.m.wikimedia.org/wiki/File:ResNet50.png. Accessed 23 Nov 2023

47. Gentry, K.E., Lewis, R.N., Glanz, H., Simõesm, P.I., Nyári, Ã.S., Reichert, M.S.: Bioacoustics in cognitive research: applications, considerations, and recommendations. Wiley Interdisc. Rev. Cogn. Sci. **11**(5), e1538 (2020). https://doi.org/10.1002/WCS.1538

48. Shorten, C., Khoshgoftaar, T.M.: A survey on image data augmentation for deep learning. J. Big Data **6**(1), 60 (2019). https://doi.org/10.1186/s40537-019-0197-0

49. Perez, L., Wang, J.: The effectiveness of data augmentation in image classification using deep learning (2017). https://doi.org/10.48550/arXiv.1712.04621. arXiv:1712.04621

50. Kiyokawa, Y., Mishima, S., Toizumi, T., Sagi, K., Kondo, R., Nomura, T.: Sound event detection with ResNet and self-mask module for DCASE 2019 task 4. Technical report (2019)

51. Park, D.S., et al.: SpecAugment: a simple data augmentation method for automatic speech recognition. In: Interspeech 2019 (ISCA), pp. 2613–2617 (2019). https://doi.org/10.21437/Interspeech.2019-2680

52. Park, D.S., et al.: Specaugment on large scale datasets. In: ICASSP 2020 - 2020 IEEE International Conference on Acoustics, Speech and Signal Processing (ICASSP), pp. 6879–6883 (2020). https://doi.org/10.1109/ICASSP40776.2020.9053205

53. Ioffe, S., Szegedy, C.: Batch normalization: accelerating deep network training by reducing internal covariate shift (2015). https://doi.org/10.48550/arXiv.1502.03167. arXiv:1502.03167

54. Bai, Y.: RELU-function and derived function review. In: 2022 International Conference on Science and Technology Ethics and Human Future (STEHF 2022), SHS Web of Conferences, vol. 144, p. 02006 (2022). https://doi.org/10.1051/shsconf/202214402006

55. Aggarwal, C.C.: Neural Networks and Deep Learning: A Textbook. Springer, Cham (2018). https://doi.org/10.1007/978-3-319-94463-0

56. Resolve: RESOLVE and CVEDIA Announce WildEyes AI, a New Technology to Save Wild Elephants and Prevent Human-Elephant Conflict (2020). https://www.resolve.ngo/blog/WildEyes-AI-Helping-to-Save-Wild-Elephants-and-Prevent-Human-Elephant-Conflict.htm

57. Pérez-Granados, C.: BirdNET: applications, performance, pitfalls and future opportunities. Ibis **165**(3), 1068–1075 (2023). https://doi.org/10.1111/ibi.13193
58. Yeoh, A.: Musk's Starlink lands in Malaysia. The Star (2023). https://www.thestar.com.my/news/nation/2023/07/26/musks-starlink-lands-in-malaysia

Author Index

J. Ren et al. (Eds.): BICS 2023, LNAI 14374, pp. 401–402, 2024.
https://doi.org/10.1007/978-981-97-1417-9

Printed in the United States
by Baker & Taylor Publisher Services